T0360528

Emergence of The Quantum from The Classical

Mathematical Aspects of Quantum Processes

Emergence of The Quantum from The Classical

Mathematical Aspects of Quantum Processes

Maurice de Gosson
University of Vienna, Austria

 **World Scientific**

W JERSEY · LONDON · SINGAPORE · BEIJING · SHANGHAI · HONG KONG · TAIPEI · CHENNAI · TOKYO

Published by

World Scientific Publishing Europe Ltd.
57 Shelton Street, Covent Garden, London WC2H 9HE
Head office: 5 Toh Tuck Link, Singapore 596224
USA office: 27 Warren Street, Suite 401-402, Hackensack, NJ 07601

Library of Congress Cataloging-in-Publication Data
Names: Gosson, Maurice de, author.
Title: Emergence of the quantum from the classical : mathematical aspects of quantum processes /
    by Maurice de Gosson (University of Vienna, Austria).
Description: New Jersey : World Scientific, 2017.
Identifiers: LCCN 2017024951 | ISBN 9781786344144 (hc : alk. paper)
Subjects: LCSH: Quantum theory. | Hamiltonian systems. | Mechanics.
Classification: LCC QC174.12 .G6738 2017 | DDC 530.12--dc23
LC record available at https://lccn.loc.gov/2017024951

British Library Cataloguing-in-Publication Data
A catalogue record for this book is available from the British Library.

Desk Editors: V. Vishnu Mohan/Jennifer Bough/Koe Shi Ying

Typeset by Stallion Press
Email: enquiries@stallionpress.com

Printed in Singapore

To my wife, Charlyne, who is all my reasons

Contents

Preface

The notion of "emergent quantum mechanics" has gained increased popularity among scientists during the two last decades. To witness, the brilliant book *On the Emergence Theme of Physics* by Robert Carroll (World Scientific, 2010) or the cycle of conferences *Emergent Quantum Mechanics* organized by my friend Gerhard Grössing in Vienna since 2011. The present work aims at being another modest contribution to this fascinating and still mysterious topic.

My book *Principles of Newtonian and Quantum Mechanics* (World Scientific 2001, 2017) was a first attempt towards showing that quantum mechanics can truly be seen as emerging from classical mechanics in its Hamiltonian formulation, thus confirming George Mackey's view according to which quantum mechanics is a refinement of Hamiltonian mechanics. This was done using the mathematical theory of the metaplectic group, which led me to introduce the notion of *Metatron* in physics; the Metatron is related to Bohm's quantum potential in a subtle way. In the present book my aim is to substantially extend the scope of the study by including many other topics which demonstrate emergence phenomena. For instance, following previous work of mine, I introduce the notion of "quantum blob"; quantum blobs are phase space ellipsoids with minimum symplectic capacity; they are geometric manifestations of the indeterminacy principle of both quantum and classical mechanics. In quantum mechanics they can be used to coarse grain phase space in a symplectic invariant way making them compatible with the strong uncertainty inequalities of Robertson and Schrödinger. Another topic I have taken up is the dependence of quantum states on the value of Planck's constant. If, as ongoing cosmological experiments tend to prove, Planck's "constant" h has fluctuations, there would be drastic consequences. For instance, if h decreases some quantum states would become classical states (and could thus no longer be entangled with other states); if conversely the value of h increases, there will be a transition of some classical states into quantum states: "emergence" is here a two-way highway!

I want to thank my old friend Basil Hiley for many discussions of the "Implicate Order" and its physical and ontological interpretations. I also thank him for having pushed me to write this book: without his insistence, it would just have remained an emergent project!

Maurice de Gosson
Vienna

Acknowledgment

This work has been supported by the grants P20442-N13 and P27773-N13 of the Austrian Research Agency FWF.

Introduction

"Some of you, I am sure, will call this mysticism. So with all due acknowledgement to the fact that physical theory is at all times relative, in that it depends on certain basic assumptions, we may, or so I believe, assert that physical theory in its present stage strongly suggests the indestructibility of Mind by Time."

Erwin Schrödinger, Mind and Matter (1958)

The aim of this book is to introduce the physics community to some new ideas and concepts related to the notion of "emergence" and to show how they can be used with profit as well in quantum and in classical mechanics.

We will try to convey to the reader our belief that new concepts from symplectic topology might really be the proper setup for a better understanding of the twilight zone between "quantum" and "classical" properties. Indeed, the borderline between true quantum effects and classical phenomena is a murky area, which is far from being fully understood. For instance, there has been during the last years an increasing interest in precise forms of both classical and quantum uncertainty principles. This interest has been triggered not only by progress in quantum optics, but also in quantum information science, and in time-frequency and signal analysis. A new mathematical discipline, *symplectic geometry*, has emerged from the study of the properties of Hamiltonian flows, and in turn, these mathematical developments have led to the discovery of unsuspected properties of Hamiltonian systems; the *principle of the symplectic camel* is one outstanding example which will be discussed. One of the themes of the present book is that quantum mechanics can be seen to emerge from classical mechanics in its Hamiltonian formulation. The study of symplectic geometry therefore is (as we have emphasized elsewhere) essential for the understanding of these developments. We will see that symplectic geometry, and its topological extension, symplectic topology allow us to state and prove a classical multi-dimensional uncertainty principle, formally similar to the quantum uncertainty relations. In fact, one recurring theme of the present book is that the "uncertainty principle" in its strong form due to Robertson and Schrödinger can be expressed in terms of a class of symplectic invariants whose definition is made possible by Gromov's non-squeezing theorem. These invariants are the *symplectic capacities* of subsets of phase space. Symplectic capacities can be viewed as a measure of uncertainty, not related to volume, but rather to area, to which they reduce when the phase space is a plane; they are moreover intensive quantities, that is, they do not depend on the number of degrees of freedom. Our everyday world is ruled by Euclidean geometry (and by its extension to curved spaces, Riemannian geometry); we can measure distances in this geometry, and hence velocities when time is taken into account. Far away from our daily experience, and much more subtle, is the mechanical phase space world, in which all the phenomena are expressed in terms of a simultaneous measure of positions and momenta. A thorough understanding of *this* world requires recourse to a somewhat counter-intuitive geometry, the symplectic geometry of Hamiltonian mechanics. Symplectic geometry is in fact highly counter-intuitive; the notion of length does not make sense there, while the notion of area does. This "areal" nature of symplectic geometry, which was not realized until very recently, has led to unexpected mathematical developments, starting in the mid-1980s with Gromov's discovery of a "non-squeezing" phenomenon which is reminiscent of the quantum uncertainty principle — but in a totally classical setting! Gromov's discovery has been followed by a constellation of related results, which have considerably stimulated and reinvigorated symplectic topology, i.e., the study of global topological properties that are invariant under the action of symplectic mappings (or "canonical transformations" as they are called in physics). Unfortunately these mathematical developments have taken place almost unnoticed by the general physics community. Another somewhat related topic is studied in the last part of this book. It is the dependence of the notion of quantum state on the value of Planck's constant. While it is easy to see using elementary quantum mechanics that, say, a doubling of Planck's constant would lead to immediately observable effects, it is less clear what happens if we allow infinitesimal variations of h, as those who have probably taken place since the first nanoseconds following the Big Bang. We will see that the tiniest variation can have drastic consequences: some quantum states becoming classical states (or the other way round); pure states can become mixed states and so on. The analysis of the situation (which is not yet fully understood at the time of writing this book) could shed a new light on how our Universe was, and what it will become.

The book is planned as follows:

- Chapters 1–3 are of a rather classical nature. In Chapter 1 we review the basics of Hamiltonian mechanics in its Hamiltonian formulation, with a strong emphasis on the symplectic character of Hamiltonian flows. We take the opportunity to review a topic which is generally not very well-known by physicists, namely the non-squeezing theorem of Gromov, which shows that Hamiltonian flows are much more "rigid" than was believed before. Gromov's theorem (alias the property of the *symplectic camel*) will give us the opportunity later in this book to state the indeterminacy principle of quantum mechanics in a totally symplectically invariant way. Chapter 2 is devoted to a very classical topic: Hamilton–Jacobi theory. It is a powerful tool which plays an important role in the understanding of the Schrödinger equation. It also allows in a very natural way to explain the notion of action and its relationship with the Lagrangian manifolds of symplectic geometry. Last, but not least, it allows the introduction of the notion of density of trajectories, which is closely related to the Van Vleck determinant. In Chapter 3 we make a transition from the classical to the quantum worlds. We begin by discussing Louis de Broglie's matter waves, which can be used to define a short-time propagator. using the latter we justify, using Hamilton–Jacobi theory, the guiding equation for matter waves: Schrödinger's equation. We finally derive and discuss Bohm's equations for quantum motion, and emphasize the role played by the *quantum potential*. The latter allows us to show that Bohm's quantum trajectories are essentially of Hamiltonian nature.
- Chapter 4 ("The Metatron") is the cornerstone of the understanding of the essential links between classical mechanics in its Hamiltonian formulation, and quantum mechanics. We begin by discussing the metaplectic representation of the symplectic group, and show that the metaplectic group is a group of unitary operators covering the symplectic group (it is in fact the double cover of the symplectic group). This leads us to

lift the Hamiltonian flows determined by quadratic Hamiltonian functions to flows of unitary operators in the metaplectic group; it turns out that these lifted flows are quantum, more precisely they satisfy the Schrödinger equation corresponding to the quantized version of the classical quadratic Hamiltonian. The metaplectic representation thus leads in a rigorous way to Schrödinger's equation (a fact ignored by most physicists); we have called the entity corresponding to this lifted flow *Metatron*. A tool of choice is introduced: the Weyl quantization, which we also use in the subsequent chapters. We thereafter generalize this theory to the case of arbitrary Hamiltonian flows, corresponding to Hamiltonian functions which are not necessarily quadratic functions of the position and momentum coordinates. For this we use recent results from symplectic topology, especially the theory of Hamiltonian symplectomorphisms originating from the work of Gromov and his followers. Using these tools we show that there is a one-to-one correspondence between Hamiltonian flows and quantum flows determined by the corresponding Schrödinger equation. This result really shows that quantum mechanics "emerges" from Hamiltonian mechanics by giving a physical interpretation of the quantum flow.

• In Chapter 5 we address the "Uncertainty principle of quantum mechanics" using the tools from symplectic geometry and symplectic topology developed in the previous chapters. We show the uncertainty principle in its strong form (the Robertson–Schrödinger inequalities, which are a precise form of the Heisenberg inequalities) has a simple interpretation in terms of the topological notion of symplectic capacity, whose definition is made possible using the principle of the symplectic camel. This leads us to introduce the notion of "quantum blob" which is a phase space ellipsoid of minimum uncertainty; quantum blobs have the advantage on previous phase space coarse-grainings used in quantum statistical mechanics to be *symplectic invariants*. We also study some related topics, as Hardy's uncertainty principle, who can be used to show that Wigner functions cannot be arbitrarily sharply concentrated in phase space.

• In Chapters 6 and 7 we study pure and mixed states, and study the influence (mathematical and physical) of possible changes in the value of Planck's constant. Chapter 6 is devoted to a *rigorous* presentation of the theory of the density matrix. We give the mathematical definition of trace class operators and of the associated Hilbert–Schmidt machinery. We discuss in detail the notion of trace, and emphasize that some "trace formulas" often used by physicists are incorrect. This naturally leads us to the phase space representation of density matrices in terms of Wigner functions, which are studied in some detail, both from the algebraic and the functional point of view. In Chapter 7 we apply these notions to the study of the consequences of a change in the value of Planck's constant on the density matrix. An essential theoretical tool is here the Kastler–Loupias–Miracle-Sole theorem, of which we give a new and relatively simple proof. We thereafter apply these results to the study of some particular cases; we emphasize that very little is actually known today about what happens to quantum states under changes of Planck's constant and that it is, at the moment of writing this book, unclear whether Planck's constant has varied since the Big Bang (although related cosmological measurements of the fine-structure constant seem to indicate it might very well be the case). This last chapter thus opens the gate to future research.

We have added, for the reader's conveniences four appendices: Appendix A develops the results in symplectic geometry we did not incorporate in Chapters 1, 2, and 5; in Appendix B we prove some results from the theory of the metaplectic group that are needed in Chapter 4. In Appendix C we review recent results about the Born–Jordan approach to quantization following previous work of ours and in Appendix D we prove the composition formula for Weyl operators.

Chapter 1

Hamiltonian Mechanics

Hamiltonian mechanics is perhaps the most powerful *classical* theory ever; it allows the prediction of the motion of celestial bodies, of aeroplanes, and of particles in fluids. While Hamilton's equations of motion are easily derived, in the simplest cases, from Newton's second law, Hamiltonian mechanics is more than just a fancy way of doing Newtonian mechanics. Hamiltonian mechanics could already be found in disguise in the work of Lagrange in celestial mechanics. Namely, Lagrange discovered that the equations expressing the perturbation of elliptical planetary motion due to interactions could be written down as a simple system of partial differential equations (known today as Hamilton's equations, but Hamilton was only six years old when Lagrange made his discovery!). It is however undoubtedly Hamilton who realized, some twenty four years later, the theoretical importance of Lagrange's discovery and exploited it fully. We mention in passing that the notation H for a Hamiltonian function was proposed by Lagrange to honor Christiaan Huygens, and *not* Hamilton!

Like the movement of a symphony, a Hamiltonian flow involves a total ordering which implies the whole movement: Past, present, and future are actively present in any one movement. When we are listening to music we are actually directly perceiving an implicate order. This order is active because it is continuously flowing in emotional responses which are inseparable from the flow itself. Similarly, the solutions of Hamilton's equations are uniquely determined for all bounded times and all locations close to the original one, exactly as in the symphony metaphor: if we observe during a tiny time interval the motion of a particle moving under the influence of a Hamiltonian flow, we see an unfoldment of the totality of the flow, which is uniquely determined by the past — and the future!

There are nowadays many texts, at various levels, presenting Hamiltonian mechanics from the symplectic point of view; for instance the books by Arnold [5], Abraham *et al.* [2], Guillemin and Sternberg [139, 140] are classical references; for good introductions to symplectic geometry from the modern point of view see [23].

1.1. Hamilton's Equations

1.1.1. The origins; examples

Most physical systems can be studied by using two specific theories originating from Newtonian mechanics, and having overlapping — but not identical — domains of validity. The first of these theories is "Lagrangian mechanics", which essentially uses variational principles (e.g., the "least action principle"); it will not be discussed at all in this book; we refer to Souriau [268] (especially p. 140) for an analysis of some of the drawbacks of the Lagrangian approach. The second theory, "Hamiltonian mechanics", is based on Hamilton's equations of motion

$$\dot{\mathbf{r}} = \nabla_{\mathbf{p}} H(\mathbf{r}, \mathbf{p}, t), \quad \dot{\mathbf{p}} = -\nabla_{\mathbf{r}} H(\mathbf{r}, \mathbf{p}, t), \tag{1.1}$$

where the *Hamiltonian function*

$$H = \frac{1}{2m}(\mathbf{p} - \mathbf{A}(\mathbf{r}, t))^2 + V(\mathbf{r}, t) \tag{1.2}$$

is associated with the "vector" and "scalar" potentials $\mathbf{A}$ and V (both possibly depending on time t). We are writing $\mathbf{r} = (x, y, z)$ for the position vector, and $\mathbf{p} = (p_x, p_y, p_z)$ for the momentum vector. For $\mathbf{A} = 0$, Hamilton's equations are simply

$$\dot{\mathbf{r}} = \frac{\mathbf{p}}{m}, \quad \dot{\mathbf{p}} = -\nabla_{\mathbf{r}} V(\mathbf{r}, t)$$

and are immediately seen to be equivalent to Newton's second law for a particle moving in a scalar potential.

The most familiar example where one has a non-zero vector potential is of course the case of a charged particle in an electromagnetic field; V is then the Coulomb potential $-e^2/|\mathbf{r}|$ whereas $\mathbf{A}$ is related to the magnetic field $\mathbf{B}$ by the familiar formula $\mathbf{B} = \nabla_{\mathbf{r}} \times \mathbf{A}$ in a convenient choice of units.

Neglecting spin effects the Hamiltonian function is here

$$H = \frac{1}{2m} \left(\mathbf{p} - \frac{e}{c}\mathbf{A} \right)^2 - \frac{e^2}{|\mathbf{r}|}. \tag{1.3}$$

An interesting particular case is provided by the electron moving in a uniform magnetic field.

Hydrogen atom in a uniform magnetic field

Consider a hydrogen atom placed in a magnetic field $\mathbf{B}$. Suppose this atom is prepared in a very highly excited but still bound state, near the ionization threshold. The electron can then be viewed, to a good approximation, as a free particle, except for the presence of the magnetic field: since r is large, we can neglect the Coulomb potential $-e^2/|\mathbf{r}|$ in (1.3), and thus assume that the Hamiltonian is

$$H = \frac{1}{2m} \left(\mathbf{p} - \frac{e}{c}\mathbf{A} \right)^2.$$

Suppose now that the magnetic field is uniform in space, and that its direction is the z-axis: $\mathbf{B} = (0, 0, B_z)$. The coordinates of $\mathbf{A}$ will then satisfy the equations

$$\frac{\partial A_y}{\partial z} - \frac{\partial A_z}{\partial y} = \frac{\partial A_x}{\partial z} - \frac{\partial A_z}{\partial x} = 0, \quad \frac{\partial A_y}{\partial x} - \frac{\partial A_x}{\partial y} = B_z$$

which have (among others) the solutions:

$$A_x = -\tfrac{1}{2}B_z y, \quad A_y = \tfrac{1}{2}B_z x, \quad A_z = 0.$$

With that choice, the vector fields $\mathbf{A}$ and $\mathbf{B}$ are related by the simple formula $\mathbf{A} = \tfrac{1}{2}(\mathbf{r} \times \mathbf{B})$. It is customary to call this gauge the *symmetric gauge*. In that gauge the Hamiltonian is

$$H_{sym} = \frac{\mathbf{p}^2}{2m} + \frac{e^2 B_z^2}{8mc^2}(x^2 + y^2) - \frac{eB_z}{2mc}L_z, \tag{1.4}$$

where the quantity $L_z = xp_y - yp_x$ is the angular momentum in the z-direction. The term $\omega_L = eB_z/2mc$ is called the *Larmor frequency*; it is one-half of the cyclotron frequency $\omega_c = eB_z/mc$. The second term on the right-hand side of (1.4) is the "diamagnetic term", and the third, the "paramagnetic term".

There are however other interesting situations with nontrivial vector potential $\mathbf{A}$, one example being the Hamiltonian function of the Coriolis force in a geocentric frame.

Coriolis force

Let us denote by $\mathbf{R}$ the rotation vector at a point O situated on the surface of the Earth: $\mathbf{R}$ is a vector which is parallel to the axis of rotation and pointing out of the ground at O in the Northern hemisphere, and into the ground in the Southern hemisphere. Its length is the angular velocity calculated in a heliocentric reference frame. At a point at latitude ϕ, this vector is $\mathbf{R} = (0, R\cos\phi, R\sin\phi)$ $(R = |\mathbf{R}|)$ where the coordinates are calculated in a "lab frame" with origin O on the Earth; the z-axis is the vertical, the y-axis points in the direction of the North, and the x-axis towards the East. Suppose now that an observer placed at O wants to calculate the total force exerted on a nearby point with mass m and velocity v. He finds that the force is

$$\mathbf{F} = m\mathbf{g} - \mathbf{F}_C \quad \text{with} \quad \mathbf{F}_C = 2m\,(\mathbf{R} \times \mathbf{v}). \tag{1.5}$$

The velocity-dependent term $\mathbf{F}_C$ is called the *Coriolis force*. The functions

$$\mathbf{A} = m\,(\mathbf{R} \times \mathbf{r}), \quad V = -m\mathbf{g} \cdot \mathbf{r} \tag{1.6}$$

are vector and scalar potentials, and the associated Hamiltonian is therefore

$$H_C = \frac{1}{2m}\,(\mathbf{p} - m\,(\mathbf{R} \times \mathbf{r}))^2 - m\mathbf{g} \cdot \mathbf{r}. \tag{1.7}$$

For a detailed study of this example, see [100, Chapter 2].

1.1.2. Systems with many degrees of freedom

A typical example of a system with many degrees of freedom consists of N identical particles with mass m in three-dimensional space interacting with each other via a pair potential V, and also interacting with some external potential U; the Hamiltonian function is here

$$H = \sum_{j=1}^{N} \frac{\mathbf{p}_j^2}{2m} + \sum_{j=1}^{N} V(\mathbf{r}_j) + \sum_{k<j} U(\mathbf{r}_j - \mathbf{r}_k),$$

where $\mathbf{r}_j = (x_j, y_j, z_j)$ is the coordinate vector of the jth particle and $\mathbf{p}_j = (p_{x,j}, p_{y,j}, p_{z,j})$ its momentum vector; more generally one can consider systems with Hamiltonian

$$H = \sum_{j=1}^{N} \frac{\mathbf{p}_j^2}{2m} + V(\mathbf{r}),$$

where V is an arbitrary function of the position coordinates.

Some notation

It will be convenient to use the shorthand notation $x = (x_1, x_2, \ldots, x_n)$ and $p = (p_1, p_2, \ldots, p_n)$ for position and momentum vectors; we will set $z = (x, p)$ and view z as an element of the phase space $\mathbb{R}_z^{2n} \equiv \mathbb{R}_x^n \times \mathbb{R}_p^n$. The integer n denotes the number of degrees of freedom. For instance, if the physical system under consideration consists of N independent particles with no further constraints, then $n = 3N$. When performing calculations, position and momentum vectors will be written as column vectors

$$\begin{pmatrix} x_1 \\ \vdots \\ x_n \end{pmatrix} \quad \text{and} \quad \begin{pmatrix} p_1 \\ \vdots \\ p_n \end{pmatrix}$$

and the corresponding phase space vector is thus $\begin{pmatrix} x \\ p \end{pmatrix}$. The set of all position vectors (the configuration space) is denoted by $\mathbb{R}_x^n$ (or simply by $\mathbb{R}^n$), similarly the set of all momentum vectors is $\mathbb{R}_p^n$. The generalized gradients in x, p, z are defined by

$$\nabla_x = \begin{pmatrix} \partial_{x_1} \\ \vdots \\ \partial_{x_n} \end{pmatrix}, \quad \nabla_p = \begin{pmatrix} \partial_{p_1} \\ \vdots \\ \partial_{p_n} \end{pmatrix}$$

and $s\nabla_z = \begin{pmatrix} \nabla_x \\ \nabla_p \end{pmatrix}$; when appearing in the text these gradients will be written as row vectors.

The notion of phase space is primordial in Hamiltonian mechanics; its importance comes from the fact that being even-dimensional it can be equipped with a symplectic structure (in fact, infinitely many), and thus opens the way to the symplectic formulation of Hamiltonian mechanics, which we will discuss in detail. Hamilton's equations can be defined quite abstractly and independently of any physical background: let us call a *Hamiltonian function* any twice continuously differentiable function of the phase space variables $z = (x, p)$. By definition the Hamilton equations associated with H are the system of $2n$ first-order differential equations

$$\dot{x}_j = \frac{dx_j}{dt} = \frac{\partial H}{\partial p_j}(x_1, \ldots, x_n; p_1, \ldots, p_n),$$

$$\dot{p}_j = \frac{dp_j}{dt} = -\frac{\partial H}{\partial x_j}(x_1, \ldots, x_n; p_1, \ldots, p_n),$$

which we can write for short as

$$\dot{x} = \nabla_p H(x, p), \quad \dot{p} = -\nabla_x H(x, p).$$

The study of this system is a purely mathematical problem belonging to the theory of differential equations; notice that the equations above form an autonomous system of differential equations in the sense that the right-hand sides do not explicitly depend on time t. Obvious physical examples (particles moving in a time-dependent potential) suggest that we extend the definitions above by allowing the Hamiltonian function to depend on time; this leads to the consideration of non-autonomous systems of the type

$$\dot{x}_j = \frac{\partial H}{\partial p_j}(x_1, \ldots, x_n; p_1, \ldots, p_n; t),$$

$$\dot{p}_j = -\frac{\partial H}{\partial x_j}(x_1, \ldots, x_n; p_1, \ldots, p_n; t),$$

that is

$$\dot{x} = \nabla_p H(x, p, t), \quad \dot{p} = -\nabla_x H(x, p, t).$$

The study of such systems also fits very well in the general theory of (ordinary) differential equations. Actually, there are conceptual advantages to including time-dependency in the theory from the very beginning. We will therefore call a Hamiltonian function a function H defined on the time-dependent phase space

$$\mathbb{R}_z^{2n} \times \mathbb{R}_t \equiv \mathbb{R}_x^n \times \mathbb{R}_p^n \times \mathbb{R}_t$$

which is twice continuously differentiable in the x, p variables; one usually asks that H should depend on the time variable t in a continuously differentiable way.

When the Hamiltonian function has the familiar form "kinetic energy plus potential function"

$$H(z,t) = \sum_{j=1}^{n} \frac{p_j^2}{2m_j} + V(x,t),$$

the Hamilton equations reduce to the Newton equations

$$\dot{x}_j = \frac{1}{m_j} p_j, \quad \dot{p}_j = -\frac{\partial V}{\partial x_j}(x_1, \ldots, x_n; t).$$

As we said above, the study of Hamilton's equation can be viewed as a subfield of the theory of ordinary differential equations (autonomous, or not). But of course this does not mean that they are easy to solve in practice outside a few particular examples: to obtain explicit "closed formulas" is usually really challenging, and most of the time even impossible (there are however efficient approximate numerical methods, e.g., the method of symplectic integrators). One important exactly solvable case is that when the Hamiltonian function is quadratic.

1.2. Quadratic Hamiltonians

The consideration of such Hamiltonians has, in addition to its intrinsic importance, undeniable pedagogical virtues: it will allow us to introduce in a natural way the notion of symplectic transformation.

1.2.1. Hamilton's equations and symplectic matrices

We will assume that the Hamiltonian function is a quadratic polynomial in the position and momentum variables, i.e., of the form

$$H(z) = \sum_{j,k} a_{jk} x_j x_k + b_{jk} p_j p_k + c_{jk} x_j p_k.$$

Such a Hamiltonian can always be written in compact form as

$$H(z) = \tfrac{1}{2} z^T M z = \tfrac{1}{2} M z \cdot z, \tag{1.8}$$

where M is a real symmetric $2n \times 2n$ matrix (M is the Hessian matrix of H, that is the matrix of second derivatives).

The archetypical example is of course the one-dimensional harmonic oscillator: here $n = 1$ and the Hamiltonian is

$$H(z) = \frac{1}{2m}(p^2 + m^2\omega^2 x^2);$$

it represents a spring-mass system, or a pendulum having small oscillations. The associated Hamilton equations are

$$\dot{x} = p/m, \quad \dot{p} = -m\omega^2 x$$

and their solutions are easily found:

$$x(t) = A\sin(\omega t + \varphi), \quad p(t) = Am\omega\cos(\omega t + \varphi).$$

A less trivial example is provided by the electron in a uniform magnetic field studied above: the Hamiltonian function (1.4) can be written as

$$H_{sym} = \sum_{j=1}^{2}\left(\frac{p_j^2}{2m} + \frac{e^2 B_z^2}{8mc^2}x_j^2\right) - \frac{eB_z}{2mc}(x_1 p_2 - x_2 p_1). \qquad (1.9)$$

Quadratic Hamiltonian functions play a privileged role in both classical and quantum mechanics. The main reason is that the associated Hamilton equations are (at least in the time-independent case) exactly solvable. Many properties of Hamiltonian systems can therefore be tested on them; in addition they play an important role in classical mechanics near equilibrium, where one can often replace the true flow by its linearization. But there is a deeper reason for which we study them here; it is because they play a fundamental role in the theory of the metaplectic representation of the symplectic group which we will study later on; quadratic Hamiltonians open the door to quantum mechanics. Let us begin by introducing some notation. The standard symplectic matrix is, by definition, the $2n \times 2n$ real matrix

$$J = \begin{pmatrix} 0_{n\times n} & I_{n\times n} \\ -I_{n\times n} & 0_{n\times n} \end{pmatrix},$$

where $0_{n\times n}$ and $I_{n\times n}$ are, respectively, the zero and unity $n \times n$ matrices (we will usually drop the subscripts $n \times n$ when the context is clear).

Note that $\det J = 1$ and that

$$J^2 = -I, \quad J^T = J^{-1} = -J$$

(the superscript ^{T} denotes transposition). The property $J^2 = -I$ shows that in a sense J is the matrix equivalent of the imaginary unit i (this analogy will be made precise later).

We will say that a real $2n \times 2n$ matrix S is symplectic if it satisfies the conditions

$$S^T J S = S J S^T = J. \tag{1.10}$$

We mention that the conditions in (1.10) are redundant: to prove that a matrix S is symplectic it suffices in fact to show that we have $S^T J S = J$ or, alternatively, $S J S^T = J$ because both conditions are equivalent (the proof of this property is left as an exercise for the reader).

The first (and perhaps the most obvious) reason for introducing the matrix J is that it allows us to write Hamilton's equations (for any Hamiltonian function, not just those that are quadratic) in the compact form

$$\dot{z} = J \nabla_z H(z, t). \tag{1.11}$$

We will also use the notation and call

$$X_H = J \nabla_z H = (\nabla_p H, -\nabla_x H)$$

the *Hamiltonian vector field*. Returning to quadratic Hamiltonians (1.8) we have $\nabla_z H = M z$ and hence Hamilton's equation (1.11) becomes the linear system

$$\dot{z} = J M z \tag{1.12}$$

which we can immediately solve, finding the solution

$$z(t) = e^{t J M} z(0). \tag{1.13}$$

Let us write $S_t = e^{t J M}$; we claim that S_t is a symplectic matrix for every value of time t. Consider in fact the matrix function

$$F(t) = S_t^T J S_t = e^{-t M J} J e^{t J M}.$$

Differentiation with respect to t yields

$$\frac{d}{dt}F(t) = (-e^{-tMJ}MJ)Je^{tJM} + e^{-tMJ}J(JM)e^{tJM}$$

$$= e^{-tMJ}Me^{tJM} - e^{-tMJ}Me^{tJM}$$

$$= 0$$

hence the function F is constant, and in particular equal to $F(0) = J$. We thus have $S_t^T J S_t = J$ for all t which proves that S_t is always symplectic.

1.2.2. The symplectic group

The set of all symplectic matrices is denoted by $\mathrm{Sp}(2n, \mathbb{R})$. It is easy to see that $\mathrm{Sp}(2n, \mathbb{R})$ is a group for matrix multiplication: if S and S' are two symplectic matrices, then

$$(SS')^T JSS' = (S')^T S^T JSS' = (S')^T JS' = J \qquad (1.14)$$

and similarly $SS'J(SS')^T = J$; also, since $J^{-1} = -J$,

$$(S^{-1})^T JS^{-1} = -(SJS^T)^{-1} - J^{-1} = J$$

and $S^{-1}J(S^{-1})^T = J$. Since the identity matrix clearly is symplectic as well it follows that $\mathrm{Sp}(2n, \mathbb{R})$ is indeed a group: it is the *standard symplectic group*.

The notion of symplectic matrix is closely related to that of symplectic form. By definition, the standard symplectic form on the phase space $\mathbb{R}_z^{2n}$ is the antisymmetric bilinear form $\sigma : \mathbb{R}_z^{2n} \times \mathbb{R}_z^{2n} \to \mathbb{R}$ associating with every pair $z = (x, p)$; $z' = (x', p')$ of phase space vectors the number

$$\sigma(z, z') = p \cdot x' - p' \cdot x = Jz \cdot z'. \qquad (1.15)$$

Notice that in the case $n = 1$ the symplectic form reduces to minus the determinant: $\sigma(z, z') = -\det(z, z')$. In the general case, we have

$$\sigma(z, z') = -\sum_{j=1}^{n} \begin{vmatrix} x_j & x_j' \\ p_j & p_j' \end{vmatrix},$$

which shows that $\sigma(z, z')$ is (up to the sign) the algebraic sum of the oriented areas of the parallelograms spanned by the projections of the vectors z, z' on the planes of conjugate coordinates x_j, p_j. It is clear that definition (1.10) of a symplectic matrix can be restated as

$$S \in \mathrm{Sp}(2n, \mathbb{R}) \iff \sigma(Sz, Sz') = \sigma(z, z') \tag{1.16}$$

for all $z, z' \in \mathbb{R}_z^{2n}$.

Assume now that we write the matrix S in block form

$$S = \begin{pmatrix} A & B \\ C & D \end{pmatrix}, \tag{1.17}$$

where A, B, C, D are matrices of size n. It is a simple exercise in matrix algebra to show that condition (1.10) is equivalent to the following constraints on the blocks A, B, C, D:

$$A^T C = C^T A, \quad B^T D = D^T B, \quad \text{and} \quad A^T D - C^T B = I. \tag{1.18}$$

Notice that the first two conditions mean that both products $A^T C$ and $B^T D$ *are symmetric.* These conditions reduce to the identity $ad - bc = 1$ when $n = 1$: in this case, A, B, C, D are the numbers a, b, c, d so that $A^T C = ac$ and $B^T D = bd$; the condition $A^T D - C^T B = I$ reduces to $ad - bc = 1$. The conditions (1.18) are equivalent to the following two sets of conditions:

$$AB^T = BA^T, \quad DC^T = CD^T, \quad DA^T - CB^T = I \tag{1.19}$$

and

$$AC^T = CA^T, \quad DB^T = BD^T, \quad AD^T - BC^T = I, \tag{1.20}$$

which are easily verified by applying (1.10) to S^{-1} and S^T (see [108] for details).

Let us next give a brief review of the Lie algebra $\mathfrak{sp}(2n, \mathbb{R})$ of the symplectic group $\mathrm{Sp}(2n, \mathbb{R})$. We have actually already encountered this Lie algebra in disguise during our discussion of the Hamilton equations determined by a quadratic Hamiltonian function

$$H(z) = \tfrac{1}{2} z^T M z.$$

We have seen that the solutions of these Hamilton equations are given by $z(t) = S_t z(0)$ where $S_t = e^{tJM}$ is in $\mathrm{Sp}(2n, \mathbb{R})$; it follows that JM belongs to the Lie algebra of the symplectic group. The following result generalizes this observation.

Proposition 1. *The Lie algebra* $\mathfrak{sp}(2n, \mathbb{R})$ *of* $\mathrm{Sp}(2n, \mathbb{R})$ *consists of all* $X \in \mathcal{M}(2n, \mathbb{R})$ *such that*

$$XJ + JX^T = 0 \quad (equivalently \ X^T J + JX = 0). \qquad (1.21)$$

Proof. Let (S_t) be a differentiable one-parameter subgroup of $\mathrm{Sp}(2n, \mathbb{R})$ and X a $2n \times 2n$ real matrix such that $S_t = \exp(tX)$. Since S_t is symplectic, we have $S_t J (S_t)^T = J$, i.e.,

$$\exp(tX) J \exp(tX^T) = J.$$

Differentiating both sides of this equality with respect to t and then setting $t = 0$ we get $XJ + JX^T = 0$, and applying the same argument to the transpose S_t^T we get $X^T J + JX = 0$ as well. Suppose conversely that X is such that $XJ + JX^T = 0$ and let us show that $X \in \mathfrak{sp}(2n, \mathbb{R})$. For this it suffices to prove that $S_t = \exp(tX)$ is in $\mathrm{Sp}(2n, \mathbb{R})$ for every t. The condition $X^T J + JX = 0$ is equivalent to $X^T = JXJ$ hence $S_t^T = \exp(tJXJ)$; since $J^2 = -I$ we have $(JXJ)^k = (-1)^{k+1} JX^k J$ and hence

$$\exp(tJXJ) = -\sum_{k=0}^{\infty} \frac{(-t)^k}{k!} (JXJ)^k = -Je^{-tX}J.$$

It follows that $S_t^T J S_t = (-Je^{-tX}J)Je^{tX} = J$ so that $S_t \in \mathrm{Sp}(2n, \mathbb{R})$ as claimed. $\qquad \square$

Note that if one writes $X \in \mathfrak{sp}(2n, \mathbb{R})$ in block matrix form then it has the form

$$X = \begin{pmatrix} U & V \\ W & -V^T \end{pmatrix},$$

where U, V, and W are $n \times n$ matrices such that

$$V = V^T \quad \text{and} \quad W = W^T.$$

In particular $\mathfrak{sp}(2, \mathbb{R})$ consists of all 2×2 matrices with vanishing trace:

$$X \in \mathfrak{sp}(2, \mathbb{R}) \Longleftrightarrow \mathrm{Tr} X = 0.$$

One should be careful to note that the exponential mapping

$$\exp : \mathfrak{sp}(2n, \mathbb{R}) \longrightarrow \mathrm{Sp}(2n, \mathbb{R})$$

is neither surjective nor injective; for instance it is not hard to prove that if $S \in \mathrm{Sp}(2, \mathbb{R})$ can be written in the form $S = \exp X$ with $X \in \mathfrak{sp}(2)$ then we must have $\mathrm{Tr} S \geq -2$ (see [108, p. 37]). However, when conditions of positivity and symmetry are imposed, one has a much better situation. In fact, denoting by $\mathrm{Sym}(2n, \mathbb{R})$ the set of real symmetric $2n \times 2n$ matrices and by $\mathrm{Sym}_+(2n, \mathbb{R})$ the subset of $\mathrm{Sym}(2n, \mathbb{R})$ consisting of positive definite matrices, we have the following propositions.

Proposition 2. *We have* $S \in \mathrm{Sp}(2n, \mathbb{R}) \cap \mathrm{Sym}_+(2n, \mathbb{R})$ *if and only if* $S = \exp X$ *with* $X \in \mathfrak{sp}(2n)$ *and* $X = X^T$. *The exponential mapping is a diffeomorphism*

$$\exp : \mathfrak{sp}(2n, \mathbb{R}) \cap \mathrm{Sym}(2n, \mathbb{R}) \longrightarrow \mathrm{Sp}(2n, \mathbb{R}) \cap \mathrm{Sym}_+(2n, \mathbb{R}).$$

Proof. If $X \in \mathfrak{sp}(2n, \mathbb{R})$ and $X = X^T$ then S is both symplectic and symmetric positive definite. Assume conversely that S is symplectic and symmetric positive definite. The exponential mapping is a diffeomorphism

$$\exp : \mathrm{Sym}(2n, \mathbb{R}) \longrightarrow \mathrm{Sym}_+(2n, \mathbb{R})$$

and hence there exists a unique $X \in \mathrm{Sym}(2n, \mathbb{R})$ such that $S = \exp X$. Let us show that $X \in \mathfrak{sp}(2n, \mathbb{R})$. Since $S = S^T$ we have $SJS = J$ and hence $S = -JS^{-1}J$. Because $-J = J^{-1}$, it follows that

$$\exp X = J^{-1}(\exp(-X))J = \exp(-J^{-1}XJ),$$

and $J^{-1}XJ$ being symmetric, we conclude that $X = J^{-1}XJ$; that is $JX = -XJ$, showing that $X \in \mathfrak{sp}(2n, \mathbb{R})$. $\quad\square$

1.2.3. Free symplectic matrices and generators

We will say that a symplectic matrix S is free if and only if given the initial and final positions x' and x, the equation $(x, p) = S(x', p')$ *uniquely* determines the initial and final momenta p', p. Writing S as a block matrix

$$S = \begin{pmatrix} A & B \\ C & D \end{pmatrix}$$

(each block of dimension n), S is free if and only if its right upper-corner is invertible, i.e., if

$$\det B \neq 0. \tag{1.22}$$

This is easily seen by rewriting the equation $(x, p) = S(x', p')$ as a system

$$Ax' + Bp' = x, \quad Cx' + Dp' = p;$$

we leave the details to the reader as an elementary exercise. The symplectic matrix

$$J = \begin{pmatrix} 0 & I \\ -I & 0 \end{pmatrix}$$

is obviously free, whereas the identity matrix is not.

We next introduce the fundamental notion of the *generating function* of a free symplectic matrix. We will say that a function W on $\mathbb{R}_x^n \times \mathbb{R}_x^n$ generates the symplectic matrix S if we have

$$(x, p) = S(x', p') \iff \begin{cases} p = \nabla_x W(x, x'), \\ p' = -\nabla_{x'} W(x, x'). \end{cases}$$

To illustrate this notion, let us discuss two elementary examples. Consider first the free particle Hamiltonian $H = p^2/2m$ $(n = 1)$. The solutions of Hamilton's equations are

$$x(t) = x(0) + \frac{p(0)}{m} t, \quad p(t) = p(0),$$

and hence $(x(t), p(t)) = S_t(x(0), p(0))$ where

$$S_t = \begin{pmatrix} 1 & t/m \\ 0 & 1 \end{pmatrix};$$

this symplectic matrix is free for $t \neq 0$. One checks directly that a (time-depending) generating function is

$$W(x, x'; t) = m \frac{(x - x')^2}{2t}.$$

Consider next the one-dimensional harmonic oscillator Hamiltonian; we choose for convenience the mass equal to one, hence

$$H = \tfrac{1}{2}(p^2 + \omega^2 x^2).$$

Here the solutions to Hamilton's equations are $(x(t), p(t)) = S_t(x(0), p(0))$ where

$$S_t = \begin{pmatrix} \cos \omega t & \sin \omega t \\ -\sin \omega t & \cos \omega t \end{pmatrix}$$

and a (time-depending) generating function is

$$W(x, x'; t) = \frac{\omega}{2 \sin \omega t} \left((x^2 + x'^2) \cos \omega t - 2xx' \right) \tag{1.23}$$

for $t \neq k\pi/\omega$ (k an integer).

More generally, we have the following proposition.

Proposition 3. *Let* $S = \begin{pmatrix} A & B \\ C & D \end{pmatrix}$ *be a free* $2n \times 2n$ *symplectic matrix:* $\det B \neq 0$.

(i) *The generating function of* S *is*

$$W(x, x') = \tfrac{1}{2} DB^{-1} x \cdot x - (B^T)^{-1} x \cdot x' + \tfrac{1}{2} B^{-1} Ax' \cdot x'; \tag{1.24}$$

the matrices DB^{-1} *and* $B^{-1}A$ *are symmetric.*

(ii) *Conversely, every quadratic form*

$$W(x, x') = \tfrac{1}{2} Px \cdot x - Lx \cdot x' + \tfrac{1}{2} Qx' \cdot x', \tag{1.25}$$

where P *and* Q *are symmetric and* $\det L \neq 0$, *is the generating function of a symplectic matrix* $S = S_W$, *namely*

$$S_W = \begin{pmatrix} L^{-1}Q & L^{-1} \\ PL^{-1}Q - L^T & PL^{-1} \end{pmatrix}. \tag{1.26}$$

Proof. (i) To prove that the generating function is given by (1.24) it suffices to check that if $(x, p) = S(x', p')$ then

$$p = \nabla_x W(x, x'), \ p' = -\nabla_{x'} W(x, x');$$

we omit the straightforward calculations here. That the matrices DB^{-1} and $B^{-1}A$ are symmetric follows from conditions (1.18) and (1.19).

(ii) Using the expression (1.25) for W, we see that

$$(x, p) = S_W(x', p') \Longleftrightarrow \begin{cases} p = Px - L^T x', \\ p' = Lx - Qx', \end{cases}$$

and since $\det L \neq 0$, we can explicitly solve these equations in (x, p); this yields (1.26) after a few straightforward calculations. That the matrix S_W is symplectic immediately follows using any of the equivalent conditions (1.18), (1.19), or (1.20). $\qquad\square$

There is thus a one-to-one correspondence between the quadratic forms W, and free symplectic matrices. In fact, to every such quadratic form W one can associate the free symplectic matrix and, conversely, every free symplectic matrix can be written in this form, and thus determines W.

Notation 4. We will use the shorthand notation $W = (P, L, Q)$ for quadratic forms (1.25).

The following consequence of Proposition 3 will be useful in our study of the metaplectic group.

Proposition 5. *The inverse $(S_W)^{-1}$ is the free symplectic matrix given by*

$$(S_W)^{-1} = S_{W^*}, \quad W^*(x, x') = -W(x', x). \tag{1.27}$$

Proof. To prove (1.27), it suffices to note that the inverse of the symplectic matrix S_W is

$$(S_W)^{-1} = \begin{pmatrix} (L^{-1})^T P & -(L^{-1})^T \\ -Q(L^{-1})P + L^T & Q(L^{-1})^T \end{pmatrix},$$

which shows that the inverse matrix $(S_W)^{-1}$ is associated with the quadratic form W^* obtained from W by changing the triple of matrices (P, L, Q) into the new triple $(-Q, -L^T, -P)$. □

We next study the relationship between free symplectic matrices and the generators of the symplectic group.

Lemma 6. *Let S_W and $S_{W'}$ be two free symplectic matrices, respectively, associated with $W = (P, L, Q)$ and $W' = (P', L', Q')$. Their product $S_W S_{W'}$ is a free symplectic matrix $S_{W''}$ if and only if*

$$\det(P' + Q) \neq 0, \tag{1.28}$$

in which case we have $W'' = (P'', L'', Q'')$ with

$$\begin{cases} P'' = P - L^T (P' + Q)^{-1} L, \\ L'' = L'(P' + Q)^{-1} L, \\ Q'' = Q' - L'(P' + Q)^{-1} L'^T. \end{cases} \tag{1.29}$$

Proof. In view of (1.26), the product $S_W S_{W'}$ is given by

$$\begin{pmatrix} L^{-1}Q & L^{-1} \\ PL^{-1}Q - L^T & PL^{-1} \end{pmatrix} \begin{pmatrix} L'^{-1}Q' & L'^{-1} \\ P'L'^{-1}Q' - L'^T & P'L'^{-1} \end{pmatrix},$$

and performing the matrix multiplication, the right upper-corner of that product is $L^{-1}(P' + Q)L'^{-1}$ which is invertible if and only if (1.28) holds. If it holds, set

$$S_{W''} = \begin{pmatrix} L''^{-1}Q'' & L''^{-1} \\ P''L''^{-1}Q'' - L''^T & P''L''^{-1} \end{pmatrix}$$

and solve successively for P'', L'', Q''. $\qquad\square$

Let P and L be two $n \times n$ matrices such that P is symmetric and L is invertible. It immediately follows from the conditions (1.18) that the matrices

$$V_P = \begin{pmatrix} I & 0 \\ -P & I \end{pmatrix}, \quad M_L = \begin{pmatrix} L^{-1} & 0 \\ 0 & L^T \end{pmatrix} \tag{1.30}$$

are symplectic. The set of all matrices V_P, M_L, together with the matrix

$$J = \begin{pmatrix} 0 & I \\ -I & 0 \end{pmatrix},$$

generates the symplectic group $\mathrm{Sp}(2n, \mathbb{R})$. To our knowledge, there are at least four proofs of this fact. One can either use a topological argument (see for instance [96] or [283]) or elementary linear algebra, as in the first Chapter of Guillemin and Sternberg [140] (but the calculations are then rather complicated). One can also use methods from the theory of Lie groups (see for instance [211]). We are going to present a fourth method, which consists of using the properties of free symplectic matrices we have developed above. This approach has the advantage of giving a rather straightforward factorization of an arbitrary symplectic matrix. We begin with the following two preparatory results.

Lemma 7. *Every free symplectic matrix S_W can be (uniquely) written as a product*

$$S_W = V_{-P} M_L J V_{-Q}, \tag{1.31}$$

where V_{-P}, V_{-Q}, and M_L are defined by (1.30).

Proof. Performing the product on the right-hand side of (1.31), we get

$$S_W = \begin{pmatrix} L^{-1}Q & L^{-1} \\ PL^{-1}Q - L^T & PL^{-1} \end{pmatrix}$$

(cf. formula (1.26)). Writing S_W in the usual block-matrix form $\begin{pmatrix} A & B \\ C & D \end{pmatrix}$, we get the following equations for P, L and Q:

$$\begin{cases} A = L^{-1}Q, \ B = L^{-1}, \\ C = PL^{-1}Q - L^T, \ D = PL^{-1}. \end{cases}$$

Since B is invertible, we get $L = B^{-1}$, $P = DB^{-1}$, $Q = B^{-1}A$. $\qquad\square$

The following important lemma is the key to our derivation.

Lemma 8. *Every symplectic matrix is the product of two free symplectic matrices.*

Proof. The proof makes use of the properties of the action of the symplectic group on Lagrangian planes, which we do not study here (see [108, 112]). $\qquad\square$

Combining the two lemmas above we get the following proposition.

Proposition 9. *The matrices V_P, M_L and J generate the symplectic group $\mathrm{Sp}(2n, \mathbb{R})$.*

Proof. In view of Lemma 8, every $S \in \mathrm{Sp}(2n, \mathbb{R})$ can be written as a product $S_W S_{W'}$ and hence, by formula (1.31) in Lemma 7,

$$S = V_{-P} M_L J V_{-(P'+Q)} M_{L'} J V_{-Q'}, \qquad (1.32)$$

which shows that S is a product of matrices of the type V_P, M_L and J. $\qquad\square$

The following consequence is obvious.

Corollary 10. *Every $S \in \mathrm{Sp}(2n, \mathbb{R})$ has determinant $\det S = 1$.*

Proof. The determinants of V_P, M_L, and J are trivially equal to one. $\square$

1.3. The Hamiltonian Flow

We now study in detail the symplectic properties of the Hamilton equations determined by an arbitrary Hamiltonian function H.

1.3.1. Hamiltonian vector fields

Recall that the letter z denotes the phase space variable (x, p). We assume that the Hamiltonian function H is defined on the time-dependent phase space $\mathbb{R}_z^{2n} \times \mathbb{R}_t$. Hamilton's equations can be written in the compact form

$$\dot{z} = X_H(z, t), \tag{1.33}$$

where X_H is the Hamiltonian vector field

$$X_H = (\nabla_p H, -\nabla_x H). \tag{1.34}$$

If H is time-independent, then Hamilton's equations form an autonomous system of differential equations, for which we can define the notion of *flow*. By definition, the flow (f_t) determined by the vector field X_H is the collection of mappings f_t which takes an "initial" point $z_0 = (x_0, p_0)$ to the point $z_t = (x_t, p_t)$ after time t, along the trajectory of X_H through z_0, i.e.,

$$\frac{d}{dt} f_t(z_0) = X_H(f_t(z_0)). \tag{1.35}$$

It is customary to call the trajectory $t \mapsto f_t(z_0)$ the *orbit* of z_0. The mappings f_t obviously satisfy the one-parameter group property:

$$f_t \circ f_{t'} = f_{t+t'}, \quad (f_t)^{-1} = f_{-t}, f_0 = I_d. \tag{1.36}$$

When H depends explicitly on time t, Hamilton's equations no longer form an autonomous system, and the mappings f_t — which are defined by the same formula (1.35) as above — no longer satisfy the one-parameter group property (1.36). This is due to the fact that X_H is no longer a "true" vector field, but is rather a family of vector fields continuously depending on the parameter t. It is then advantageous to modify the definition of the notion of flow in the following way: given "initial" and "final" times t' and t, we denote by $f_{t,t'}$ the mapping that takes a point $z' = (x', p')$ at time t' to the point $z = (x, p)$ at time t along the trajectory determined by Hamilton's equations, equivalently:

$$\frac{d}{dt} f_{t,t'}(z') = X_H(f_{t,t'}(z')) \tag{1.37}$$

(cf. formula (1.35)). This amounts to finding the value $z(t) = (x(t), p(t))$ at time t of the solution to Hamilton's equations with initial value $z' = (x', p')$

at time t'. The family $(f_{t,t'})$ of phase space transformations satisfies

$$f_{t,t'} \circ f_{t',t''} = f_{t,t''}, \quad (f_{t,t'})^{-1} = f_{t',t}, \quad f_{t,t} = I_{\mathrm{d}}, \qquad (1.38)$$

which express *causality* in classical mechanics. When the initial time t' is 0, we will write f_t instead of $f_{t,0}$ and call (f_t) the "time-dependent flow", but one must then be careful to remember that in general $f_t \circ f_{t'} \neq f_{t+t'}$.

We will often omit the composition sign $\circ$ and write $f_{t,t'} f_{t',t''}$, $f_t f_{t'}$, etc. instead of $f_{t,t'} \circ f_{t',t''}$, $f_t \circ f_{t'}$, etc.

The solutions of Hamilton's equations determine curves

$$t \longmapsto (x(t), p(t), t)$$

in the time-dependent phase space $\mathbb{R}_z^{2n} \times \mathbb{R}_t$. Hamilton's equations are trivially equivalent to the system of differential equations

$$\frac{d}{dt}(x, p, t) = \tilde{X}_H(x, p, t), \qquad (1.39)$$

where

$$\tilde{X}_H = (\nabla_p H, -\nabla_x H, 1)$$

is the *suspended Hamiltonian vector field*. The point in introducing the redundant variable t in (1.39) is that $\tilde{X}_H$ is now a "true" vector field, however not on $\mathbb{R}_z^{2n}$ but on $\mathbb{R}_z^{2n} \times \mathbb{R}_t$, to which we can apply the standard theory and terminology of autonomous systems, whereas, as observed above, the ordinary Hamilton vector field $X_H = (\nabla_p H, -\nabla_x H)$ is in general, strictly speaking, a family of vector fields indexed by time t.

The projections of the integral curves $t \mapsto (x(t), p(t), t)$ of $\tilde{X}_H$ on the ordinary phase space $\mathbb{R}_x^n \times \mathbb{R}_p^n$ are just the integral curves of X_H; the mappings $\tilde{f}_t$ defined by

$$\frac{d}{dt} \tilde{f}_t(z, t) = \tilde{X}_H(\tilde{f}_t(z, t))$$

automatically satisfy the one-parameter group property

$$\tilde{f}_t \circ \tilde{f}_{t'} = \tilde{f}_{t+t'}, \quad (\tilde{f}_t^{-1}) = \tilde{f}_{-t}. \qquad (1.40)$$

We will call the family $(\tilde{f}_t)$ the *suspended flow* determined by the Hamiltonian function H. There is a simple relation between the time-dependent flow $(f_{t,t'})$ defined above and the suspended flow, namely

$$(f_{t,t'}(x',p'),t) = \tilde{f}_{t-t'}(x',p',t'), \tag{1.41}$$

which is easily checked.

There are natural questions regarding the uniqueness and, indeed, existence of any solution of Hamilton's equations

$$\dot{z} = J\nabla_z H(z)$$

with an initial condition $z(0) = z_0$. Since Hamilton's equations form a differential system, these questions can be answered using the standard theory of differential equations. Below we give a sufficient (but not necessary) criterion when the Hamiltonian function is of the physical type "kinetic energy plus potential", but here is a very general result.

Proposition 11. *Suppose that the Hamilton vector field* $X_H = J\nabla_z H$ *vanishes outside some compact subset* Ω *of phase space* $\mathbb{R}_z^{2n}$. *Then the vector field* X_H *is complete, i.e., each solution of Hamilton's equations* $\dot{z} = X_H(z)$, $z(0) = z_0$ *is defined (and is unique) for each* $z_0 \in \Omega$, *and this solution exists for all times* t.

For a proof, see for instance [2, §4.1]. This result corresponds to the notion of well-defined Hamiltonian dynamics persisting eternally. Physically, this makes sense for a bounded Universe, since momenta cannot become infinite in view of special relativity. Mathematically, one can always reduce the situation to that of a bounded domain Ω using the following trick: let φ be a smooth function equal to one on Ω and equal to a constant outside an arbitrarily small neighborhood $\tilde{\Omega}$ of Ω. Then the vector field $X_H^\varphi = J\nabla_z(\varphi H)$ satisfies the hypotheses of the proposition above on Ω.

Here is another existence result which is sufficient for many applications to physics. We assume that the Hamiltonian is time-independent and of the usual type

$$H(x,p) = \sum_{j=1}^n \frac{p_j^2}{2m_j} + V(x),$$

where V is an infinitely differentiable potential.

Proposition 12. *If there exist constants* a *and* $b \geq 0$ *such that*

$$V(x) \geq a - b|x|^2,$$

then the solutions of Hamilton's equations

$$\frac{dx_j}{dt} = \frac{p_j}{m_j}, \quad \frac{dp_j}{dt} = -\frac{\partial V}{\partial x_j}(x)$$

$$x(0) = x_0, \quad p(0) = p_0$$

$(1 \leq j \leq n)$ *exist for all times (and are unique for each choice of initial condition (x_0, p_0)).*

Let us give a proof in the particular case $b = 0$; for a general proof, see [2, §4.1]. We thus assume that the potential is bounded from below: $V(x) \geq a$. Since Hamilton's equations are insensitive to the addition of a constant to the Hamiltonian, we may assume $a = 0$, and rescaling if necessary the momentum and position coordinates, it is no restriction neither to assume $m_j = 1$ for $1 \leq j \leq n$. In view of the local existence theory for ordinary differential equations, it suffices to show that the solutions $t \mapsto z(t)$ remain in bounded sets for finite times. For notational simplicity we moreover assume $n = 1$ (the argument is easily generalized to an arbitrary number of degrees of freedom). Let $t \mapsto (x(t), p(t))$ be a solution of the Hamilton equations

$$\frac{dx}{dt} = p, \quad \frac{dp}{dt} = -\frac{\partial V}{\partial x}(x)$$

and let $E = H(z(t))$ be the energy; since $H \geq V$ we have $E \geq V(x(t))$. In view of the triangle inequality we have

$$|x_j(t)| \leq |x(0)| + |x(t) - x(0)| \leq |x(0)| + \int_0^t \left| \frac{d}{dt} x(s) \right| ds;$$

since $\dot{x}(s) = p(s)$ and

$$p(t) = \sqrt{2(E - V(x(t)))} \leq \sqrt{2E}, \tag{1.42}$$

we have

$$|x(t)| \leq |x(0)| + \int_0^t |p(s)| ds$$

$$\leq |x(0)| + \int_0^t \sqrt{2(E - V(x(s)))} ds$$

so that

$$|x(t)| \leq |x(0)| + t\sqrt{2E}. \tag{1.43}$$

The inequalities (1.42) and (1.43) show that for t in any finite time-interval $[0, T]$ the functions $t \mapsto x(t)$ and $t \mapsto p(t) = \dot{x}(t)$, and hence $t \mapsto z(t)$, stay in a bounded set. Thus, the solutions cannot "go to infinity in a finite time".

1.3.2. The effect of a gauge transformation

Assume that the Hamiltonian function is of the particular type

$$H = \sum_{j=1}^{n} \frac{1}{2m_j} (p_j - A_j(x, t))^2 + V(x, t); \tag{1.44}$$

such Hamiltonians can be viewed as a generalization of the electromagnetic Hamiltonian (1.2), where $A = (A_1, \ldots, A_n)$ can be viewed as a generalized vector potential. The datum of a pair (A, V) is called a choice of gauge. We will call a gauge transformation any mapping

$$(A, V) \longmapsto \left(A + \nabla_x \chi, V - \frac{\partial \chi}{\partial t} \right), \tag{1.45}$$

where $\chi = \chi(x, t)$ is a smooth function of the variables x and t. A gauge transformation takes the Hamiltonian function (1.44) into the new function

$$H^{\chi} = \sum_{j=1}^{n} \frac{1}{2m_j} (p_j - A_j - \nabla_x \chi)^2 + V(x, t) - \frac{\partial \chi}{\partial t}. \tag{1.46}$$

While the Hamiltonian flows (f_t) and (f_t^{χ}) of H and H^{χ} are not identical, we observe that a gauge transformation does not affect the motion in configuration space.

Proposition 13. *The motion of a system in a gauge (A, V) is determined by the system of second-order differential equations*

$$\ddot{x}_j + \frac{1}{m_j} \left(\frac{\partial A_j}{\partial t} (x, t) + \frac{\partial V}{\partial x_j} (x, t) \right) = 0 \tag{1.47}$$

$(1 \leq j \leq n)$. *That system is invariant under every gauge transformation* $(A, V) \mapsto (A', V')$.

Proof. (Cf. [100, Proposition 21]). The Hamilton equations for H^χ are

$$\dot{x}_j = \frac{1}{m_j}(p_j - A_j) \quad \text{and} \quad \dot{p}_j = \frac{1}{m_j}(p_j - A_j)\frac{\partial A_j}{\partial x_j} - \frac{\partial V}{\partial x_j}.$$

Differentiating $\dot{x}_j$ with respect to t and then inserting the value of $\dot{p}_j$ given by the second set of equations, we get (1.47). That equation does not depend on the choice of gauge, for example, if we replace A_j by $A_j' = A_j + \frac{\partial \chi}{\partial x_j}$ and V by $V' = V - \frac{\partial \chi}{\partial t}$ we have

$$\frac{\partial A_j'}{\partial t} + \frac{\partial V'}{\partial x_j} = \frac{\partial A_j}{\partial t} + \frac{\partial V}{\partial x_j}$$

so that the left-hand side of (1.47) does not change. $\quad\square$

1.3.3. The symplectic character of Hamiltonian flows

We have been discussing the notion of Hamiltonian flow in a rather general way, and almost every definition and property we have given can to a great extent be generalized to any vector field on a Euclidean (or Riemannian) space. We will now demonstrate a very particular property, which is characteristic of Hamiltonian flows and systems. Let us introduce some notation and terminology.

We will call a *symplectomorphism* (the term canonical transformation is also used in the literature) any diffeomorphism f of the phase space $\mathbb{R}^{2n}_z$ whose Jacobian matrix is symplectic at every point where it is defined. That is, f is an infinitely differentiable and invertible mapping $\mathbb{R}^{2n}_z \to \mathbb{R}^{2n}_z$ whose inverse f^{-1} is also infinitely differentiable, and such that, for every z,

$$Df(z)^T J Df(z) = Df(z) J Df(z)^T = J,$$

where $Df(z)$ is the Jacobian matrix of f calculated at the point z. Recall that the Jacobian matrix $Df(z)$ is the $2n \times 2n$ matrix

$$Df(z') = \begin{pmatrix} \dfrac{\partial x}{\partial x'}(z') & \dfrac{\partial x}{\partial p'}(z') \\ \dfrac{\partial p}{\partial x'}(z') & \dfrac{\partial p}{\partial p'}(z') \end{pmatrix}, \tag{1.48}$$

where $(x, p) = f(x', p')$ and

$$\frac{\partial x}{\partial x'} = \frac{\partial(x_1, \ldots, x_n)}{\partial(x_1', \ldots, x_n')}$$

is the matrix of derivatives of $x = x(x', p')$ with respect to the variables $x' = (x_1', \ldots, x_n')$ and similar definitions for the other entries of (1.48).

Lemma 14. *Let $(f_{t,t'})$ be the time-dependent Hamiltonian flow determined by a Hamiltonian function H. The Jacobian matrix $S_{t,t'}(z') = Df_{t,t'}(z')$ satisfies the "variational equation"*

$$\frac{d}{dt}S_{t,t'}(z') = JH''(f_{t,t'}(z'),t)S_{t,t'}(z'), \ S_{t,t} = I, \qquad (1.49)$$

where $H''(f_{t,t'}(z'),t)$ is the Hessian matrix of H in the variables x,p calculated at time t at the point $z = f_{t,t'}(z')$.

Proof. Let us first prove that $S_{t,t'}(z')$ satisfies the variational equation (1.49). The time-derivative of the Jacobian matrix $S_{t,t'}(z')$ is

$$\frac{d}{dt}S_{t,t'}(z') = \frac{d}{dt}(Df_{t,t'}(z')) = D\left(\frac{d}{dt}f_{t,t'}(z')\right),$$

i.e.,

$$\frac{d}{dt}S_{t,t'}(z') = D(X_H(f_t^H(z'))),$$

where X_H is the Hamilton vector field. Using the fact that $X_H = J\nabla_z H$ together with the chain rule, we have

$$D(X_H(f_{t,t'}(z'))) = D(J\nabla_z H)(f_{t,t'}(z'))$$
$$= JD(\nabla_z H)(f_{t,t'}(z'))$$
$$= JH''(f_{t,t'}(z'))Df_{t,t'}(z'),$$

and hence $S_{t,t'}(z')$ satisfies the variational equation (1.49).   □

Proposition 15. *The Jacobian matrix $S_{t,t'}(z') = Df_{t,t'}(z')$ is symplectic for every z'. The time-dependent Hamiltonian flow $(f_{t,t'})$ thus consists of symplectomorphisms.*

Proof. Setting for simplicity $S_{t,t'} = S_{t,t'}(z')$ we consider the matrix $A_{t,t'} = S_{t,t'}^T J S_{t,t'}$. Differentiating $A_{t,t'}$ with respect to t and using the product rule for matrices we get, using formula (1.49),

$$\frac{dA_t}{dt} = \frac{d(S_{t,t'})^T}{dt} JS_{t,t'} + (S_{t,t'})^T J \frac{dS_{t,t'}}{dt}$$
$$= (S_{t,t'})^T H''(z) S_{t,t'} - (S_{t,t'})^T H''(z) S_{t,t'}$$
$$= 0.$$

It follows that the matrix $S_{t,t'}^T JS_{t,t'}$ is constant in t, hence, in particular, $S_{t,t}^T JS_{t,t} = J$ (because $S_{t,t}$ is the identity) so that $S_{t,t'}$ is symplectic for all $t \in \mathbb{R}$ as claimed. □

There are several alternative proofs of the fact that the matrices $S_{t,t'}(z')$ are symplectic; most of them use the language and notation of differential geometry (see for instance [2] or [23]). The advantage of the proof above is its relative simplicity, and the fact that it works both in the time-dependent and time-independent cases. It moreover demonstrates that the Jacobians of a Hamiltonian flow satisfy a first-order (nonlinear) matrix equation.

1.4. Liouville's Theorem and the Symplectic Camel

Liouville's theorem is a venerable topic from the theory of Hamiltonian systems (for applications to specific problems, see for instance [5, 92, 230]); we will see that it can be considerably improved as soon as $n > 1$ using results from modern symplectic geometry and topology.

1.4.1. Liouville's theorem

Suppose we are studying a physical system $\mathcal{S}$ consisting of a very large number N of particles, so large that it would be a hopeless task to keep track of each particle individually. What one can do, however, is to make measurements about average properties of certain quantities or characteristics associated with the system $\mathcal{S}$. This leads to the consideration of "particle densities" and probabilities, which one studies by using statistical methods. It turns out that the same procedure is used for the experimental study of a small number of non-interacting particles (or even of a single particle). This is because the only physical knowledge we can have of any "real" system is the result of measurements, which are essentially intervals of numbers. What one does in these cases, is to examine the properties of statistical ensembles which are large numbers of ideally identical systems, and to treat the data thus obtained again by probabilistic and statistical methods. For instance, suppose we want to describe the motion of a single particle under the action of some field. To obtain maximum precision, we must perform a great number of measurements of position and velocity, on similar particles, and this under conditions being ideally kept identical. We can then represent the results of our position and velocity measurements as a swarm of points in phase space, to which one can apply statistical methods. Furthermore, if the number of observations is very large, we can approximate this swarm of points with a "fluid" in phase space. We can therefore speak about the average density of that fluid: it is the average number of points per unit volume in phase space. We thus picture the fluid as a continuous system, i.e., we identify the swarm of points with a fluid having a continuously differentiable density $\rho(x, p, t)$ at the point (x, p) at time t.

Now, a fundamental postulate of classical statistical mechanics is that along each trajectory from $t \mapsto z(t)$ followed by a "particle" of the "fluid", the density function ρ satisfies Liouville's condition

$$\frac{d}{dt}\rho(z(t), t) = 0. \tag{1.50}$$

The idea of introducing probability densities in phase space is due to Josiah Willard Gibbs. He called the condition above the *principle of density in phase*. Following this principle, classical statistical mechanics becomes a theory in which the motion of particles (or systems) is deterministic, but unpredictable individually: the particles move in phase space as if they constituted an incompressible fluid of varying density. Note that conservation of volume does not, however, mean conservation of shape; this was called by Gibbs the *principle of extension in phase*. Liouville's condition can be motivated and justified heuristically in two different ways. We begin with a classical "particle counting" argument. Consider at time $t = 0$ a small volume Ω in phase space surrounding some given point-like particle. The boundary $\partial\Omega$ of Ω is formed by some surface of neighboring particles. In the course of time, the measure of the volume will remain constant in view of Liouville's theorem, although the volume Ω itself will be moved and distorted. Now, any particle inside Ω must remain inside Ω: if some particle were to cross the boundary of Ω it would occupy at some time the same position in phase space as one of the particles defining $\partial\Omega$. Since the subsequent motion of any particle is entirely and uniquely determined by its location in phase space at a given time, the two particles would then travel together from there on, but this is absurd, and the particle can thus never leave Ω. Reversing the argument, we also find that no particle can ever enter Ω, so that the total number of particles within Ω must remain constant. Summarizing, both the measure of the volume and the number of particles are constant, and so is thus the density, as claimed. Consequently, Liouville's condition (1.50) must hold.

A second possible interpretation of Liouville's condition is of a probabilistic nature. Assume that the total mass

$$m \equiv \int \rho(z,t) d^n z$$

is non-zero and finite. Choosing proper units we can choose $m = 1$ and thus assume that the normalization condition

$$\int \rho(z,t) \, d^n z = 1 \tag{1.51}$$

holds for all t. This allows us to view ρ as a probability density. In fact, if we consider, as in the argument above, an "infinitesimal volume" Ω with measure ΔV around the point z, then $\rho(z,0)\Delta V$ will be the probability of finding a given particle inside Ω. Assuming that the time-evolution is governed by some Hamiltonian function H with flow (f_t), the probability of finding that particle in the image Ω_t of Ω by the flow (f_t) is then

$$\rho(z(t),t)\Delta V_t = \rho(z,0)\Delta V \tag{1.52}$$

and since $\Delta V_t = \Delta V$ because volume is preserved, we must have

$$\rho(z(t),t) = \rho(z,0) \tag{1.53}$$

which is again Liouville's condition. We see that, either way, Liouville's condition (1.50) appears to be a *conservation law*: no particles in phase space can be created or destroyed in classical statistical mechanics.

1.4.2. Mathematical statement of Liouville's theorem

Let us now do some rigorous work. We will follow closely Arnold [5].

We begin by making the following observation: let M be any square matrix with dimension, say, m. For t small we have

$$\det(I + tM) = 1 + t \operatorname{Tr} M + O(t^2), \tag{1.54}$$

where $\operatorname{Tr} M$ is the trace of M (it is the sum of the diagonal elements of M) and the remainder $O(t^2)$ is a function going to zero at least as fast as t^2. This formula is easily obtained by expanding the determinant in the usual way: this leads to 1 plus m terms in t, and the remaining terms then contain powers t^k with $k \geq 2$. Suppose now that we have a system of m ordinary equations $\dot{y} = g(y,t)$ ($y = (y_1, \ldots, y_m)$) where $g = (g_1, \ldots, g_m)$ is a vector-valued function on $\mathbb{R}^m$; following the general principles expressed earlier, this system determines a time-dependent flow $(f_{t,t'})$. We will need the following general technical lemma.

Lemma 16. *Let $\Omega_{t'}$ be a region with finite volume $V(t') = \mathrm{Vol}(\Omega_{t'})$ in $\mathbb{R}^m$. Let $\Omega_t = \varphi_{t,t'}(\Omega_{t'})$ be the image of $\Omega_{t'}$ by the flow-mapping $\varphi_{t,t'}$ determined by the system $\dot{y} = g(y,t)$. Then*

$$\dot{V}(t) = \int_{\Omega_t} \mathrm{div}\, g(y,t) d^m y, \tag{1.55}$$

where $\dot{V}(t)$ is the derivative $dV(t)/dt$ calculated at time t'.

Proof. We have

$$V(t) = \int_{\Omega_t} d^m y = \int_{\Omega_{t'}} \det\left(\nabla_y f_{t,t'}(y)\right) d^m y, \tag{1.56}$$

where the second equality follows from the formula for the change of variables in a multiple integral. We next observe that a Taylor expansion of the function $t \rightarrow \varphi_{t,t'}(y)$ at t' yields, taking the relations $\dot{y} = g(y,t)$ and $\varphi_{t',t'}(y) = y$ into account,

$$f_{t,t'}(y) = y + g(y,t')(t - t') + O((t - t')^2).$$

It follows that

$$\nabla_y f_{t,t'}(y) = I + \nabla_y g(y,t')(t - t') + O((t - t')^2),$$

and hence, using formula (1.54),

$$\begin{aligned}
\det\left(\nabla_y f_{t,t'}(y)\right) &= \det[I + \nabla_y g(y,t')(t - t') + O((t - t')^2)] \\
&= 1 + (t - t')\mathrm{Tr}\, \nabla_y g(y,t') + O((t - t')^2) \\
&= 1 + (t - t')\mathrm{div}\, g(y,t') + O((t - t')^2).
\end{aligned}$$

Insertion of this expression in the right-hand side of (1.56) yields

$$V(t) = \int_{\Omega_{t'}} d^m y + (t - t') \int_{\Omega_{t'}} \mathrm{div}\, g(y,t') d^m y + O((t - t')^2) \int_{\Omega_{t'}} d^m y.$$

The first integral on the right-hand side being just $V(t')$ and this can be rewritten

$$\frac{V(t) - V(t')}{t - t'} = \int_{\Omega_{t'}} \operatorname{div} g(y, t') d^m y + V(t') O(t - t')$$

hence formula (1.55) letting $t \to t'$. □

Using this lemma, we obtain Liouville's theorem and equation (which are actually both due to Gibbs!).

Proposition 17. *Let H be a time-dependent Hamiltonian function and $(f_{t,t'})$ its time-dependent flow. Let $\Omega_{t'}$ be a measurable region in phase space at time t'. Then*

$$\operatorname{Vol}[f_{t,t'}(\Omega_{t'})] = \operatorname{Vol}(\Omega_{t'}) \tag{1.57}$$

for all times t for which $f_{t,t'}$ is defined.

Proof. The time-dependent flow is determined by Hamilton's equations $\dot{z} = X_H(z, t)$. Replacing y with z and $g(y, t)$ with $X_H(z, t)$ (and setting $m = 2n$) we are thus in the situation described by the lemma above. Setting $V(t) = \operatorname{Vol}[f_{t,t'}(\Omega_{t'})]$ it follows that formula (1.55) becomes

$$\dot{V}(t') = \int_{\Omega_{t'}} \operatorname{div} X_H(z, t') d^m z.$$

We now observe that in view of the definition $X_H = (\nabla_p H, -\nabla_x H)$ of the Hamilton vector field we have

$$\operatorname{div} X_H(z, t') = \nabla_x(\nabla_p H) - \nabla_p(\nabla_x H) = 0$$

and hence $\dot{V}(t') = 0$. Since the choice of the time t' is arbitrary we thus have $\dot{V}(t) = 0$ for all t and the volume $V(t)$ is thus constant, which proves formula (1.57). □

Closely related to the Liouville–Gibbs theorem is Liouville's equation, which describes the time evolution of densities. Suppose we have a phase space distribution ρ determining the probability that a Hamiltonian system will be found in some infinitesimal phase space volume (see the discussion in the previous section). The result is elegantly expressed if one introduces the

Poisson bracket notation: if $F = F(z,t)$, $G = G(z,t)$ are two differentiable functions defined on extended phase space their Poisson bracket is

$$\{F, G\} = \nabla_x F \cdot \nabla_p G - \nabla_p F \cdot \nabla_x G. \tag{1.58}$$

Equivalently

$$\{F, G\} = \sum_{j=1}^{n} \frac{\partial F}{\partial x_j} \frac{\partial G}{\partial p_j} - \frac{\partial F}{\partial p_j} \frac{\partial G}{\partial x_j}.$$

Proposition 18. *Let $\rho' = \rho'(z)$ be the datum of a probability density on $\mathbb{R}^{2n}$ and at time t' and set*

$$\rho(z, t) = \rho'(f_{t,t'}^{-1}(z)) = \rho'(f_{t',t}(z)), \tag{1.59}$$

where $(f_{t,t'})$ is the time-dependent flow of some Hamiltonian function H. The function ρ satisfies the Liouville equation

$$\frac{\partial \rho}{\partial t}(z, t) + \{\rho(z, t), H(z, t)\} = 0. \tag{1.60}$$

Proof. We first observe that formula (1.59) is equivalent to

$$\rho(f_{t,t'}(z'), t) = \rho'(z'). \tag{1.61}$$

We have, using the chain rule,

$$\frac{d}{dt}\rho(f_{t,t'}(z'), t) = \nabla_z \rho(f_{t,t'}(z'), t) \frac{d}{dt} f_{t,t'}(z') + \frac{\partial \rho}{\partial t}(f_{t,t'}(z'), t)$$

$$= \nabla_z \rho(f_{t,t'}(z'), t) X_H(f_{t,t'}(z'), t) + \frac{\partial \rho}{\partial t}(f_{t,t'}(z'), t).$$

Since $X_H = (\nabla_p H, -\nabla_x H)$ we have

$$\nabla_z \rho(f_{t,t'}(z'), t) X_H(f_{t,t'}(z'), t)$$

$$= \nabla_x \rho(f_{t,t'}(z'), t) \nabla_p H(f_{t,t'}(z'), t) - \nabla_p \rho(f_{t,t'}(z'), t) \nabla_x H(f_{t,t'}(z'), t),$$

that is

$$\nabla_z \rho(f_{t,t'}(z'), t) X_H(f_{t,t'}(z'), t) = \{\rho(f_{t,t'}(z'), t), H(f_{t,t'}(z'), t)\}$$

and hence

$$\frac{d}{dt}\rho(f_{t,t'}(z'),t) = \{\rho(f_{t,t'}(z'),t), H(f_{t,t'}(z'),t)\} + \frac{\partial}{\partial t}\rho(f_{t,t'}(z'),t).$$

One then concludes setting $z = f_{t,t'}(z')$ and noting that we have

$$\frac{d}{dt}\rho(f_{t,t'}(z'),t) = 0$$

in view of formula (1.61). $\qquad\qquad\qquad\qquad\qquad\qquad\qquad\qquad\qquad\qquad$ □

1.4.3. The symplectic camel

Assume again that we are dealing with a physical system $\mathcal{S}$ consisting of a very large number of point-like particles forming a "cloud" filling a subset Ω of phase space $\mathbb{R}^{2n}_z$. Suppose that this cloud is, at time $t = 0$, spherical so Ω is a phase space ball $B^{2n}(r) : |z - z_0| \leq r$. The orthogonal projection of that ball on any plane of coordinates will always be a circle with area πr^2. As time evolves, the cloud of points will distort and may take after a while a very different shape, while keeping constant volume. In fact, since conservation of volume has nothing to do with conservation of shape, one might very well envisage that the ball $B^{2n}(r)$ can be stretched in all directions by the Hamiltonian flow (f_t), and eventually get very thinly spread out over huge regions of phase space, so that the projections on any plane could *a priori* become arbitrarily small after some time t. This possibility is perfectly consistent with a rather subtle result due to Katok [171], which can be stated as follows: consider two bounded domains Ω and Ω' in $\mathbb{R}^{2n}$ which are both diffeomorphic to the ball $B^{2n}(r)$ and have the same volume. Then, for every $\varepsilon > 0$ there exist a Hamiltonian function H and a time t such that $\mathrm{Vol}(f_t(\Omega)\Delta\Omega') < \varepsilon$. Here $f_t(\Omega)\Delta\Omega'$ denotes the symmetric difference of the two sets $f_t(\Omega)$ and Ω': it is the set of all points that are in $f_t(\Omega)$ or Ω', but not in both. Katok's lemma thus shows that up to sets of arbitrarily small measure ε any kind of phase space spreading is *a priori* possible for a volume-preserving flow.

However (and this was unknown until Gromov's breakthrough in 1985), the orthogonal projection of the set $f_t(B^{2n}(r))$ on any plane of *conjugate coordinates* x_j, p_j will never decrease below its original value! That projection will in fact be a surface in this plane, and that surface will have an area of at least πr^2. The condition that the plane is a plane of conjugate coordinates is essential: had we chosen a plane of coordinates x_j, p_k with $j \neq k$ or any plane of coordinates x_j, x_k or p_j, p_k then the projection could become arbitrarily small. The property just described is a consequence (in fact, an equivalent statement) of the "symplectic non-squeezing theorem".

Gromov's non-squeezing theorem is very surprising and has many direct and indirect consequences. For instance, not so long ago many people used to believe that whatever could be done by a volume-preserving diffeomorphism could be done (at least to an arbitrarily good approximation) by symplectomorphisms (i.e., canonical transformations). This belief was an extrapolation from the case $n = 1$ where both notions in fact coincide. The first step towards a better understanding of the peculiarities of symplectomorphisms was formulated in the early 1970s under the name of "Gromov's alternative": the group $\text{Symp}(2n, \mathbb{R})$ of symplectomorphisms is either C^0-closed in the group of all diffeomorphisms, or its closure in the C^0-topology is the group of volume-preserving diffeomorphisms. One consequence of the symplectic non-squeezing theorem is that it is the first possibility that holds true: in general we cannot approximate a volume-preserving transformation by a sequence of symplectomorphisms; this shows that symplectomorphisms really are in a sense very different in nature (and properties) from general Liouville-type mappings.

Let us denote by $Z_j^{2n}(R)$ the cylinder in $\mathbb{R}^{2n}$ with radius R and based on the plane of x_j, p_j coordinates:

$$Z_j^{2n}(R) = \{(x, p) : x_j^2 + p_j^2 \leq R^2\}.$$

What Gromov proved in 1985, using complex analysis (the theory of pseudo-holomorphic curves) is the following crucial result, which is the act of birth of modern symplectic topology.

Proposition 19 (Gromov). *If there exists a symplectomorphism f in $\mathbb{R}^{2n}$ sending the ball $B^{2n}(r)$ in some cylinder $Z_j^{2n}(R)$, then we must have $r \leq R$.*

A word of explanation for the word "some" is the statement of Gromov's theorem. Let $Z_j^{2n}(R)$ and $Z_k^{2n}(R)$ be two cylinders with $j \neq k$. A phase space point $z = (x, p)$ is in $Z_j^{2n}(R)$ if and only if $x_j^2 + p_j^2 \leq R^2$ and in $Z_k^{2n}(R)$ if and only if $x_k^2 + p_k^2 \leq R^2$; we can switch from one condition to the other by the coordinate permutation $\tau_{j,k}$ swapping the pairs (x_j, p_j) and (x_k, p_k) and leaving all other pairs (x_ℓ, p_ℓ) invariant. Since $\tau_{j,k}$ obviously leaves $\sigma(z, z')$ unchanged, it is a (linear) symplectomorphism. Thus, if f sends $B^{2n}(r)$ in $Z_j^{2n}(R)$ then $\tau_{j,k} \circ f$ sends $B^{2n}(r)$ in $Z_k^{2n}(R)$ so that there exists a symplectomorphism sending $B^{2n}(r)$ in $Z_j^{2n}(R)$ if and only if there exists one sending $B^{2n}(r)$ in $Z_k^{2n}(R)$. The property of Hamiltonian flows described above immediately follows: assume that the orthogonal projection of $f_t(B^{2n}(r))$ on the x_j, p_j plane has an area $\leq \pi r^2$. Then any cylinder based on that plane and having a radius $\geq r$ would contain $f_t(B^{2n}(r))$.

It is essential for the non-squeezing theorem to hold that the considered cylinder is based on a x_j, p_j plane (or, more generally, on a symplectic plane). For instance, if we replace the cylinder $Z_j^{2n}(R)$ by the cylinder $Z_{12}^{2n}(R) : x_1^2 + x_2^2 \leq R^2$ based on the x_1, x_2 plane, it is immediate to check that the linear symplectomorphism f defined by $f(x, p) = (\lambda x, \lambda^{-1} p)$ sends $B^{2n}(r)$ into $Z_{12}^{2n}(R)$ as soon as $\lambda \leq r/R$. Also, one can always "squeeze" a large ball into a big cylinder using volume-preserving diffeomorphisms that are not canonical. Here is an example in the case $n = 2$ that is very easy to generalize to higher dimensions: let the linear mapping f be defined by the matrix

$$
M = \begin{pmatrix} \lambda & 0 & 0 & 0 \\ 0 & \lambda^{-1} & 0 & 0 \\ 0 & 0 & \lambda & 0 \\ 0 & 0 & 0 & \lambda^{-1} \end{pmatrix}.
$$

Clearly $\det M = 1$ and f is hence volume-preserving; f is however not symplectic if $\lambda \neq 1$. Choosing again $\lambda \leq r/R$, the mapping f sends $B^{2n}(R)$ into $Z_1^{2n}(r)$.

Gromov's theorem actually holds when $Z_j^{2n}(R)$ is replaced by any cylinder with radius R based on a symplectic plane, i.e., a two-dimensional subspace $\mathcal{P}$ of $\mathbb{R}^{2n}$ such that the restriction of σ to $\mathcal{P}$ is non-degenerate (equivalently, $\mathcal{P}$ has a basis $\{e, f\}$ such that $\sigma(e, f) \neq 0$). The planes $\mathcal{P}_j$ of coordinates x_j, p_j are of course symplectic, and given an arbitrary symplectic plane $\mathcal{P}$ it is easy to construct a linear symplectomorphism S_j such that $S_j(\mathcal{P}) = \mathcal{P}_j$. It follows that a symplectomorphism f sends $B^{2n}(r)$ in the cylinder $Z_j^{2n}(R)$ if and only if $f \circ S_j$ sends $B^{2n}(r)$ in the cylinder $Z^{2n}(R)$ with the same radius based on $\mathcal{P}$.

1.5. Geometric Optics and Hamiltonian Mechanics

Geometric optics (also called ray optics) views light as having a corpuscular nature; its propagation can be defined in terms of straight lines (rays), which are the trajectories of these corpuscles. The concern of geometric optics is the location and direction of these rays. Geometric optics can be viewed as the short-wave limit of physical optics, where interference and other wave phenomena can be neglected. A particularly simple theory of geometric optics is the paraxial linear approximation, leading to Gaussian optics. For a modern detailed study of optics we refer to [5, 139, 140] which give quite rigorous accounts; Holm [162] (Chapter 1) might be more accessible to physicists less interested in mathematical rigor. Also see the review paper [209] by Masoliver and Ros, where a succinct but well-written account of the optical-mechanical analogy and its evolution is given.

1.5.1. The optical Hamiltonian

Let us begin with a discussion of the optical-mechanical analogy which historically goes back to William Rowan Hamilton [144] who, by a real *tour de force*, proposed an analogy between the trajectory of material particles and the path of light rays. As a special illuminating case, we study the paraxial approach which is valid when the angle θ between the ray and optical axis is sufficiently small, justifying the linear approximations $\sin\theta \approx \theta$ and $\cos\theta \approx 1$. Roughly speaking, paraxial optics applies when the transversal size is small compared with the longitudinal size of the objects, images and constructive parameters of the optical system; a rule of thumb in optics is that the paraxial approximation is valid when the angle θ is less than $30°$. For an image forming system, paraxial optics corresponds to a "perfect" system. We will see that the mathematical study of both geometric and paraxial optics provides an easy application of elementary symplectic geometry.

We consider a three-dimensional optical medium (air, vacuum, glass, etc.) in which the speed of light is a function $\mathbf{v} = \mathbf{v}(\mathbf{r})$ of position; here $\mathbf{r} = (x, y, z)$. The *index of refraction* (or refractive index) at a point $M(\mathbf{r})$ of that medium is the quotient

$$n(\mathbf{r}) = \frac{c}{v(\mathbf{r})}, \quad v(\mathbf{r}) = |\mathbf{v}(\mathbf{r})|,$$

where c is the speed of light in vacuum; we thus always have $n \geq 1$. Suppose now that a light ray originates at an initial point M of the medium and moves in a direction specified by a unit vector $\mathbf{u}$. The ray arrives at a final point M' in a direction specified by a final unit vector $\mathbf{u}'$. The "imaging problem" of geometric optics is the problem of the determination of the relation between the pairs $(M, \mathbf{u})$ and $(M', \mathbf{u}')$. Assuming that the coordinates x, y, z can be parameterized in a piecewise continuously differentiable way:

$$x = x(t), \quad y = y(t), \quad z = z(t), \tag{1.62}$$

the *optical length* along a ray from M to M' is the line integral

$$L(M, M') = \int_{t_M}^{t_{M'}} n(x, y, z)\sqrt{1 + \dot{x}^2 + \dot{y}^2 + \dot{z}^2}\, dt, \tag{1.63}$$

where M and M' have respective coordinates $(x(t_M), y(t_M), z(t_M))$ and $(x(t_{M'}), y(t_{M'}), z(t_{M'}))$. Now, a basic law of optics is Fermat's principle which says that the optical path minimizes or maximizes the time of propagation of light between two points. Thus, among all curves (1.62), the choices that correspond to optical paths (i.e., possible light rays) for fixed initial and end points M and M' are those which maximize or minimize the integral (1.63). One can show, using standard variational methods, that Fermat's principle is equivalent to the Euler–Lagrange equations

$$\frac{d}{dt}\left(\frac{\partial L}{\partial \dot{x}}\right) - \frac{\partial L}{\partial x} = 0,$$

$$\frac{d}{dt}\left(\frac{\partial L}{\partial \dot{y}}\right) - \frac{\partial L}{\partial y} = 0, \tag{1.64}$$

$$\frac{d}{dt}\left(\frac{\partial L}{\partial \dot{z}}\right) - \frac{\partial L}{\partial z} = 0$$

(L is called the *optical Lagrangian* when $\dot{x}$ and $\dot{y}$ are viewed as independent variables). The conjugate momenta are defined by the formulas

$$p_x = \frac{\partial L}{\partial \dot{x}} = \frac{n(x, y, z)\dot{x}}{\sqrt{1 + \dot{x}^2 + \dot{y}^2 + \dot{z}^2}},$$

$$p_y = \frac{\partial L}{\partial \dot{y}} = \frac{n(x, y, z)\dot{y}}{\sqrt{1 + \dot{x}^2 + \dot{y}^2 + \dot{z}^2}}, \tag{1.65}$$

$$p_z = \frac{\partial L}{\partial \dot{z}} = \frac{n(x, y, z)\dot{z}}{\sqrt{1 + \dot{x}^2 + \dot{y}^2 + \dot{z}^2}};$$

setting $\mathbf{p}^2 = p_x^2 + p_y^2 + p_z^2$, $\mathbf{r}^2 = x^2 + y^2 + z^2$ and $\dot{\mathbf{r}}^2 = \dot{x}^2 + \dot{y}^2 + \dot{z}^2$ we thus have

$$\mathbf{p}^2 - n(\mathbf{r})^2 = -\frac{n(\mathbf{r})^2}{1 + \dot{\mathbf{r}}^2}, \tag{1.66}$$

and hence we can rewrite the formulas (1.65) as

$$\mathbf{p} = \sqrt{n(\mathbf{r})^2 - \mathbf{p}^2}\,\dot{\mathbf{r}}. \tag{1.67}$$

It follows that the velocity vector is given by

$$\mathbf{v} = \dot{\mathbf{r}} = \frac{\mathbf{p}}{\sqrt{n(\mathbf{r})^2 - \mathbf{p}^2}}. \tag{1.68}$$

The *optical Hamiltonian* H is obtained from the Lagrangian L by a Legendre transform:

$$H = \mathbf{p} \cdot \dot{\mathbf{r}} - L; \tag{1.69}$$

using formulas (1.68) this yields the explicit expression

$$H = -\sqrt{n(\mathbf{r})^2 - \mathbf{p}^2}. \tag{1.70}$$

Euler–Lagrange's equations (1.64) are then equivalent to Hamilton's equations

$$\dot{\mathbf{r}} = \nabla_{\mathbf{p}} H, \quad \dot{\mathbf{p}} = -\nabla_{\mathbf{r}} H. \tag{1.71}$$

When the index of refraction is constant, $n(\mathbf{r}) = n$, these equations reduce to

$$\dot{\mathbf{r}} = \frac{\mathbf{p}}{\sqrt{n(\mathbf{r})^2 - \mathbf{p}^2}}, \quad \dot{\mathbf{p}} = \mathbf{0}$$

and their solutions are

$$\mathbf{r}(t) = \mathbf{r}(0) + \mathbf{v}(0)t, \quad \mathbf{p}(t) = \mathbf{p}(0),$$

where the initial velocity vector $\mathbf{v}(0)$ is given by (1.68). We see that the light rays are *straight lines*.

1.5.2. Paraxial optics

We now consider a simple optical system consisting of refracting surfaces separated by regions where the refraction index remains constant. A typical example is a "cascade" of lenses separated by vacuum, or air and the surface of the sea. A light ray entering such a device will propagate along a broken line, as it is being refracted by the various surfaces separating the media with constant index. We next make the following three simplifying restrictions: first, we exclusively consider optical systems where all the refracting surfaces are rotationally symmetric about the optical axis. This hypothesis is for instance satisfied by a sequence of ordinary parallel round lenses. Moreover, we suppose that all rays lie in a plane containing the optical axis. Finally, and *this* is indeed a serious restriction, we only study light rays that travel at small inclinations around the optical axis. More specifically, we assume that the angles that the rays form with that axis are so small that we can disregard their squares in all our calculations, thus neglecting all terms of order two, or higher, which appear in the expansions of the trigonometric functions of these angles. Such rays are called *paraxial rays* in optics. Using the small angle approximation $\sin \theta \approx \theta$, we replace the exact version of Snell's law of refraction

$$n \sin \theta = n' \sin \theta' \qquad (1.72)$$

(which is deduced from Fermat's principle) by its linear approximation

$$n\theta = n'\theta'. \qquad (1.73)$$

Here n and n' are the respective refraction indices of two adjacent regions; the speed of light in these regions is thus $v = c/n$ and $v' = c/n'$. The angles θ and θ' are the angles of the light ray with the normal to the surface separating these two regions.

Let us now choose an origin O and a length unit on the optical axis, which we identify with the z-axis. By definition, a "reference line" $z = z_0$ is a plane orthogonal to the optical axis and passing through z; it is thus a plane which is parallel to the x, y-axis. We next introduce coordinates on each reference line. This can be done by specifying a ray by two numbers as it passes through the line z. These numbers are the height q of the point above the optical axis where the ray hits the line z, and the quantity $p_x = nq$, where n is the index of refraction at that point; q is the angle of the ray with the optical axis. It will become clear in a moment why we are choosing nq, and not q as a variable. Suppose now that we pick one reference line z' at the entrance of the optical system, and another, z, at the exit. We will call these particular reference lines the "input line" and the "output line". We can thus specify the ray by the two coordinates (x', p'_x) when it enters the system by the input line, and by (x, p_x) when it leaves it by the output line. We now want to know what type of dependence can be expected between (x', p'_x) and (x, p_x). It is actually not difficult to see that the relation must be linear, because we are using the first-order formulation (1.73) of Snell's law. Thus, the new coordinates (x, p_x) are related to the old coordinates (x', p'_x) by a formula of the type

$$\begin{pmatrix} x \\ p_x \end{pmatrix} = \begin{pmatrix} A & B \\ C & D \end{pmatrix} \begin{pmatrix} x' \\ p'_x \end{pmatrix}, \tag{1.74}$$

where A, B, C and D are some real numbers depending on both the optical system, and on the reference lines z and z' that are being used. The 2×2 matrix appearing in (1.74) is called the optical (or ray-transfer) matrix of the system relative to the reference lines z and z'. The terminology "ABCD matrix" also appears in the literature. Of course, the choice of two reference lines, one at the "input", the other at the "output", is arbitrary. In fact, by viewing the optical system as a juxtaposition of adjacent subsystems, we are actually free to choose as many intermediate reference lines as we like, and we can describe the light ray when it passes through each of these planes. If there are, for instance, three lines z, z' and z'', and if we denote by

$$\begin{pmatrix} A & B \\ C & D \end{pmatrix}, \quad \begin{pmatrix} A' & B' \\ C' & D' \end{pmatrix}, \quad \text{and} \quad \begin{pmatrix} A'' & B'' \\ C'' & D'' \end{pmatrix} \tag{1.75}$$

the optical matrices relative to (z', z), (z', z''), and (z'', z), respectively, then these matrices are related by the product formula

$$\begin{pmatrix} A & B \\ C & D \end{pmatrix} = \begin{pmatrix} A' & B' \\ C' & D' \end{pmatrix} \begin{pmatrix} A'' & B'' \\ C'' & D'' \end{pmatrix} \tag{1.76}$$

(to the first subsystem corresponds the first matrix on the right). It follows that the most general optical matrix can be reduced to the calculation of products of matrices

$$\begin{pmatrix} A_1 & B_1 \\ C_1 & D_1 \end{pmatrix} \begin{pmatrix} A_2 & B_2 \\ C_2 & D_2 \end{pmatrix} \cdots \begin{pmatrix} A_N & B_N \\ C_N & D_N \end{pmatrix}$$

corresponding to arbitrarily small parts of the optical system under consideration.

Let us determine the optical matrices in the following elementary cases (for explicit calculations see for instance the first chapter of Guillemin and Sternberg [140]).

Propagation through free space: A light ray travels in a straight line between two reference lines z, z' in the same medium with index n. The effect of propagation is here to translate the location of the ray in proportion to the angle at which it travels leaving this angle unchanged. The optical matrix is

$$u_d = \begin{pmatrix} 1 & d \\ 0 & 1 \end{pmatrix} \quad \text{with } d = (z - z')/n \qquad (1.77)$$

(the number d is called the "reduced distance").

Refraction by a surface: A light ray is refracted by a curve $z = f(x)$ separating two regions with constant indices n' and n. Assume that the right and left reference planes are "infinitely close" to the surface, so that the free propagation effects in the case above can be neglected. Rotational symmetry requires that $f(-x) = f(x)$ hence $f'(0) = 0$ and by Taylor's formula, $f(x) = f(0) + \frac{1}{2}f''(0)x^2 + O(x^3)$. Neglecting terms $O(x^3)$ we may thus assume that the curve is a parabola

$$z = z' + \tfrac{1}{2}kx^2,$$

where k is a constant associated with the curvature of the surface at its intersection with the optical axis. The position of the ray remains unchanged but the angles on each side of the surface satisfy the approximate Snell's law (1.73). Suppose the initial medium has index of refraction n and the final medium has index of refraction n'. The optical matrix is then

$$V_{-P} = \begin{pmatrix} 1 & 0 \\ -P & 1 \end{pmatrix} \quad \text{with } P = \frac{n' - n}{k}; \qquad (1.78)$$

the number P is called the "lens power" in optics.

Passage through a thin lens: A thin lens is obtained by using two close spherical surfaces; the inside of the lens has refractive index n' and the lens is imbedded in a medium with index n. Representing the optical matrices of the surfaces on the left and the right by two matrices of the type (1.78) and neglecting the travel of the light ray between these two surfaces, the optical matrix of the thin lens is

$$V_{-P}V_{-P'} = \begin{pmatrix} 1 & 0 \\ -(P + P') & 1 \end{pmatrix} = V_{-(P+P')}.$$

This formula is often called the "lens-maker's equation" because it tells which combination of refractive index and radii of curvature will give a lens of a desired focal length.

We note that all the elementary matrices described above have determinant one; hence the matrix associated with an arbitrary optical system will also have determinant one, because it can be written as a product of matrices of these two types. It is not difficult to prove that, conversely, every unimodular matrix can be factorized as a product of matrices (1.77) and (1.78); since the proof follows from a more general situation, we defer it. It follows that the unimodular group $S\ell(2, \mathbb{R})$ is the natural reservoir for all optical matrices.

1.5.3. Optical matrices in higher dimensions

We are now dropping the assumption of rotational symmetry (but we still assume that the angles of the rays with the optical axis are small). To see what happens we closely follow the argument in the first chapter of Guillemin and Sternberg [140]. In specifying a light ray we now need twice as many variables: the coordinates x and y which are needed to indicate where the ray intersects a plane transverse to the optical axis (again identified with the z-axis) and the angles θ_x and θ_y which govern the direction of the ray. Now, in three-dimensional space $\mathbb{R}^3_{x,y,z}$ a direction is specified by a normalized vector $\mathbf{v} = (v_x, v_y, v_z)$, $|\mathbf{v}| = 1$. If that vector is close to the positive z-axis, it will have the form $\mathbf{v} = (\theta_x, \theta_y, v_z)$ with $v_z \approx 1 - \frac{1}{2}(\theta_x^2 + \theta_y^2) \approx 1$ since θ_x and θ_y are assumed to be small. Assuming that the medium is isotropic we set $p_x = n\theta_x$ and $p_y = n\theta_y$. Considering two planes z and z' the ray will now correspond to the vectors

$$
u = \begin{pmatrix} x \\ y \\ p_x \\ p_y \end{pmatrix}, \quad u' = \begin{pmatrix} x' \\ y' \\ p'_x \\ p'_y \end{pmatrix},
$$

respectively. Let us find the relationship between these two vectors. Since we are ignoring quadratic and higher order terms, this relation will be linear; that is, there exists a 4×4 matrix S such that $u' = Su$. As in the rotationally symmetric case previously discussed, every optical system can be viewed as a sequence of free propagations and of refractions, so it suffices to consider these two cases. Assume first that a light ray travels in a straight line between two reference planes z, z' in the same medium with index n. The effect of propagation is now to translate the location of the ray in proportion to the angles θ_x and θ_y at which it travels leaving these angles unchanged. Ignoring quadratic and higher order terms in v_x and v_y this is the same as moving a distance $z - z'$ along the line supporting $\mathbf{v}$, that is $x - x' = (z - z')v_x$ and $y - y' = (z - z')v_y$. The optical matrix is thus here

$$U_d = \begin{pmatrix} I & dI \\ 0 & I \end{pmatrix} \quad \text{with } d = (z - z')/n, \qquad (1.79)$$

where I is the 2×2 identity matrix. Next consider the case of refraction by a surface. It is sufficient to assume that the surface is given by the formula

$$z' = z + \tfrac{1}{2}M\mathbf{r} \cdot \mathbf{r},$$

where $\mathbf{r} = (x, y)$ and $M = M^T$ is a 2×2 matrix. One shows that in this case the optical matrix is

$$V_{-P} = \begin{pmatrix} I & 0 \\ -P & I \end{pmatrix} \quad \text{with } P = -(n - n')M; \qquad (1.80)$$

the matrix P is symmetric since M is.

It turns out that the concept of generating function has a straightforward interpretation in terms of the mechanical-optical analogy. The optical length of a ray proceeding from x' to x in a medium of index n is here $L = n\ell$ where ℓ is the actual length. Write

$$L(x, x'; z, z') = n(z - z') + \Delta L(x, x'; z, z'), \qquad (1.81)$$

where $n(z - z')$ is the optical length of a ray proceeding exactly along the optical axis; the term $\Delta L(x, x'; z, z')$ measures the deviation of the "true" optical length to that of a perfectly coaxial ray. In the paraxial approximation, we have

$$\Delta L(x, x'; z, z') = n\frac{(x - x')^2}{2(z - z')} \qquad (1.82)$$

which is immediately identifiable with the generating function

$$W(x, x'; z, z') = m\frac{(x - x')^2}{2(z - z')} \qquad (1.83)$$

of the free particle. This is perfectly in accordance with the assumptions made for the paraxial approximation where p_x is viewed as small; this allows us to replace the optical Hamiltonian

$$H = -\sqrt{n^2 - p_x^2} = -n\left(1 - \frac{p_x^2}{2n^2}\right) + O(p_x^2)$$

with its linear approximation

$$H \approx \frac{p_x^2}{2n} - n.$$

Replacing the refractive index n with mass m and disregarding the constant $-n$ this is just the free particle Hamiltonian function.

1.5.4. The passage from ray optics to wave optics

We have seen that the motion of the light corpuscles is governed by symplectic geometry. Various experiences show that light actually also has a wave-like behavior; this leads us to postulate that there is an optical wavefunction ψ. We write such a wavefunction in polar form

$$\psi(x,t) = a(x,t)e^{i\Phi(x,t)} \tag{1.84}$$

which we propose to determine. We argue as follows: as light propagates from the point x' to the point x, it will undergo both a phase change and attenuation. The phase change is simply

$$\Delta\Phi = \frac{2\pi L}{\lambda}, \tag{1.85}$$

where the quantity L is the optical length of the trajectory from x' to x. Using the expression (1.81), we get

$$\Delta\Phi = \frac{2\pi n}{\lambda} + \frac{2\pi \Delta L}{\lambda}.$$

In view of the expression (1.82) for the eikonal, it follows that the light will contribute along the optical path leading from (x', z') to (x, z) the quantity

$$K \exp\left[i\left(\Phi(x,t) + \frac{2\pi n}{\lambda}(z - z') + \frac{2\pi n}{\lambda}\frac{(x - x')^2}{2(z - z')}\right)\right],$$

where K is an attenuation factor, to be determined. We begin by noting that the net contribution at x of all these terms is obtained by integrating over x'. Assume from now on that the index of refraction n is constant in position and time (it is, for instance, equal to one in vacuum); then K does not depend on x or x', and we thus have:

$$\psi(x, z) = K \int \exp\left[i\left(\Phi(x,t) + \frac{2\pi n}{\lambda}(z - z') + \frac{2\pi n}{\lambda}\frac{(x - x')^2}{2(z - z')}\right)\right]$$
$$\times \psi(x', z')\, dx'. \tag{1.86}$$

To calculate K, we remark that since the total intensity of light must be the same on the z-plane as that on the initial ($z' = 0$)-line, we must have

$$\int |a(x, z)|^2\, dx = \int |a(x', z')|^2\, dx'$$

and this condition leads, after some calculations, to

$$|K|^2 = \frac{n}{\lambda(z - z')}$$

and we can fix the argument of K by requiring that $\psi(x, z) \to \psi(x, z')$ as $z \to z'$. Using for instance the method of stationary phase, or the theory of Fresnel integrals, this finally yields the value

$$K = e^{i\alpha(z)}\left(\frac{n}{\lambda z}\right)^{1/2}, \quad \arg z = \begin{cases} 0 & \text{for } z > 0, \\ 1 & \text{for } z < 0, \end{cases}$$

where $\alpha(z)$ is some undetermined continuous function, vanishing at $z = 0$. The phase factor $e^{i\alpha(z)}$ can be determined by the following argument: the optical matrices S_z satisfy the group property $S_z S_{z'} = S_{z+z'}$, hence we must also have $\widehat{S}_z \widehat{S}_{z'} = \widehat{S}_{z+z'}$ where $\widehat{S}_z$ is the operator taking $\psi(x', z')$ to $\psi(x, z)$, and it is not difficult to show that this is only possible if one chooses $\alpha(z) = 0$, mod 2π. Neglecting the term $\exp(2\pi i n z/\lambda)$, we define

$$\psi(x, z) = \left(\frac{n}{i\lambda}\right)^{1/2} \int_{-\infty}^{\infty} \exp\left[2\pi i n \frac{(x - x')^2}{2\lambda z}\right] \psi_0(x')dx' \tag{1.87}$$

and a straightforward computation of partial derivatives shows that it satisfies the partial differential equation

$$i\lambda \frac{\partial \psi}{\partial z} = -\frac{\lambda^2}{4\pi} \frac{\partial^2 \psi}{\partial x^2}.$$

Formally, this is exactly Schrödinger's equation for a free particle with unit mass if one replaces λ by Planck's constant h!

Formula (1.87) can be written

$$\psi(x, z) = \int_{-\infty}^{\infty} G(x, x', z)\psi_0(x')\, dx'$$

where the kernel

$$G(x, x', z) = \left(\frac{n}{i\lambda}\right)^{1/2} \exp\left[2\pi i n \frac{(x - x')^2}{2\lambda z}\right]$$

is viewed as a "point source" of particles emanating from x' at time 0.

This last step, the passage from ray optics to wave optics, *is essential*; as we will see it is totally similar — both formally, and in spirit — to the passage from Hamiltonian mechanics to quantum mechanics. But we still have a long way to go in order to demonstrate this rigorously.

Chapter 2

Hamilton–Jacobi Theory

The Hamilton–Jacobi equation is an unavoidable topic in any discussion of Hamiltonian mechanics; it moreover has the advantage of immediately providing the link between Hamiltonian mechanics and wave optics discovered by Hamilton.

From a historical perspective, the fundamentals of the Hamilton–Jacobi equation first appeared in the studies of Hamilton in the 1820s for problems in wave optics; in 1834 Hamilton noticed the optical–mechanical analogy the theory provided, and in 1837 Carl Gustav Jacob Jacobi applied the method to general problems in the calculus of variations. From a historical point of view Hamilton–Jacobi theory is one of the oldest methods of resolution of Hamilton's equations; its interest in mathematics and physics has been revived since the 1980s with the introduction of the theory of viscosity solutions in hydrodynamics. It is encountered in problems of mechanics, geometry, optics, front propagation, computer vision, optimal control, and differential games. We will mainly focus on the theoretical aspects of the method, and on the related mathematical objects (generating functions) which are conceptually more important than the applications themselves.

2.1. The Action Integral

We begin by reviewing the properties of an integral symplectic invariant which is identified with *action* in physics. We are following here the discussion in [113].

2.1.1. Symplectic invariance property

In what follows $\gamma(t)$, $0 \leq t \leq 2\pi$, is a loop in phase space: $\gamma(t) = (x(t), p(t))$ where $x(0) = x(2\pi)$, $p(0) = p(2\pi)$; the functions $x(t)$ and $p(t)$ are supposed to be continuously differentiable. By definition, the first Poincaré invariant associated with $\gamma(t)$ is the integral

$$I(\gamma) = \oint_\gamma p\,dx = \int_0^{2\pi} p(t)\dot{x}(t)\,dt. \tag{2.1}$$

The fundamental property is that $I(\gamma)$ is a symplectic invariant. This means that if we replace the loop $\gamma(t)$ by a new loop $S\gamma(t)$ where S is a symplectic matrix, the integral (2.1) will keep the same value $I(S\gamma) = I(\gamma)$, that is

$$\oint_\gamma p\,dx = \oint_{S\gamma} p\,dx. \tag{2.2}$$

The proof is not very difficult if we carefully use the relations characterizing symplectic matrices. We will first need a differentiation rule for vector-valued functions, generalizing the product formula from elementary calculus. Suppose that $u(t) = (u_1(t), \ldots, u_n(t))$ and $v(t) = (v_1(t), \ldots, v_n(t))$ are vectors depending on the variable t and such that each component $u_j(t)$, $v_j(t)$ is differentiable. Let M be a symmetric matrix of size n and consider the real-valued function $Mv(t) \cdot u(t)$. Its derivative is given by the formula

$$\frac{d}{dt}[Mv(t) \cdot u(t)] = Mv(t) \cdot \dot{u}(t) + M\dot{v}(t) \cdot u(t) \tag{2.3}$$

as is easily verified. Let us now go back to the proof of the symplectic invariance of the first Poincaré invariant. Writing the symplectic matrix S in block form $\begin{pmatrix} A & B \\ C & D \end{pmatrix}$, the loop $S\gamma(t)$ is parameterized by

$$S\gamma(t) = (Ax(t) + Bp(t), Cx(t) + Dp(t)), \quad 0 \leq t \leq 2\pi.$$

We thus have, by definition of the Poincaré invariant,

$$I(S\gamma) = \int_0^{2\pi} (A\dot{x}(t) + B\dot{p}(t))(Cx(t) + Dp(t))\,dt;$$

expanding the product in the integrand, we have $I(S\gamma) = I_1 + I_2$ where

$$I_1 = \int_0^{2\pi} C^T A\dot{x}(t) \cdot x(t)dt + \int_0^{2\pi} D^T B\dot{p}(t) \cdot p(t)dt,$$

$$I_2 = \int_0^{2\pi} C^T B\dot{p}(t) \cdot x(t)dt + \int_0^{2\pi} D^T A\dot{x}(t) \cdot p(t)dt.$$

We claim that $I_1 = 0$. Recall that the matrix S is symplectic if and only if its blocks A, B, C, D satisfy the set of equivalent conditions

$$\begin{cases} A^T C, \ D^T B \text{ symmetric}, \ A^T D - C^T B = I, \\ AB^T, \ CD^T \text{ symmetric}, \ AD^T - BC^T = I, \qquad (2.4) \\ DC^T, \ AB^T \text{ symmetric}, \ DA^T - CB^T = I. \end{cases}$$

Applying the differentiation formula (2.3) with $u = v = x$, we have

$$\int_0^{2\pi} C^T A\dot{x}(t) \cdot x(t)dt = \frac{1}{2} \int_{0.}^{2\pi} \frac{d}{dt}(C^T Ax(t) \cdot x(t))dt = 0$$

because $x(0) = x(2\pi)$. Likewise, applying (2.3) with $u = v = p$ we get

$$\int_0^{2\pi} D^T B\dot{p}(t) \cdot p(t)dt = 0$$

hence $I_1 = 0$ as claimed. We next consider the term I_2. Rewriting the integrand of the second integral as

$$C^T B\dot{p}(t) \cdot x(t) = B^T Cx(t) \cdot \dot{p}(t),$$

we have

$$I_2 = \int_0^{2\pi} B^T Cx(t) \cdot \dot{p}(t)dt + \int_0^{2\pi} D^T A\dot{x}(t) \cdot p(t)dt,$$

that is, since $D^T A = I + B^T C$ by transposition of the third equality in (1.18),

$$I_2 = \int_0^{2\pi} p(t) \cdot \dot{x}(t)dt + \int_0^{2\pi} \left[p(t) \cdot B^T CA\dot{x}(t) + \dot{p}(t) \cdot B^T CAx(t) \right]dt.$$

Using again the rule (2.3) and noting that the first integral is precisely $I(\gamma)$ we get, $D^T A$ being symmetric,

$$I_2 = I(\gamma) + \int_0^{2\pi} \frac{d}{dt} \left[B^T C A x(t) \cdot p(t) \right] dt.$$

The equality $I(S\gamma) = I(\gamma)$ follows noting that the integral on the right-hand side is

$$B^T C A x(2\pi) \cdot p(2\pi) - B^T C A x(0) \cdot p(0) = 0$$

since $(x(2\pi), p(2\pi)) = (x(0), p(0))$.

The observant reader will have noticed that we really needed all of the properties of a symplectic matrix contained in the set of conditions (1.18); this shows that the symplectic invariance of the first Poincaré invariant is a characteristic property of symplectic matrices.

2.1.2. Poincaré–Cartan and Helmholtz

Given an arbitrary Hamiltonian function H on the extended phase space, the associated Hamilton–Jacobi equation is the partial differential equation with unknown $\Phi = \Phi(x, t)$:

$$\frac{\partial \Phi}{\partial t} + H(x, \nabla_x \Phi, t) = 0. \tag{2.5}$$

This equation is usually nonlinear; for instance if H is of the type

$$H(z, t) = \sum_{j=1}^{n} \frac{p_j^2}{2m_j} + V(x, t),$$

the associated Hamilton–Jacobi equation is explicitly

$$\frac{\partial \Phi}{\partial t} + \sum_{j=1}^{n} \frac{1}{2m_j} \left(\frac{\partial \Phi}{\partial x_j} \right)^2 + V(x, t) = 0.$$

Traditionally, the interest of this equation comes from the fact that the knowledge of a sufficiently general solution Φ yields the solutions of Hamilton's equations for H. At first sight it may seem strange that one replaces a system of ordinary differential equations by a nonlinear partial differential equation, but this procedure is often the only available method! For the sake of completeness, here is a description of the method.

Proposition 20 (Jacobi). *Let* $\Phi = \Phi(x, t, \alpha)$ *be a solution of*

$$\frac{\partial \Phi}{\partial t} + H(x, \nabla_x \Phi, t) = 0 \tag{2.6}$$

depending on n non-additive constants of integration $\alpha_1, \ldots, \alpha_n$ such that the Hessian matrix

$$\Phi''_{x,\alpha}(x, t, \alpha) = \left(\frac{\partial^2 \Phi}{\partial x_j \partial \alpha_k}(x, t, \alpha) \right)_{1 \le j, k \le n}$$

is invertible:

$$\det \Phi''_{x,\alpha}(x, t, \alpha) \ne 0. \tag{2.7}$$

Let $\beta_1, \ldots, \beta_n$ be n other arbitrary constants and set $\alpha = (\alpha_1, \ldots, \alpha_n)$, $\beta = (\beta_1, \ldots, \beta_n)$. The functions $x(t)$ and $p(t)$ determined by the implicit equations

$$\nabla_\alpha \Phi(x, t, \alpha) = \beta, \;\; p = \nabla_x \Phi(x, t, \alpha) \tag{2.8}$$

are solutions of Hamilton's equations for H.

Proof. Condition (2.7) implies, using the implicit function theorem, that the equation $\nabla_\alpha \Phi(x, t, \alpha) = \beta$ has a unique solution $x(t) = (x_1(t), \ldots, x_n(t))$ for each t. Inserting $x(t)$ in the formula $p = \nabla_x \Phi(x, t, \alpha)$ we also get a function $p(t) = \nabla_x \Phi(x(t), t, \alpha)$. Let us show that the functions $x(t), p(t)$ are solutions of Hamilton's equations for H. Differentiating equation (2.6) with respect to α yields, using the chain rule,

$$\frac{\partial^2 \Phi}{\partial \alpha_j \partial t} + \sum_{k=1}^n \frac{\partial H}{\partial p_k} \frac{\partial^2 \Phi}{\partial \alpha_j \partial x_k} = 0 \tag{2.9}$$

for $1 \le j \le n$. Similarly, differentiating the first equation (2.8) with respect to t yields

$$\frac{\partial^2 \Phi}{\partial t \partial \alpha_j} + \sum_{k=1}^n \frac{\partial^2 \Phi}{\partial x_k \partial \alpha_j} \dot{x}_k = 0 \tag{2.10}$$

for $1 \le j \le n$. Subtracting (2.10) from (2.9) we get

$$\sum_{k=1}^n \frac{\partial^2 \Phi}{\partial x_k \partial \alpha_j} \left(\frac{\partial H}{\partial p_k} - \dot{x}_k \right) = 0;$$

in matrix notation

$$\Phi''_{x,\alpha}(x,t,\alpha)(\nabla_p H(x,p,t) - \dot{x}) = 0.$$

Since the matrix $\Phi''_{x,\alpha}(x,t,\alpha)$ is assumed to be non-singular, this proves that

$$\dot{x}_k = \frac{\partial H}{\partial p_k} \quad \text{for } 1 \le k \le n \tag{2.11}$$

hence we have the first half of Hamilton's equations. To get the second half we differentiate the Hamilton–Jacobi equation (2.6) with respect to x; this yields

$$\frac{\partial H}{\partial x_j} + \frac{\partial^2 \Phi}{\partial x_j \partial t} + \sum_{k=1}^{n} \frac{\partial H}{\partial p_k} \frac{\partial^2 \Phi}{\partial x_k \partial x_j} = 0; \tag{2.12}$$

we next differentiate $p_j = \partial \Phi / \partial x_j$ with respect to t:

$$\dot{p}_j = \frac{\partial^2 \Phi}{\partial t \partial x_j} + \sum_{k=1}^{n} \frac{\partial^2 \Phi}{\partial x_k \partial x_j} \dot{x}_k, \tag{2.13}$$

that is, taking (2.11) into account,

$$\dot{p}_j = \frac{\partial^2 \Phi}{\partial t \partial x_j} + \sum_{k=1}^{n} \frac{\partial^2 \Phi}{\partial x_k \partial x_j} \frac{\partial H}{\partial p_k}. \tag{2.14}$$

Comparison of (2.12) and (2.14) immediately yields,

$$\dot{p}_k = -\frac{\partial H}{\partial x_k} \quad \text{for } 1 \le k \le n,$$

which is the second half of Hamilton's equations.   □

The Poincaré–Cartan invariant associated with a Hamiltonian function H (time-dependent or not) is by definition the one-differential form

$$\alpha_H = pdx - H(z,t)dt,$$

where pdx is the first Poincaré invariant

$$pdx = \sum_{j=1}^{n} p_j dx_j$$

discussed above. The subject of integral invariants was initiated in a systematic way by the mathematician Elie Cartan in his celebrated *Leçons sur les invariants intégraux* (1922).

We assume, as usual, that the solutions of the Hamilton equations for H are defined for all times. Let now γ be a closed curve in the extended phase space $\mathbb{R}^2_{x,p} \times \mathbb{R}_t$. We denote by $T_H(\gamma)$ the *trajectory tube* generated by the action of the Hamiltonian flow: it is the surface swept out by γ under the action of the flow of the suspended vector field $\tilde{X}_H = (\nabla_p H, -\nabla_x H, 1)$. That is, each point (x_0, p_0, t_0) of γ describes the curve

$$\tilde{f}_t(x_0, p_0, t_0) = (f_{t+t_0, t_0}(x_0, p_0), t + t_0)$$

in $\mathbb{R}^2_{x,p} \times \mathbb{R}_t$ parameterized by t. The suspended vector field being tangent to $T_H(\gamma)$, at each point (x, p, t) of $T_H(\gamma)$ we must have

$$\tilde{X}_H(x, p, t) \cdot \mathbf{n}(x, p, t) = 0, \qquad (2.15)$$

where $\mathbf{n}(x, p, t)$ is the (outwardly oriented) normal vector to $T_H(\gamma)$ at the point (x, p, t).

Lemma 21. *The integral of the Poincaré–Cartan form along any closed curve λ lying on (i.e., not encircling) a trajectory tube $T_H(\gamma)$ is equal to zero:*

$$\int_\lambda p \, dx - H dt = 0. \qquad (2.16)$$

Proof. Let us give the proof in the case $n = 1$. Setting $u = (x, p, t)$, we have

$$\oint_\lambda p \, dx - H dt = \int_\lambda (p, 0, -H) \cdot du. \qquad (2.17)$$

Let us denote by D the surface on the tube $T_H(\gamma)$ encircled by λ. By Gauss's integral formula we have

$$\oint_\lambda p \, dx - H dt = \iint_D [\nabla_u \times (p, 0, -H)] \cdot \mathbf{n} \, dS, \qquad (2.18)$$

where dS is the area element and $\mathbf{n}$ the normal vector. Now

$$\nabla_u \times (p, 0, -H) = \tilde{X}_H \qquad (2.19)$$

and formula (2.16) follows, using (2.15). In the case of general dimension n one applies the multi-dimensional Stokes' theorem to the chain $\sigma = \gamma_1 - \gamma_2$, leading to

$$\oint_{\gamma_1} \alpha_H - \oint_{\gamma_2} \alpha_H = \oint_\sigma \alpha_H = \int_D d\alpha_H,$$

where D is the piece of surface of $T_H(\gamma)$ bounded by γ, and one then shows that the exterior derivative $d\alpha_H$ vanishes on $T_H(\gamma)$. For a detailed proof in this case see [5, Chapter 9] or [100, Chapter 4]. □

This result allows us to give a simple proof of Helmholtz's classical theorem on the conservation of vorticity.

Proposition 22 (Helmholtz). *Let γ_1 and γ_2 be two arbitrary closed curves encircling the same tube of trajectories $T_H(\gamma)$.*

(i) *The integrals of the Poincaré–Cartan form along γ_1 and γ_2 are equal:*

$$\oint_{\gamma_1} \alpha_H = \oint_{\gamma_2} \alpha_H. \tag{2.20}$$

(ii) *If in particular γ_1 and γ_2 lie in the parallel planes $t = t_1$ and $t = t_2$, then*

$$\oint_{\gamma_1} p dx = \oint_{\gamma_2} p dx. \tag{2.21}$$

Proof. (i) Let us give the curves γ_1 and γ_2 the same orientation, and apply the following "surgery" to the piece of trajectory tube $T(\gamma_1, \gamma_2)$ limited by them: choose a point (z_1, t_1) on γ_1, and a point (z_2, t_2) on γ_2. Let now γ be a curve in $T(\gamma_1, \gamma_2)$ joining these two points, and define the chain

$$\sigma = \gamma_1 + \gamma - \gamma_2 - \gamma.$$

It is a loop encircling $T(\gamma_1, \gamma_2)$ as follows: it starts from (z_1, t_1) and runs (in the positive direction) along γ_1; once it has returned to (z_1, t_1), it runs along γ until it reaches the point (z_2, t_2) and then runs (now in the negative direction) around γ_2 until it is back to (z_2, t_2); finally it runs back to (z_1, t_1) along γ. Obviously σ is contractible to a point, and hence formula (2.16) applies, and the integral of the Poincaré–Cartan form along σ vanishes. Since the contributions from γ and $-\gamma$ cancel, we get

$$\oint_\sigma \alpha_H = \oint_{\gamma_1} \alpha_H + \oint_{-\gamma_2} \alpha_H = 0,$$

which proves formula (2.20). The statement (ii) follows from the statement (i) since $dt = 0$ on planes with constant t. □

Helmholtz's theorem can be rephrased by saying that the Poincaré–Cartan form is an *integral invariant*. The study of integral invariants is an important and rich topic; see [5] for a detailed discussion.

2.2. Generating Functions and Action

2.2.1. Free symplectomorphisms

Assume that you throw a piece of chalk from a point A to reach some other point B. You decide you want to reach B after a given time $t-t'$. Then there will be only one possible trajectory, and both the initial and final momenta are uniquely determined (at least if the time interval $t - t'$ is small enough: for large $t - t'$ uniqueness is not preserved, because of exotic possibilities, such as the piece of chalk making one, or several, turns around the earth). This can be stated in terms of the flow $(f_{t,t'})$ determined by the Hamiltonian H of the piece of chalk by saying that $f_{t,t'}$ is a free *symplectomorphism*. Furthermore this is a generalization to the nonlinear case of the notion of free symplectic matrix introduced in Chapter 1, motivating the following definition.

A symplectomorphism $f_{t,t'}$ is *free* if, given initial and final positions x' and x, the equation

$$(x, p) = f_{t,t'}(x', p')$$

uniquely determines the initial and final momenta p', p.

Recall that in the linear case a symplectic matrix

$$S = \begin{pmatrix} A & B \\ C & D \end{pmatrix}$$

is free if $\det B \neq 0$ and that (Proposition 2) a generating function for S is the quadratic form

$$W(x, x') = \tfrac{1}{2} DB^{-1} x^2 - B^{-1} x \cdot x' + \tfrac{1}{2} B^{-1} A x'^2. \tag{2.22}$$

We also showed that if, conversely, W is a quadratic form of the type

$$W(x, x') = \tfrac{1}{2} P x^2 - L x \cdot x' + \tfrac{1}{2} Q x'^2 \tag{2.23}$$

with $P = P^T$, $Q = Q^T$, $\det L \neq 0$, then the matrix

$$S_W = \begin{pmatrix} L^{-1}Q & L^{-1} \\ PL^{-1}Q - L^T & PL^{-1} \end{pmatrix} \tag{2.24}$$

is a free symplectic matrix whose generating function is given precisely by (2.23).

Here is a useful criterion for the general (nonlinear) case.

Lemma 23. *The symplectomorphism $f_{t,t'}$ is free if and only if $(x, p) = f_{t,t'}(x', p')$ is such that the Jacobian matrix*

$$\frac{\partial(x, x')}{\partial(p', x')} = \frac{\partial(x_1, \ldots, x_n; x'_1, \ldots, x'_n)}{\partial(p'_1, \ldots, p'_n; x'_1, \ldots, x'_n)}$$

is invertible. (In particular, if $f_{t,t'}$ is a symplectic matrix $S = \begin{pmatrix} A & B \\ C & D \end{pmatrix}$ then it is free if and only if $\det B \neq 0$).

Proof. We begin by noting that

$$\det \frac{\partial(x, x')}{\partial(p', x')} = \det \frac{\partial x}{\partial p'}. \tag{2.25}$$

This identity follows from the definition of the Jacobian matrix: we have

$$\frac{\partial(x, x')}{\partial(p', x')} = \begin{pmatrix} \dfrac{\partial x}{\partial p'} & \dfrac{\partial x}{\partial x'} \\ 0 & I_{n \times n} \end{pmatrix}$$

and hence

$$\det \frac{\partial(x, x')}{\partial(p', x')} = \det \left(\frac{\partial x}{\partial p'} \right) \det I_{n \times n} = \det \left(\frac{\partial x}{\partial p'} \right).$$

It thus suffices to prove that f is free if and only if we have

$$\det \frac{\partial x}{\partial p'} \neq 0. \tag{2.26}$$

Now, in view of the implicit function theorem, the equation $x = x(x', p')$ can be solved (locally) in p' if and only if the Jacobian matrix $\partial x / \partial p'$ is non-singular, that is if (2.26) holds. When $f_{t,t'} = S$ we have $x = Ax' + Bp'$ and $p = Cx' + Dp'$ hence $\partial x / \partial p' = B$ and S is thus free if and only if $\det B \neq 0$, as claimed. $\qquad \square$

Let $(f_{t,t'})$ be the time-dependent flow determined by a Hamiltonian

$$H = \sum_{j=1}^{n} \frac{1}{2m_j} \left(p_j - A_j(x,t)\right)^2 + U(x,t). \tag{2.27}$$

We are going to show that $f_{t,t'}$ is always a free symplectomorphism for sufficiently short time intervals $t - t'$ (but different from zero: the identity mapping is not free!).

Proposition 24. *For every $z_0 = (x_0, p_0)$ there exists a number $\varepsilon > 0$ such that $f_{t,t'}$ is a free symplectomorphism near z_0 for $0 < |t - t'| < \varepsilon$.*

Proof. Set $z = f_{t,t'}(z')$. The first-order Taylor expansion of z at $t = t'$ is

$$z = z' + (t - t')X_H(z') + O((t - t')^2).$$

Denoting by M the mass matrix (i.e., the diagonal matrix with non-zero elements the masses m_j), the Hamilton vector field can be concisely expressed as

$$X_H = (M^{-1}(p - A),\ M^{-1}(p - A)\nabla_x A - \nabla_x U)$$

and hence

$$\begin{aligned}
\frac{\partial x}{\partial p'} &= (t - t')M^{-1} + O\left((t - t')^2\right) \\
&= (t - t')M^{-1}\left(I_{n \times n} + O_{n \times n}(t - t')\right),
\end{aligned}$$

where $O\left((t - t')^k\right)$ $(k = 1, 2)$ is an $n \times n$ matrix whose entries go to zero at least as fast as $(t - t')^k$ when $t \to t'$. It follows that when $t - t' \to 0$ we have

$$\det \frac{\partial x}{\partial p'} = (t - t')^n[\det M^{-1} + O((t - t')^n)]$$

hence $\partial x/\partial p'$ will be invertible near x_0 if $|t - t'| \neq 0$ is sufficiently small. In view of Lemma 23, this means that $f_{t,t'}$ is free at x_0 for those values of t, t'. $\qquad\square$

We point out that the conclusions of the proposition above remain valid for the following weaker condition on the Hamiltonian function H:

$$\det\left(\frac{\partial^2 H}{\partial p_j \partial p_k}\right)_{1 \leq j,k \leq n} \neq 0. \tag{2.28}$$

This is easy to verify by reproducing the proof above. Hamiltonian functions verifying (2.28) are said to be *convex*.

It turns out that free symplectomorphisms are "generated" by functions defined on twice the configuration space. This important property, closely related to the Hamilton–Jacobi equation, is studied in the next section.

We are going to show that to every free symplectomorphism $f_{t,t'}$ we can associate, as in the linear case, a generating function $W(x, x'; t, t')$, which is a function such that:

$$(x, p) = f_{t,t'}(x', p') \iff \begin{cases} p = \nabla_x W(x, x'; t, t'), \\ p' = -\nabla_{x'} W(x, x'; t, t'). \end{cases} \tag{2.29}$$

Proposition 25. (i) *The function W defined by*

$$W(x, x'; t, t') = \int_{x',t'}^{x,t} pdx - Hds \tag{2.30}$$

(*where the integral is calculated along the phase space trajectory leading from (x', p', t') to (x, p, t)) is a free generating function for the symplectomorphism $f_{t,t'}$.*

Proof. Let $W_{t,t'}$ be defined by formula (2.30). We have to show that if $(x, p) = f_{t,t'}(x', p')$ then

$$p = \nabla_x W(x, x'; t, t'), \quad p' = -\nabla_{x'} W_{t,t'} W(x, x'; t, t'). \tag{2.31}$$

It suffices in fact to prove the first of these identities: since $(f_{t,t'})^{-1} = f_{t',t}$ we have

$$W(x, x'; t, t') = -W(x', x; t', t) \tag{2.32}$$

and the second formula in (2.31) will hence automatically follow from the first. We set out to prove that

$$p_j = \frac{\partial W}{\partial x_j}(x, x'; t, t') \tag{2.33}$$

for $1 \leq j \leq n$. Assuming for notational simplicity that $n = 1$, and giving an increment Δx to x, we set out to evaluate the difference

$$W(x + \Delta x, x'; t, t') - W(x, x'; t, t').$$

Let γ_1 and γ_2 be the trajectories joining, respectively, (x', p', t') to (x, p, t) and $(x', p' + \Delta p', t')$ to $(x + \Delta x, p + \Delta p, t)$. The two vectors $p' + \Delta p'$ and $p + \Delta p$ are the new initial and final momenta corresponding to the new final position $x + \Delta x$. We have, by definition of $W_{t,t'}$:

$$W(x, x'; t, t') = \int_{\gamma_1} \alpha_H$$

and

$$W(x + \Delta x, x'; t, t') = \int_{\gamma_2} \alpha_H.$$

Let now μ be an arbitrary curve joining the point (x', p') to the point $(x', p' + \Delta p')$ while keeping time constant and equal to t', and $\mu_{t,t'}$ the image of that curve by $f_{t,t'}$: $\mu_{t,t'}$ is thus a curve joining (x, p) to $(x + \Delta x, p + \Delta p)$ in the hyperplane *time* $= t$. In view of Helmholtz's theorem we have

$$\int_{\gamma_2} \alpha_H - \int_{\gamma_1} \alpha_H = \int_{\mu_{t,t'}} \alpha_H - \int_{\mu} \alpha_H$$

and hence, since $dt = 0$ on μ and $\mu_{t,t'}$ and $dx = 0$ on μ:

$$W_{t,t'}(x + \Delta x, x') - W_{t,t'}(x, x') = \int_{\mu_{t,t'}} p \, dx.$$

The choice of the curve μ being arbitrary, we can assume that $\mu_{t,t'}$ is the line segment

$$x(s) = x + s\Delta x, \quad p(s) = p + s\Delta x.$$

$(0 \leq s \leq 1)$ in which case the integral along $\mu_{t,t'}$ becomes

$$\int_{\mu_{t,t'}} p \, dx = p\Delta x + \frac{1}{2}(\Delta x)^2$$

and hence

$$\frac{W_{t,t'}(x + \Delta x, x') - W_{t,t'}(x, x')}{\Delta x} = p + \frac{1}{2}\Delta x$$

from which follows that

$$p = \frac{\partial W}{\partial x}(x, x'; t, t')$$

letting $\Delta x \to 0$; formula (2.33) follows. $\square$

One of the main appeals of generating functions is that they allow the Hamilton equations to be solved in an elegant way. Let H again be a Hamiltonian function

$$H = \sum_{j=1}^{n} \frac{1}{2m_j}(p_j - A_j)^2 + U \qquad (2.34)$$

and $W = W(x, x'; t, t')$ the generating function that it determines. We are going to show that W allows us to solve explicitly the Hamilton equations

$$\dot{x} = \nabla_p H(x, p, t), \quad \dot{p} = -\nabla_x H(x, p, t).$$

Proposition 26. *Let $W(x, x'; t, t')$ be the generating function determined by a Hamiltonian (2.34). The equations*

$$\begin{cases} p = \nabla_x W(x, x'; t, t') \\ p' = -\nabla_{x'} W(x, x'; t, t') \end{cases} \qquad (2.35)$$

determine the flow $(f_{t,t'})$ for all t, t' for which the generating function exists. In fact, for given (x', p') and t' the two functions

$$t \longmapsto x(x', p', t), \quad t \longmapsto p(x', p', t)$$

implicitly defined by these equations are the solutions of Hamilton's equations for H with initial data $x(t') = x'$ and $p(t') = p'$.

Proof. Differentiating the second equation (2.35) with respect to time yields

$$\frac{dp'}{dt} = -W''_{x,x'}\dot{x} - \nabla_{x'}\left(\frac{\partial W}{\partial t}\right),$$

where $W''_{x,x'}$ is the matrix of mixed second derivatives:

$$W''_{x,x'} = \left(-\frac{\partial^2 W}{\partial x_i \partial x'_j}\right)_{1 \le i, j \le n}.$$

Since $dp'/dt = 0$ this implies

$$-W''_{x,x'}\dot{x}(t) + \nabla_{x'}[H(x, \nabla_x W, t)] = 0.$$

Using the chain rule we get

$$\nabla_{x'}(H(x, p, t)) = \left(\frac{\partial p}{\partial x'}\right)^T \nabla_p H(x, p, t) = W''_{x,x'}\nabla_p H(x, p, t)$$

and hence

$$W''_{x,x'}(\dot{x} - \nabla_p H(x, p, t)) = 0.$$

Since the matrix $W''_{x,x'}$ is non-singular, this is equivalent to

$$\dot{x} = \nabla_p H(x, p, t)$$

which is the Hamilton equation for x. Similarly, differentiating the first equation (2.35) with respect to t and taking Hamilton–Jacobi's equation into account we get

$$\dot{p} = \nabla_x \left(\frac{\partial W}{\partial t}\right) + W''_{x,x'}\dot{x}$$

$$= -\nabla_x(H(x, \nabla_x W)) + W''_{x,x'}\dot{x}$$

$$= -(\nabla_x H)(x, \nabla_x W, t) + W''_{x,x'}(\nabla_p H(x, \nabla_x W, t) - \dot{x})$$

$$= -(\nabla_x H)(x, \nabla_x W, t)$$

$$= -(\nabla_x H)(x, p, t)$$

which ends the proof of Proposition 26. $\qquad\qquad\square$

We call an affine symplectomorphism any mapping $\mathbb{R}^{2n} \to \mathbb{R}^{2n}$ of the type $T(z)S$ where $S \in \mathrm{Sp}(2n, \mathbb{R})$ and $T(z)$ is the phase space translation $z' \mapsto z' + z$. Noticing that

$$ST(z) = T(Sz)S, \quad T(z)S = ST(S^{-1}z)$$

we have

$$(T(z)S)(T(z')S') = T(z + Sz')SS'$$

hence the product of two affine symplectomorphisms is also an affine symplectomorphism. Since

$$(T(z)S)^{-1} = S^{-1}T(-z) = T(-S^{-1}z)S^{-1},$$

the set of all such mappings forms a group; that group is denoted by $\mathrm{ASp}(2n,\mathbb{R})$ and is called the *affine* (or sometimes *inhomogeneous*) *symplectic group*. We will use the notation $\langle S, z \rangle = T(z)S$.

The following result characterizes the generating functions of the free elements of $\mathrm{ISp}(2n,\mathbb{R})$.

Proposition 27. *An affine symplectomorphism $\langle S, z_0 \rangle$, $z_0 = (x_0, p_0)$, is free if and only if S is a free symplectic matrix. A free generating function of $\langle S, z_0 \rangle$ is the inhomogeneous quadratic polynomial*

$$W_{z_0}(x, x') = W(x - x_0, x') + p_0 x, \tag{2.36}$$

where W is a free generating function for S. Conversely, if W is the generating function of a symplectic matrix S, then any polynomial

$$W_{z_0}(x, x') = W(x, x') + \alpha x + \alpha' x' \tag{2.37}$$

$(\alpha, \alpha' \in \mathbb{R}^n_x)$ is a generating function of an affine symplectic transformation, the translation vector $z_0 = (x_0, p_0)$ being given by

$$(x_0, p_0) = (B\alpha, D a + \beta) \tag{2.38}$$

when $S = \begin{pmatrix} A & B \\ C & D \end{pmatrix}$.

Proof. Let W_{z_0} be defined by (2.36), and set $(x', p') = S(x'', p'')$, $(x, p) = T(z_0)(x', p')$. We have

$$p dx - p' dx' = (p dx - p'' dx'') + (p'' dx'' - p' dx')$$

$$= p dx - (p - p_0) d(x - x_0) + dW(x'', x')$$

$$= d(p_0 x + W(x - x_0, x'))$$

which shows that W_{z_0} is a generating function. Finally, formula (2.38) is obtained by a direct computation, expanding $W(x - x_0, x')$. $\qquad\square$

2.2.2. Hamilton–Jacobi equation and generating functions

It turns out that the family of generating functions $W(x, x'; t, t')$ provides us with a solution of Hamilton–Jacobi's equation.

Proposition 28. *For fixed x' and t', the function $\Phi_{x',t'} : (x, t) \mapsto W(x, x'; t, t')$ defined above is a solution of Hamilton–Jacobi's equation: we have*

$$\frac{\partial}{\partial t} \Phi_{x',t'}(x, t) + H(x, \nabla_x \Phi_{x',t'}, t) = 0 \tag{2.39}$$

for $|t - t'|$ sufficiently small.

Proof. Fixing the starting point x' and the initial time t', x will depend only on t. Writing $\Phi = \Phi_{x',t'}$ we have, by the chain rule:

$$\frac{d\Phi}{dt}(x, t) = \nabla_x \Phi(x, t) \cdot \dot{x} + \frac{\partial \Phi}{\partial t}(x, t), \tag{2.40}$$

that is, since $p = \nabla_x \Phi(x, t)$:

$$\frac{d\Phi}{dt}(x, t) = p \cdot \dot{x} + \frac{\partial \Phi}{\partial t}(x, t). \tag{2.41}$$

On the other hand, by definition of Φ:

$$\Phi(x, t) = \int_{t'}^{t} (p(s) \cdot \dot{x}(s) - H(x(s), p(s), s)) ds$$

so that

$$\frac{d\Phi}{dt}(x, t) = p \cdot \dot{x} - H(x, p, t). \tag{2.42}$$

Equating the right-hand sides of (2.41) and (2.42), we finally get

$$\frac{\partial \Phi}{\partial t}(x, t) + H(x, p, t) = 0$$

which is Hamilton–Jacobi's equation for Φ since $p = \nabla_x \Phi(x, t)$. $\quad\square$

We are going to show that the generating function also allows us to solve the Cauchy problem for the Hamilton–Jacobi equation.

Proposition 29. *Suppose that there exists ε such that for $0 < |t - t'| < \varepsilon$, the mappings $f_{t,t'}$ are free symplectomorphisms. The Cauchy problem*

$$\begin{cases} \dfrac{\partial \Phi}{\partial t} + H\left(x, \nabla_x \Phi, t\right) = 0 \\ \Phi(x, t') = \Phi'(x) \end{cases} \tag{2.43}$$

has a unique solution $\Phi = \Phi(x, t)$, defined for $0 < |t - t'| < \varepsilon$, and that solution is given by the formula

$$\Phi(x, t) = \Phi'(x') + \int_{x', t'}^{x, t} p \, dx - H \, ds, \tag{2.44}$$

where the initial point x' is defined by the condition

$$(x, p) = f_{t, t'}(x', \nabla_x \Phi(x')). \tag{2.45}$$

Proof. We first note that formula (2.45) really defines x': since $f_{t,t'}$ is free, the datum of x and x' unambiguously determines p and p'; assigning x and $p' = \nabla_x \Phi(x')$ thus also unambiguously determines both p and x'. Now, it is clear that $\lim_{t' \to t} \Phi(x, t') = \Phi'(x)$ since $x' \to x$ as $t' \to t$, so that the Cauchy condition is satisfied. To prove that Φ is a solution of Hamilton–Jacobi's equation one first shows, as in the proof of Proposition 25, that

$$\Phi(x + \Delta x, t + \Delta t) - \Phi(x, t) = \int_\beta p \, dx - H \, dt,$$

where β is the line in phase space joining (x, p, t) to $(x + \Delta x, p + \Delta p, t + \Delta t)$, p and $p + \Delta p$ being determined by the relations $p = \nabla_x W(x, x'; t, t')$, and

$$p + \Delta p = \nabla_x W(x + \Delta x, x' + \Delta x'; t + \Delta t, t')$$

($\Delta x'$ is the increment of x' determined by Δx). Thus,

$$\Phi(x + \Delta x, t + \Delta t) - \Phi(x, t)$$
$$= p \Delta x + \frac{1}{2} \Delta p \Delta x - \Delta t \int_0^1 H(x + s \Delta x, p + s \Delta p, t + s \Delta t) ds$$

and hence

$$\frac{\Phi(x, t + \Delta t) - \Phi(x, t)}{\Delta t} = -\int_0^1 H(x, p + s \Delta p, t + s \Delta t) ds$$

from which follows that

$$\frac{\partial \Phi}{\partial t}(x,t) = -H(x,p,t) \tag{2.46}$$

since $\Delta p \to 0$ when $\Delta t \to 0$. Similarly,

$$\Phi(x + \Delta x, t) - \Phi(x,t) = p\Delta x + \tfrac{1}{2}\Delta p \Delta x$$

and $\Delta p \to 0$ as $\Delta x \to 0$ so that

$$\frac{\partial \Phi}{\partial x}(x,t) = p. \tag{2.47}$$

Combining (2.46) and (2.47) shows that Φ satisfies Hamilton–Jacobi's equation, as claimed. $\qquad\square$

2.2.3. Short-time approximations

It is in most cases impossible to solve explicitly Hamilton–Jacobi's Cauchy problem. We are going to see that it is however rather straightforward to obtain asymptotic solutions to that problem for short times.

Physicists working on the Feynman integral (about which we will have much to say later) commonly use the following "midpoint approximations" for the generating function W:

$$W(x,x_0;t_0+\Delta t) \approx \sum_{j=1}^{n} \frac{m_j}{2\Delta t}(x_j - x_0)^2 - \frac{1}{2}(V(x) + V(x_0))\Delta t \tag{2.48}$$

or even

$$W(x,x_0;t_0+\Delta t) \approx \sum_{j=1}^{n} \frac{m_j}{2\Delta t}(x_j - x_0)^2 - V(\tfrac{1}{2}(x + x_0))\Delta t. \tag{2.49}$$

(See Schulman's book [258] for variants of these "rules".) However, as already pointed out by Makri and Miller [200, 201], these "approximations" are wrong; they fail to be correct even to first order in Δt! Let us exemplify this strong statement with a simple example. Consider the one-dimensional harmonic oscillator with Hamiltonian function

$$H(x,p) = \frac{1}{2m}(p^2 + m^2\omega^2 x^2).$$

The corresponding generating function is, choosing $t_0 = 0$,

$$W(x, x_0; \Delta t) = \frac{m\omega}{2\sin\omega\Delta t}((x^2 + x_0^2)\cos\omega\Delta t - 2xx_0). \qquad (2.50)$$

Using the approximations $\sin\omega\Delta t = \omega\Delta t + O(\Delta t^3)$ and $\cos\omega\Delta t = 1 + O(\Delta t^2)$ valid for $\Delta t \to 0$ we get the expansion

$$W(x, x_0; \Delta t) = \frac{m}{2\Delta t}(x - x_0)^2 - \frac{m\omega^2}{6}(x^2 + xx_0 + x_0^2)\Delta t + O(\Delta t^2);$$

$$(2.51)$$

this correct expression is of course totally different from the approximations obtained by using the "rules" (2.48) or (2.49).

In fact, Makri and Miller have shown in [200, 201] that the correct asymptotic expression for the generating function is given by formula (2.55) below which we are going to prove along the lines in [100, Chapter 4]. Let us first introduce the following notation: for a continuous function $f : \mathbb{R}_x^n \to \mathbb{R}$ we denote by $\bar{f}(x, x')$ its average value on the line segment joining x_0 to x:

$$\widetilde{f}(x, x_0) = \int_0^1 f(sx + (1-s)x_0)\, ds \qquad (2.52)$$

and we set $\widetilde{f}(x) = \widetilde{f}(x, x)$. Observe that we have

$$\nabla_x \bar{f}(x, x) = \int_0^1 s\nabla_x f(x)\, ds = \frac{1}{2}\nabla_x f(x). \qquad (2.53)$$

The function $\widetilde{f}(x, x_0)$ is the solution of a certain singular partial differential equation.

Lemma 30. *For any continuous function* $f : \mathbb{R}_x^n \to \mathbb{R}$ *the equation*

$$(x - x_0) \cdot \nabla_x u + u = f \qquad (2.54)$$

has $u(x) = \widetilde{f}(x, x_0)$ *as the only smooth solution defined on all of* $\mathbb{R}_x^n$.

Proof. It is immediate to check that if u and v are two solutions of the equation (2.54) then we have

$$u(x) - v(x) = \sum_{j=1}^n \frac{k_j}{x_j}$$

so that (2.54) has at most one solution defined on all of $\mathbb{R}_x^n$. Let us show that $u = \tilde{f}$ indeed is a solution. It is of course sufficient to assume that $x_0 = 0$ and to show that $u(x) = \tilde{f}(x)$ satisfies $x \cdot \nabla_x u + u = 0$. We have

$$(x \cdot \nabla_x u + u)(x) = \int_0^1 (x \cdot (\nabla_x f)(sx)s + f(sx)) \, ds$$

$$= \sum_{j=1}^n \int_0^1 \left(sx_j \frac{\partial f}{\partial x_j}(sx) + f(sx) \right) ds$$

$$= \int_0^1 \frac{d}{ds}(sf(sx)) \, ds$$

hence our claim. $\qquad\qquad\qquad\qquad\qquad\qquad\qquad\qquad\qquad\qquad\square$

Let us now give the correct short-time approximation to the generating function.

Proposition 31. *Let H be a time-dependent Hamiltonian function of the type*

$$H(z,t) = \sum_{j=1}^n \frac{1}{2m_j} p_j^2 + V(x,t).$$

The generating function W has the following asymptotic expansion for $\Delta t \to 0$:

$$W(x,x_0;t_0 + \Delta t, t_0) = \sum_{j=1}^n \frac{m_j}{2\Delta t}(x_j - x_{0,j})^2 - \tilde{V}(x,x_0,t_0)\Delta t + O(\Delta t^2),$$

$$(2.55)$$

where $\tilde{V}(x,x_0,t_0)$ is the average value of the potential over the straight line joining x_0 at time t_0 to x at time $t_0 + \Delta t$ with constant velocity:

$$\tilde{V}(x,x_0,t_0) = \int_0^1 V(\lambda x + (1-\lambda)x_0, t_0) d\lambda. \qquad (2.56)$$

Proof. It is of course no restriction to assume that the initial time is $t_0 = 0$; we will write $W(x,x_0;\Delta t, 0) = W(x,x_0;\Delta t)$. We begin with the case $n = 1$ corresponding to a particle moving along the x-axis under the influence of a scalar potential. The Hamiltonian function is thus here

$$H = \frac{p^2}{2m} + V(x,t).$$

The generating function determined by H satisfies the Hamilton–Jacobi equation

$$\frac{\partial W}{\partial t} + \frac{1}{2m}\left(\frac{\partial W}{\partial x}\right)^2 + V = 0. \tag{2.57}$$

Let us denote by W_{free} the free-particle generating function:

$$W_{\text{free}}(x, x_0; \Delta t) = m\frac{(x - x_0)^2}{2\Delta t} \tag{2.58}$$

and look for a solution of Hamilton–Jacobi's equation of the form $W = W_{\text{free}} + R$. Inserting $W_{\text{free}} + R$ in (2.57), and expanding the squared bracket, we see that the function $R = R(x, x_0; \Delta t)$ has to satisfy the singular partial differential equation

$$\frac{\partial R}{\partial t} + \frac{1}{2m}\left(\frac{\partial R}{\partial x}\right)^2 + V + \frac{1}{\Delta t}(x - x_0)\frac{\partial R}{\partial x} = 0. \tag{2.59}$$

Expanding R to the second order in Δt

$$R = W_0 + W_1\Delta t + W_2^2\Delta t + O\left(\Delta t^3\right),$$

where the W_j $(j = 0, 1, 2)$ are smooth functions of x and x_0, we immediately see that we must have $W_0 = 0$, and W_1 and W_2 must satisfy the conditions

$$W_1 + (x - x_0)\frac{\partial W_1}{\partial x} + V = 0$$

and

$$2W_2 + (x - x_0)\frac{\partial W_2}{\partial x} = 0.$$

The general solution of the first of these equations is

$$W_1(x, x_0) = \frac{k}{x - x_0} - \frac{1}{x - x_0}\int_{x_0}^{x} V(x')dx',$$

where k is an arbitrary constant. Since we only want smooth solutions, we must choose $k = 0$ and we get $W_1 = -\widetilde{V}$ where $\widetilde{V}(x, x_0)$ is the average value of the potential V on the interval $[x_0, x]$:

$$\widetilde{V}(x, x_0) = \frac{1}{x - x_0} \int_{x_0}^{x} V(x')dx'.$$

Similarly, the only possible choice leading to a smooth W_2 is $W_2 = 0$. It follows that the generating function has the asymptotic form:

$$W = \widetilde{W} + O\left(\Delta t^3\right), \tag{2.60}$$

where

$$\widetilde{W}(x, x'; \Delta t) = m\frac{(x - x_0)^2}{2\Delta t} - \widetilde{V}(x, x_0)\Delta t. \tag{2.61}$$

The case of an arbitrary dimension n is proven similarly using Lemma 30 above. $\qquad \square$

The case where a vector potential is present is treated similarly. Consider a single particle with mass m in a gauge $(\mathbf{A}, U)$. The Hamiltonian function is thus here

$$H = \frac{1}{2m}\left(\mathbf{p} - \mathbf{A}\right)^2 + V, \tag{2.62}$$

where $\mathbf{A} = \mathbf{A}(\mathbf{r}, t)$ and $V = V(\mathbf{r}, t)$.

Proposition 32. *The generating function determined by the Hamiltonian function* (2.62) *is asymptotically given, for* $\Delta t = t - t_0 \to 0$, *by* $W = \widetilde{W} + O\left(\Delta t^2\right)$ *where*

$$\widetilde{W} = m\frac{(\mathbf{r} - \mathbf{r}_0)^2}{2\Delta t} + (\mathbf{r} - \mathbf{r}_0) \cdot \widetilde{\mathbf{A}}(\mathbf{r}, \mathbf{r}_0; t_0) - \widetilde{U}(\mathbf{r}, \mathbf{r}_0; t_0)\Delta t$$

and $\widetilde{U} = \widetilde{U}(\mathbf{r}, \mathbf{r}_0; t_0)$ *is the average on* $[\mathbf{r}, \mathbf{r}_0]$ *at time* t_0 *of the function* $\mathbf{r} \mapsto UV(\mathbf{r}, \mathbf{r}_0; t_0)$ *defined by*

$$U = \frac{1}{2m}(\widetilde{(\mathbf{r} - \mathbf{r}_0)} \times \mathbf{B})^2 - (\mathbf{r} - \mathbf{r}_0) \cdot \frac{\partial \mathbf{A}}{\partial t} + V, \tag{2.63}$$

where $\mathbf{B}$, $\partial \mathbf{A}/\partial t$, *and* V *are calculated at time* t_0.

Proof. (Cf. [100, Chapter 4, Proposition 105]). We look for a solution $W = W(\mathbf{r}, \mathbf{r}_0; t, t_0)$ of Hamilton–Jacobi's equation

$$\frac{\partial W}{\partial t} + H(\mathbf{r}, \nabla_{\mathbf{r}} W, t) = 0 \tag{2.64}$$

having an asymptotic expansion

$$W(\mathbf{r}, \mathbf{r}_0; t, t_0) \sim m \frac{(\mathbf{r} - \mathbf{r}_0)^2}{2(t - t_0)} + \sum_{j=0}^{\infty} W_j(\mathbf{r}, \mathbf{r}_0; t_0)(t - t_0)^j \tag{2.65}$$

for small values of $t - t_0$. Introducing the notation

$$\mathbf{A}_0 = \mathbf{A}(\mathbf{r}, t_0), \ \mathbf{A}_1 = \frac{\partial \mathbf{A}}{\partial t}(\mathbf{r}, t_0), \ U_0 = U(\mathbf{r}, t_0)$$

insertion of the asymptotic expansion (2.65) in Hamilton–Jacobi's equation (2.64) leads to the conditions

$$(\mathbf{r} - \mathbf{r}_0) \cdot (\nabla_{\mathbf{r}} W_0 - \mathbf{A}_0) = 0 \tag{2.66}$$

and

$$(\mathbf{r} - \mathbf{r}_0) \cdot (\nabla_{\mathbf{r}} W_1 - \mathbf{A}_1) + W_1 = -\frac{1}{2m} (\nabla_{\mathbf{r}} W_0 - \mathbf{A}_0)^2 - U_0. \tag{2.67}$$

Equation (2.66) is easily solved using Lemma 30, yielding

$$W_0 = (\mathbf{r} - \mathbf{r}_0) \cdot \widetilde{\mathbf{A}_0}, \tag{2.68}$$

where $\widetilde{\mathbf{A}}_0 = \widetilde{\mathbf{A}}(\mathbf{r}, \mathbf{r}_0; t_0)$ is the average of $\mathbf{A}_0 = \mathbf{A}(\mathbf{r}, t_0)$ on the line segment $[\mathbf{r}_0, \mathbf{r}]$. The solution of equation (2.67) is

$$W_1 = \left(-\frac{1}{2m} (\nabla_{\mathbf{r}} W_0 - \mathbf{A}_0)^2 + (\mathbf{r} - \mathbf{r}_0) \cdot \mathbf{A}_1 - U_0 \right)^{\sim}$$

$$= \left(-\frac{1}{2m} (\nabla_{\mathbf{r}} W_0 - \mathbf{A}_0)^2 \right)^{\sim} + ((\mathbf{r} - \mathbf{r}_0) \cdot \mathbf{A}_1)^{\sim} - \widetilde{U_0}$$

(we are using the notation $(f)^{\sim} = \widetilde{f}$ for typographical reasons). To complete the proof it thus suffices to show that

$$\nabla_{\mathbf{r}} W_0 - \mathbf{A}_0 = ((\mathbf{r} - \mathbf{r}_0) \times \mathbf{B})^{\sim}$$

(calculated at $(\mathbf{r}, \mathbf{r}_0; t_0)$). In view of the classical formula from vector calculus

$$\nabla_{\mathbf{r}}(\mathbf{f} \cdot \mathbf{g}) = (\mathbf{f} \cdot \nabla_{\mathbf{r}}) \cdot \mathbf{g} + (\mathbf{g} \cdot \nabla_{\mathbf{r}}) \cdot \mathbf{f} + \mathbf{f} \times (\nabla_{\mathbf{r}} \times \mathbf{g}) + \mathbf{g} \times (\nabla_{\mathbf{r}} \times \mathbf{f}),$$

we have

$$\nabla_{\mathbf{r}} W_0 - \mathbf{A}_0$$
$$= ((\mathbf{r} - \mathbf{r}_0) \cdot \nabla_{\mathbf{r}}) \widetilde{\mathbf{A}_0} + (\widetilde{\mathbf{A}_0} \cdot \nabla_{\mathbf{r}})(\mathbf{r} - \mathbf{r}_0) - \widetilde{\mathbf{A}_0} + (\mathbf{r} - \mathbf{r}_0) \times (\nabla_{\mathbf{r}} \times \widetilde{\mathbf{A}_0}),$$

where we have used the fact that

$$\nabla_{\mathbf{r}} W_0 - \mathbf{A}_0 = \nabla_{\mathbf{r}}((\mathbf{r} - \mathbf{r}_0) \cdot \widetilde{\mathbf{A}_0}) - \widetilde{\mathbf{A}_0}$$

and $\nabla_{\mathbf{r}} \times (\mathbf{r} - \mathbf{r}_0) = 0$. Simplifying, we get

$$\nabla_{\mathbf{r}} W_0 - \mathbf{A}_0 = ((\mathbf{r} - \mathbf{r}_0) \cdot \nabla_{\mathbf{r}}) \widetilde{\mathbf{A}_0} + \widetilde{\mathbf{A}_0} + (\mathbf{r} - \mathbf{r}_0) \times (\nabla_{\mathbf{r}} \times \overline{\mathbf{A}_0}),$$

that is

$$\nabla_{\mathbf{r}} W_0 - \mathbf{A}_0 = (\mathbf{r} - \mathbf{r}_0) \times (\nabla_{\mathbf{r}} \times \widetilde{\mathbf{A}_0}).$$

By definition of $\overline{\mathbf{A}_0}$ we have

$$\nabla_{\mathbf{r}} \times \widetilde{\mathbf{A}_0}(\mathbf{r}, \mathbf{r}_0; t_0) = \int_0^1 s \left(\nabla_{\mathbf{r}} \times \mathbf{A} \right) (s\mathbf{r} + (1 - s)\mathbf{r}_0; t_0) \, ds$$

hence, since $\nabla_{\mathbf{r}} \times \mathbf{A} = \mathbf{B}$:

$$\nabla_{\mathbf{r}} \times \widetilde{\mathbf{A}_0}(\mathbf{r}, \mathbf{r}_0; t_0) = \int_0^1 s\mathbf{B}(s\mathbf{r} + (1 - s)\mathbf{r}_0; t_0) ds$$

yielding

$$(\mathbf{r} - \mathbf{r}_0) \times (\nabla_{\mathbf{r}} \times \overline{\mathbf{A}_0}) = \int_0^1 s(\mathbf{r} - \mathbf{r}_0) \times \mathbf{B}(s\mathbf{r} + (1 - s)\mathbf{r}_0; t_0) \, ds$$
$$= \overline{(\mathbf{r} - \mathbf{r}_0) \times \mathbf{B}}.$$

Thus

$$\nabla_{\mathbf{r}} W_0 - \mathbf{A}_0 = ((\mathbf{r} - \mathbf{r}_0) \times \mathbf{B})^{\sim} \tag{2.69}$$

as claimed, and this completes the proof of the proposition. $\square$

The result above extends to the case of n numbers of freedom in a straightforward way; we refer to [100, Chapter 4, Proposition 107], for a statement and a proof.

2.3. The Van Vleck Density

The Van Vleck determinant was defined by John Hasbrouck Van Vleck [278, 279], who won the Nobel Prize in 1977 for his contributions to the understanding of the behavior of electrons in magnetic solids. Van Vleck's original insight went as follows (see [250, 251] for a detailed discussion): consider the Hamilton–Jacobi equation

$$\frac{\partial \Phi}{\partial t} + H(x, \nabla_x \Phi, t) = 0$$

and assume, as in Proposition 20 (Jacobi), that the solution Φ depends in addition to positions and time, on non-additive constants $\alpha_1, \ldots, \alpha_n$, and that the Hessian matrix

$$\Phi''_{x,\alpha}(x, t, \alpha) = \left(\frac{\partial^2 \Phi}{\partial x_j \partial \alpha_k}(x, t, \alpha) \right)_{1 \leq j,k \leq n}$$

is invertible. Van Vleck showed that one could quite generally derive from Φ a conserved quantity in configuration space, namely the determinant

$$D = \det \Phi''_{x,\alpha}(x, t, \alpha)$$

and that D satisfied the equation of continuity

$$\frac{\partial D}{\partial t} + \nabla_x(Dv) = 0,$$

where the velocity vector is determined by Φ. In this section we are going to retrieve rigorously this result when the solution Φ is the generating function W previously studied (in which case the constants α_j are the positions x'_j). Van Vleck's aim was to construct a semi-classical form of the quantum propagator and the determinant D he introduced is therefore often thought to be an object whose use is limited to semi-classical approximations (Wentzel–Kramers–Brillouin theory), or to non-rigorous justifications of the Feynman "integral". We will see that the Van Vleck determinant has a perfectly classical — and rigorous! — interpretation as a density of trajectories in Hamiltonian mechanics.

2.3.1. The trajectory density

Consider a system with time-dependent Hamiltonian

$$H(x, p, t) = \sum_{j=1}^{n} \frac{1}{2m_j} (p_j - A_j(x, t))^2 + V(x, t).$$

The associated Hamilton equations are

$$\dot{x}_j = \frac{1}{m_j}\,(p_j - A_j) \quad \text{and} \quad \dot{p}_j = \frac{1}{m_j}\,(p_j - A_j)\frac{\partial A_j}{\partial x_j} - \frac{\partial V}{\partial x_j}.$$

Differentiating the first equation with respect to t and then inserting the value of $\dot{p}_j$ given by the second equation, we see that the position coordinates x_j satisfy the following system of n coupled second-order differential equations:

$$\ddot{x}_j + \frac{1}{m_j}\left(\frac{\partial V}{\partial x_j}(x,t) + \frac{\partial A_j}{\partial t}(x,t)\right) = 0$$

$(1 \leq j \leq n)$. This system, which describes the motion of the particle in configuration space, has a unique solution $x(t) = (x_1(t),\ldots,x_n(t))$ for each set of initial conditions $x_j(t') = x'$, $\dot{x}_j(t') = \dot{x}'$. Recall that if the time interval $|t - t'|$ is small enough, then there will exist a unique trajectory in configuration space joining two points x', x in a time $t - t'$, and both the initial and final velocities v' and v are unambiguously determined by the datum of (x',t') and (x,t). Suppose now that we vary the initial velocity v' by a small amount. We will then obtain another trajectory from x' at time t', and passing close to x at time t. Repeating the procedure a great number of times, we will obtain a whole family of trajectories emanating from x' at time t'. These trajectories may eventually intersect, but as long as t is sufficiently close to t', they will spread and form a "fan" of non-intersecting curves in configuration space. We now ask the following question:

"What is the relation between the deviations Δx from the arrival point x corresponding to small changes $\Delta p'$ of the initial momentum p'?"

Let us pause and consider the following example: a tennis ball is being smashed by a player in the x,y-plane from a point (x',y') at time t' to a point (x,y) with an initial momentum vector $\mathbf{p}' = (p'_x, p'_y)$. The different trajectories through (x',y') will never cross each other outside that point. Let us now change slightly the initial momentum vector: we replace $\mathbf{p}'$ with $\mathbf{p}' + \Delta\mathbf{p}'$. The position vector $\mathbf{r}$ will then be changed into some new position vector $\mathbf{r}+\Delta\mathbf{r}$, the coordinate increments Δx and Δy being given by $\Delta x = (t - t')p'_x/m$ and $\Delta y = (t - t')p'_y/m$. We can evaluate quantitatively the change of trajectory by introducing the Jacobian determinant of the transformation $\mathbf{r} \mapsto \mathbf{p}'$:

$$\det \frac{\partial(p'_x, p'_y)}{\partial(x, y)} = \begin{vmatrix} \dfrac{m}{t - t'} & 0 \\ 0 & \dfrac{m}{t - t'} \end{vmatrix} = \left(\frac{m}{t - t'} \right)^2. \tag{2.70}$$

This quantity measures the rate of variation of the "number" of trajectories arriving at (x, y) at time t, when one changes the initial momentum allowing that point to be reached from (x', y') at time t'. We can thus view the determinant (2.70) as a measure of the "density of trajectories" arriving at (x, y, t) from (x', y', t'). Notice that this density becomes infinite when $t \to t'$; this can be intuitively interpreted by saying that for small $\Delta \mathbf{p}$, smaller values of $t - t'$ lead to smaller position fluctuations $\Delta \mathbf{r}$: if we diminish $t - t'$ there will be a "greater concentration" of trajectories coming from (x', y', t') in the neighborhood of (x, y, t).

In the case of a general Hamiltonian, one proceeds exactly in the same way, considering the limit, as $\Delta x \to 0$ in $\mathbb{R}^n_x$, of the determinant of the matrix

$$\frac{\Delta p'}{\Delta x} = \left(\frac{\Delta p'_i}{\Delta x_j} \right)_{1 \le i, j \le n}. \tag{2.71}$$

By definition, the limit of the determinant of the matrix (2.71) is called (when it exists) the *Van Vleck determinant*, or the *Van Vleck density of trajectories*. It is denoted by $\rho(x, x'; t, t')$. Thus, by definition:

$$\rho(x, x'; t, t') = \det \frac{\partial p'}{\partial x}(x, x'; t, t'). \tag{2.72}$$

We mention that the Jacobian matrix $\partial p' / \partial x$ is sometimes called the *Jacobi field* in the literature.

Of course ρ can take negative values, so it is not a "density" in the usual sense. Also notice that we have already encountered the Jacobian matrix $\partial p' / \partial x$ when discussing the notion of free symplectomorphism; in fact we established that

$$\det \frac{\partial x}{\partial p'} = \frac{\det \partial(x, x')}{\partial(p', x')}$$

(formula (2.25)), hence we can rewrite definition (2.72) as

$$\rho(x, x'; t, t') = \left(\det \frac{\partial(x, x')}{\partial(p', x')} \right)^{-1}. \tag{2.73}$$

It turns out that the Van Vleck determinant will exist provided that $t-t'$ is sufficiently small, because it is related to the existence of a generating function for the flow, and can be expressed in terms of that function.

Proposition 33. *There exists $\varepsilon > 0$ such that the density of trajectories is defined for all $0 < |t - t'| < \varepsilon$. In fact, $\rho(x, x'; t, t')$ is defined whenever the symplectomorphism $f_{t,t'}$ defined by $(x, p) = f_{t,t'}(x', p')$ is free. When this is the case, the function ρ is given by the formula*

$$\rho(x, x'; t, t') = \det[-W''_{xx'}(x, x'; t, t')], \tag{2.74}$$

where

$$W''_{xx'} = \left(-\frac{\partial^2 W}{\partial x_i \partial x'_j}\right)_{1 \le i,j \le n} \tag{2.75}$$

is the matrix of second derivatives in x_j, x'_j of $-W(x, x'; t, t')$.

Proof. The existence of ρ follows from the fact that there exists $\varepsilon > 0$ such that $f_{t,t'}$ is free provided that $0 < |t - t'| < \varepsilon$ (see Proposition 24). Suppose from now on that this condition holds. Then the Jacobian determinant

$$\det\left(\frac{\partial(x, x')}{\partial(p', x')}\right) = \det\left(\frac{\partial x}{\partial p'}\right)$$

is different from zero (see Lemma 23). Recalling that $p = \nabla_x W(x, x'; t, t')$ and $p' = -\nabla_{x'} W(x, x'; t, t')$ we have

$$p'_i = -\frac{\partial W}{\partial x'_i}(x, x'; t, t')$$

for $1 \le i \le n$, and thus

$$\frac{\partial p'}{\partial x} = \left(-\frac{\partial^2 W}{\partial x_i \partial x'_j}\right)_{1 \le i,j \le n}$$

from which formula (2.74) follows using definition (2.72) of the Van Vleck density. $\square$

Here is an elementary but important example. Assume that the flow $(f_{t,t'})$ is linear; it thus consists of symplectic matrices: $f_{t,t'} = S_{t,t'}$

where

$$S_{t,t'} = \begin{pmatrix} A(t,t') & B(t,t') \\ C(t,t') & D(t,t') \end{pmatrix}.$$

Provided that $|t - t'| > 0$ is small enough these symplectic matrices will be free, that is $\det B(t,t') \neq 0$, in which case the generating function $W = W(x, x'; t, t')$ is given by

$$W = \tfrac{1}{2} DB^{-1}x \cdot x - (B^T)^{-1}x \cdot x' + \tfrac{1}{2} B^{-1} Ax' \cdot x'$$

where $A = A(t,t')$, etc. An immediate calculation then yields the result

$$\rho(x, x'; t, t') = \det B(t,t')^{-1}. \tag{2.76}$$

Thus, in this case, the Van Vleck density is independent of the initial and final positions x' and x.

2.3.2. The continuity equation satisfied by Van Vleck's density

We now set out to prove that the Van Vleck density satisfies the continuity equation

$$\frac{\partial \rho}{\partial t} + \operatorname{div}(\rho v) = 0,$$

where v is the velocity expressed in terms of the initial and final points x' and x. The proof is not quite straightforward, and requires a few technical results regarding the differential equation satisfied by the Jacobian determinant of systems of (autonomous, or non-autonomous) differential equations.

Consider a differential system

$$\dot{x}(t) = f(x(t), t), \ x = (x_1, \ldots, x_n), \ f = (f_1, \ldots, f_n) \tag{2.77}$$

where the f_j are real-valued functions defined in some open set $U \subset \mathbb{R}^n$. We assume that each of the solutions $x_1, \ldots, x_n$ depends smoothly on n parameters $\alpha_1, \ldots, \alpha_n$. Setting $\alpha = (\alpha_1, \ldots, \alpha_n)$, we write the solution of the system as $x = x(\alpha, t)$. It turns out that the Jacobian determinant of the mapping $(\alpha, t) \mapsto x(\alpha, t)$ satisfies a simple differential equation.

Let us first prove the following straightforward lemma on matrices depending on a parameter.

Lemma 34. *The determinant of any invertible matrix $M(t)$ depending smoothly on t satisfies the differential equation*

$$\frac{d}{dt}\det M(t) = \det M(t)\mathrm{Tr}\left(\frac{dM(t)}{dt}M^{-1}(t)\right), \qquad (2.78)$$

where Tr *is the trace.*

Proof. Replacing, if necessary, $M(t)$ by $AM(t)$ where A is a conveniently chosen constant invertible matrix, we may assume without loss of generality that $\|M(t) - I\| < 1$, and define the logarithm of $M(t)$ by the convergent series

$$\mathrm{Log}\, M(t) = \sum_{j=0}^{\infty} (-1)^{j+1}(M(t) - I)^{j-1}.$$

Writing $M(t) = \exp(\mathrm{Log}\, M(t))$ we have

$$\det M(t) = \exp(\mathrm{Tr}(\mathrm{Log}\, M(t)))$$

and hence, differentiating both sides of this equality:

$$\frac{d}{dt}\det M(t) = \left(\frac{d}{dt}\mathrm{Tr}(\mathrm{Log}\, M(t))\right)\det M(t).$$

This yields formula (2.78), because

$$\frac{d}{dt}\mathrm{Tr}(\mathrm{Log}\, M(t)) = \mathrm{Tr}\left(\frac{d}{dt}\mathrm{Log}\, M(t)\right)$$

$$= \mathrm{Tr}\left(M^{-1}(t)\frac{dM(t)}{dt}\right)$$

$$= \mathrm{Tr}\left(\frac{dM(t)}{dt}M^{-1}(t)\right),$$

where the last equality follows from the fact that we have $\mathrm{Tr}(AB) = \mathrm{Tr}(BA)$ for all $m \times m$ matrices A, B. $\qquad\square$

Formula (2.78) is a particular case of the more general "Jacobi formula" which says that the derivative of every differentiable matrix $M(t)$

(invertible, or not) is given by

$$\frac{d}{dt}\det M(t) = \text{Tr}(M^{\text{co}}(t)^T)\frac{dM(t)}{dt}, \qquad (2.79)$$

where $M^{\text{co}}(t)$ is the cofactor matrix of $M(t)$.

Lemma 35. *Let $x = x(\alpha, t)$ be a solution of the differential system (2.77) and suppose that the Jacobian determinant*

$$Y(\alpha, t) = \det\left(\frac{\partial x(\alpha, t)}{\partial(\alpha, t)}\right) \qquad (2.80)$$

does not vanish for (α, t) in some open subset D of the product $\mathbb{R}^n \times \mathbb{R}_t$. Then Y satisfies the scalar differential equation

$$\frac{\partial Y}{\partial t}(\alpha, t) = Y(\alpha, t)\text{Tr}\left(\frac{\partial f}{\partial x}(x(\alpha, t))\right). \qquad (2.81)$$

Proof. We are following Maslov and Fedoriuk [208, p. 78]. We first note that by the chain rule we have the following identity between Jacobian matrices:

$$\frac{\partial}{\partial t}\left(\frac{\partial x(\alpha, t)}{\partial(\alpha, t)}\right) = \frac{\partial f}{\partial x}(x(\alpha, t))\frac{\partial x(\alpha, t)}{\partial(\alpha, t)}. \qquad (2.82)$$

Choosing $M(t) = \partial x(\alpha, t)/\partial(\alpha, t)$ in Lemma 34 above, we see that

$$\frac{d}{dt}Y(\alpha, t) = Y(\alpha, t)\text{Tr}\left[\frac{\partial}{\partial t}\left(\frac{\partial x(\alpha, t)}{\partial(\alpha, t)}\right)\left(\frac{\partial x(\alpha, t)}{\partial(\alpha, t)}\right)^{-1}\right]$$

which is precisely equation (2.81) in view of (2.82). $\qquad \square$

Let us now prove the main result of this section.

Proposition 36. *The function $(x, t) \mapsto \rho(x, x'; t, t')$ satisfies, for fixed values of x' and t', the equation*

$$\frac{\partial \rho}{\partial t} + \text{div}(\rho v) = 0, \qquad (2.83)$$

where v is the velocity vector at x of the trajectory passing through that point at time t, and starting from x' at time t'. Thus

$$v = \nabla_p H(x, p, t) \quad \text{if } (x, p) = f_{t, t'}(x', p').$$

Proof. Keeping x' and t' fixed, we set $\rho(x,t) = \rho(x,x';t,t')$, and consider Hamilton's equations

$$\dot{x}(t) = \nabla_p H(x(t),p(t),t), \ \ x(t') = x',$$

$$\dot{p}(t) = -\nabla_x H(x(t),p(t),t), \ \ p(t') = p',$$

where p' can be varied at will. The solution $x(t)$ of the first equation is thus parameterized only by p', since x' is fixed. We may thus apply Lemma 35 with $f = \nabla_p H$, $\alpha = p'$ and

$$Y(p',t) = \det \frac{\partial x(p',t)}{\partial(p',t)} = \begin{vmatrix} \dfrac{\partial x}{\partial p'} & \dfrac{\partial x}{\partial t} \\ 0_{1\times n} & 1 \end{vmatrix}.$$

The function Y is simply the inverse of ρ, calculated at $(x(t),t)$:

$$Y(p',t) = \frac{1}{\rho(x(t),t)}$$

and (2.81) yields

$$\frac{d}{dt}\left(\frac{1}{\rho(x(t),t)}\right) = \frac{1}{\rho(x(t),t)}\operatorname{Tr}\left[\frac{\partial}{\partial x}(\nabla_p H(x(t),p(t),t))\right],$$

that is

$$\frac{d}{dt}\rho(x(t),t) + \rho(x(t),t)\operatorname{Tr}\left[\frac{\partial}{\partial x}(\nabla_p H(x(t),p(t),t))\right] = 0. \tag{2.84}$$

Now,

$$\operatorname{Tr}\left[\frac{\partial}{\partial x}(\nabla_p H(x(t),p(t),t))\right] = \nabla_x \cdot \nabla_p H(x(t),p(t),t) \tag{2.85}$$

so that (2.84) can be rewritten

$$\frac{d}{dt}\rho(x(t),t) + \rho(x(t),t)\nabla_x \cdot \nabla_p H(x(t),p(t),t) = 0.$$

On the other hand, the total derivative of $t \mapsto \rho(x(t),t)$ is

$$\frac{d}{dt}\rho(x(t),t) = \frac{\partial \rho}{\partial t}(x(t),t) + \nabla_x \rho(x(t),t)\dot{x}(t)$$

$$= \frac{\partial \rho}{\partial t}(x(t),t) + \nabla_x \rho(x(t),t) \cdot \nabla_p H(x(t),p(t),t). \tag{2.86}$$

The continuity equation (2.83) follows. $\qquad \square$

When using (2.83) one must be careful to express the velocity components in terms of the position coordinates. For instance, if H is the free particle Hamiltonian $|p|^2/2m$, the continuity equation is

$$\frac{\partial \rho}{\partial t} + \frac{\partial}{\partial x}\left(\rho \frac{p_x}{m}\right) + \frac{\partial}{\partial y}\left(\rho \frac{p_y}{m}\right) + \frac{\partial}{\partial z}\left(\rho \frac{p_z}{m}\right) = 0$$

and the density $\rho(\mathbf{r}, \mathbf{r}'; t, t') = m^2/(t - t')^2$ that we defined solves the latter only if we use the values

$$p_x = \frac{x - x'}{t - t'}, \quad p_y = \frac{y - y'}{t - t'}, \quad p_z = \frac{z - z'}{t - t'}.$$

Chapter 3

Matter Waves, Schrödinger's Equation, and Bohm's Theory

The notion of "matter wave" goes back to Louis de Broglie who suggested in 1924 that just as light has both wave-like and particle-like properties, electrons (and more generally particles) also could have wave-like properties. Erwin Schrödinger published in 1924 a paper describing how de Broglie's matter waves evolve in time: this equation is nowadays universally referred to as "Schrödinger's equation". Later, in 1952, David Bohm pushed these ideas further, viewing de Broglie's waves as in some sense "piloting" the quantum mechanical particle.

In this chapter, we show how the notion of matter wave leads, using the theory of Van Vleck's density of trajectories, together with a short-time argument, to Schrödinger's equation. We thereafter show how Bohm's theory of quantum motion emerges.

3.1. Introductory (but Important) Remarks

Schrödinger's equation

$$i\hbar\frac{\partial}{\partial t}\psi(x,t) = \left[-\frac{\hbar^2}{2m}\nabla_x^2 + V(x,t)\right]\psi(x,t),$$

which usually comes complemented with the prescription of an initial value

$$\psi(x,t_0) = \psi(x_0)$$

at some time $t = t_0$, is the master equation of quantum mechanics. It plays a role analogous to that of Newton's second law in classical mechanics. This last statement will become more precise when we discuss later in this chapter Bohm's theory of quantum motion. Erwin Schrödinger postulated his equation in 1925 as an attempt to find a wave equation describing the evolution of Louis de Broglie's matter waves. It is often said that Schrödinger actually "guessed" his equation by a heuristic argument. While his arguments were not fully rigorous, he did not pull the equation out of thin air; he actually relied quite strongly on the Hamilton–Jacobi theory (for detailed historical expositions, see [167, 210, 230]), and as we will see, this was the right thing to do: what he did was to use an optical-mechanical analogy, very much along the lines already discussed in the first chapter of this book. More about all this in a moment, but we would like to first focus on some general aspects of Schrödinger's equation, which are, as we will see, quite surprising and unexpected. This discussion is usually not found in the literature, one of the reasons being perhaps that we have become so used to the general form of Schrödinger's equation that there seems to be no matter for further discussion.

Every researcher in the modern theory of partial differential equations and microlocal analysis knows that it is usual to classify partial (linear) differential equations in three, well studied, categories:

- *Elliptic equation*: the archetype is the Laplace equation $\Delta \psi = 0$ or its obvious extension, the Helmholtz equation

$$(\Delta + k^2)\psi = f.$$

  Elliptic equations normally correspond to stationary processes; they are mathematically very well understood and their solutions have "nice" properties; in particular they cannot have more singularities than the right-hand side f of the equation has (elliptic operators are "hypoelliptic", to use modern terminology).
- *Hyperbolic equation*: here the archetype is the wave equation

$$\frac{1}{c^2}\frac{\partial^2 \psi}{\partial t^2} - \Delta \psi = 0$$

  and its variants. Hyperbolic equations typically describe the propagation of a wave, or of a perturbation; the theory of these equations is very well understood, although there still exist a few open problems of a rather technical nature, which are being studied.

- *Parabolic equation*: this is typically the heat equation

$$k\frac{\partial\psi}{\partial t} - \Delta\psi = f$$

where k is a *real* factor. Parabolic equations describe diffusion processes; their mathematical study can sometimes be reduced to that of elliptic equations, with which they share many properties (in particular, hypoellipticity). A notable fact is that parabolic equations make a distinction between "past" and "future": as every engineering student knows from textbook examples, the heat equation can be solved for positive times only.

While one can safely say that practically every important partial differential equation from mathematical physics or engineering belongs to one of the categories above, Schrödinger's equation is an outstanding exception. Being an equation describing wave propagation, one could have expected that it should be of the hyperbolic type, but this is not the case because in hyperbolic equations the solution is differentiated twice with respect to time (which is at the origin of the fact that waves can be written as a superposition of two waves, one "incoming", the other "outgoing"); in Schrödinger's equation only the first time-derivative appears. Neither is Schrödinger's equation of the parabolic type: for an equation to qualify as being parabolic the constant k in front of the time derivative must be, as already stressed, real. In Schrödinger's equation this constant is the pure imaginary quantity $i\hbar$, and the change of k into $i\hbar$ fatally destroys the parabolic character of the equation, which can no longer be studied using standard methods. So, is Schrödinger's equation impossible to classify? No, because it has a deep meaning in functional analysis, via Marshall Stone's theorem on one-parameter groups of operators. Because of the importance of this fact, we have to explain the underlying mathematics in detail. Let $\mathcal{H}$ be an infinite-dimensional Hilbert space (which the reader can identify, if he wishes so, with $L^2(\mathbb{R}^n)$). A strongly continuous unitary one-parameter group $(U_t)_{t\in\mathbb{R}}$ on $\mathcal{H}$ is a family of unitary operators $U_t : \mathcal{H} \to \mathcal{H}$ indexed by $\mathbb{R}$ and satisfying $\lim_{t\to t_0} U_t\psi = U_{t_0}\psi$ for every $\psi \in \mathcal{H}$ (and every t_0) and such that

$$U_t U_{t'} = U_{t+t'}, \ U_0 = I_\mathrm{d}$$

for all t, t' (group property). Stone's theorem [27] from functional analysis tells us that there exists a unique (not necessarily bounded) self-adjoint operator A on $\mathcal{H}$ such that for every t we have

$$U_t = e^{iAt}$$

where A (the "infinitesimal generator") is uniquely determined by the group $(U_t)_{t \in \mathbb{R}}$. Now, let us make an innocuous change of notation: replace, in the formula above, A with $-\widehat{H}/\hbar$; there is nothing tricky here, and mathematically we do not even have to justify such a trivial change of notation. The formula above becomes

$$U_t = e^{-i\widehat{H}t/\hbar},$$

where $\widehat{H}$ is a self-adjoint operator on $\mathcal{H}$ (we will identify $\widehat{H}$ with the infinitesimal generator of the unitary group $(U_t)_{t \in \mathbb{R}}$). Letting U_t act on some $\psi_0 \in \mathcal{H}$ we set

$$\psi = U_t \psi_0 = e^{-i\widehat{H}t/\hbar} \psi_0$$

and formally differentiating this equality with respect to t we get

$$i\hbar \frac{\partial \psi}{\partial t} = \widehat{H} \psi$$

which reduces to Schrödinger's equation if the infinitesimal generator $\widehat{H}$ is the partial differential operator

$$\widehat{H} = -\frac{\hbar^2}{2m} \nabla_x^2 + V(x, t).$$

(This formal analogy has been observed by other authors, for instance Jauch [168] in his *Foundations of Quantum Mechanics*.) The main observation to keep in mind is that the self-adjointness of the operator $\widehat{H}$ is equivalent to the unitariness of the one-parameter group $(U_t)_{t \in \mathbb{R}}$ (called in quantum mechanics the "evolution group" or "quantum propagator"). This is well known, but we repeat the argument for completeness. Identifying now the Hilbert space $\mathcal{H}$ with $L^2(\mathbb{R}^n)$ it is clear that the operator $\widehat{H}$ is essentially self-adjoint, in the sense that

$$\int (\widehat{H}\psi(x))^* \phi(x) dx = \int \psi(x)^* \widehat{H}\phi(x) dx.$$

Let us show that this implies (and is equivalent to) the fact that the $\widehat{U}_t$ are unitary operators on $L^2(\mathbb{R}^n)$. Set

$$u(t) = \langle U_t\psi | U_t\psi \rangle$$

(where ψ is in, say, $\mathcal{S}(\mathbb{R}^n)$); differentiating the function u with respect to t and using the product rule we have

$$i\hbar \frac{du}{dt}(t) = \langle \widehat{H}U_t\psi | U_t\psi \rangle - \langle U_t\psi | \widehat{H}U_t\psi \rangle$$

and this zero if and only if

$$\langle \widehat{H}U_t\psi | U_t\psi \rangle = \langle U_t\psi | \widehat{H}U_t\psi \rangle,$$

that is if and only if $\widehat{H}$ is (formally) self-adjoint. But this is equivalent to saying that $\frac{du}{dt} = 0$, that is, $u(t) = u(0)$ for all t, which is in turn equivalent to the condition

$$\langle U_t\psi | U_t\psi \rangle = \langle \psi | \psi \rangle$$

showing that U_t is unitary. (The pure mathematician will object that we have been a little bit sloppy in our argumentation, because the notion of adjointness we are using (formal adjointness) is related in a subtle way to the usual notion of adjointness common from functional analysis; this slight sloppiness does however not affect in any way the correctness of the overall argumentation.)

3.2. Matter Waves and the Free Particle

Isaac Newton (and before him, Pierre Gassendi) believed that light had a corpuscular nature; however when the corpuscular theory failed to adequately explain diffraction phenomena, interference and the polarization of light, the corpuscular theory was abandoned in favor of Christiaan Huygens' wave theory. Funnily enough, as we know, both Newton and Huygens were right — but for different reasons! It was the merit of Louis de Broglie to discover that matter also displayed this kind of strange duality.

3.2.1. The free particle's phase

Consider a particle with mass m moving freely with velocity $\mathbf{v}$ in physical space $\mathbb{R}^3_\mathbf{r}$; its Hamiltonian function is thus

$$H(\mathbf{r}, \mathbf{p}) = \frac{p^2}{2m}, \ p^2 = \mathbf{p} \cdot \mathbf{p}.$$

In conformity with de Broglie's matter wave postulate, this particle is associated with a plane having phase

$$\Theta_{\mathrm{rel}}(\mathbf{r}, t) = \mathbf{k} \cdot \mathbf{r} - \omega(\mathbf{k})t + C; \tag{3.1}$$

here the wave vector $\mathbf{k}$ and the frequency $\omega(\mathbf{k})$ are defined by the *relativistic* equations

$$\mathbf{k} = \frac{m\mathbf{v}}{\hbar}, \ \omega(\mathbf{k}) = \frac{mc^2}{\hbar}.$$

In (3.1), C is an arbitrary real constant to be fixed following our needs. Expressing (3.1) in terms of the momentum $\mathbf{p} = \hbar\mathbf{k}$ and energy $E = \hbar\omega(\mathbf{k})$, we get

$$\Theta_{\mathrm{rel}}(\mathbf{r}, t) = \frac{1}{\hbar}\left(\mathbf{p} \cdot \mathbf{r} - mc^2 t\right) + C. \tag{3.2}$$

The crucial step for the derivation of Schrödinger's equation — which is a *nonrelativistic equation* — is to observe that for small values of $|\mathbf{v}|/c$ the energy has the following asymptotic expansion:

$$mc^2 = \frac{m_0 c^2}{\sqrt{1 - (v/c)^2}} = m_0 c^2 + \frac{1}{2}m_0 v^2 + O\left(\frac{v^4}{c^2}\right). \tag{3.3}$$

This allows us to rewrite (3.2) as

$$\Theta_{\mathrm{rel}}(\mathbf{r}, t) = \frac{1}{\hbar}\Phi(\mathbf{r}, t) - \frac{m_0 c^2}{\hbar}t + O\left(\frac{v^4}{c^2}\right), \tag{3.4}$$

where Φ is the function,

$$\Phi(\mathbf{r}, t) = \mathbf{p}_0 \cdot \mathbf{r} - \frac{p_0^2}{2m}t + C\hbar \tag{3.5}$$

($\mathbf{p}_0 = m_0\mathbf{v}$, $p_0 = |\mathbf{p}_0|$). When the velocity $\mathbf{v}$ is small, we can neglect the terms $O(v^4/c^2)$ in (3.4), so that $\Theta_{\text{rel}}(\mathbf{r}, t)$ is approximated by

$$\Theta'_{\text{rel}}(\mathbf{r}, t) = \frac{1}{\hbar}\Phi(\mathbf{r}, t) - \frac{m_0 c^2}{\hbar}t + C\hbar. \tag{3.6}$$

Now, there is no point in keeping the term $m_0 c^2 t/\hbar$ (its presence affects neither the phase nor the group velocities) so that we can take as definition of the phase

$$\Theta(\mathbf{r}, t) = \frac{1}{\hbar}\left(\Phi(\mathbf{r}, t) + C\right) \tag{3.7}$$

and fix the constant C by requiring that, at time $t = t_0$, the equation $\Theta = 0$ determines the phase plane $\mathbf{p} \cdot \mathbf{r} = \mathbf{p}_0 \cdot \mathbf{r}_0$. This leads to the relation

$$\Phi(\mathbf{r}, t) = \mathbf{p}_0 \cdot (\mathbf{r} - \mathbf{r}_0) - \frac{p_0^2}{2m}(t - t_0) \tag{3.8}$$

so the phase Φ is simply the gain in action when the free particle proceeds from $\mathbf{r}_0$ at time t_0 to $\mathbf{r}$ at time t with velocity $\mathbf{v}_0 = \mathbf{p}_0/m$: we have

$$\Phi(\mathbf{r}, t) = \int_{\mathbf{r}_0, t_0}^{\mathbf{r}, t} \mathbf{p}dr - Hdt,$$

where the integral is calculated along the path (here a line segment) joining the initial point $\mathbf{r}_0, t_0$ to the final point $\mathbf{r}, t$; we are using the notation $\mathbf{p}dr = p_x dx + p_y dy + p_z dz$. It follows (as can be directly seen from (3.8)) that Φ is a solution of the Hamilton–Jacobi initial value problem for the free particle Hamiltonian

$$\frac{\partial \Phi}{\partial t} + \frac{1}{2m}(\nabla_{\mathbf{r}}\Phi)^2 = 0, \quad \Phi(\mathbf{r}, t_0) = \mathbf{p}_0 \cdot (\mathbf{r} - \mathbf{r}_0).$$

This last observation provides the gateway to the derivation of Schrödinger's equation for the free particle.

3.2.2. The free particle propagator

The phase of a matter wave is defined on the time-dependent phase space $\mathbb{R}_{\mathbf{r}} \times \mathbb{R}_{\mathbf{p}} \times \mathbb{R}_t$; as is the case for energy, the phase has no absolute meaning. For instance, it depends on the momentum vector $\mathbf{p}_0$, which can take arbitrary values. For a free particle with mass m, the choice of the momentum can thus be any vector, and unless we have measured it, all the "potentialities" associated with these possible phases are present.

Since there is no reason for privileging some particular "origin" $(\mathbf{r}_0, \mathbf{p}_0, t_0)$, we write $(\mathbf{r}', \mathbf{p}', t')$ instead of $(\mathbf{r}_0, \mathbf{p}_0, t_0)$, and set

$$\Phi_{\mathbf{p}'}(\mathbf{r}, \mathbf{r}'; t, t') = \mathbf{p}' \cdot (\mathbf{r} - \mathbf{r}') - \frac{p'^2}{2m}(t - t'), \tag{3.9}$$

where $p'^2 = \mathbf{p}' \cdot \mathbf{p}'$. We next define the function

$$G_{\text{free}}(\mathbf{r}, \mathbf{r}'; t, t') = \left(\frac{1}{2\pi\hbar}\right)^3 \int e^{\frac{i}{\hbar}\Phi_{\mathbf{p}'}(\mathbf{r}, \mathbf{r}'; t, t')} d^3\mathbf{p}', \tag{3.10}$$

where $d^3\mathbf{p}' = dp'_x dp'_y dp'_z$. The integral in (3.10) is a Fresnel-type integral; it is convergent (but of course not absolutely convergent); we will calculate it in a moment. We first note that it immediately follows from (3.9) by differentiation of (3.10) under the summation sign that G satisfies the Schrödinger equation

$$i\hbar\frac{\partial}{\partial t}G_{\text{free}} = -\frac{\hbar^2}{2m}\nabla^2_{\mathbf{r}}G_{\text{free}}. \tag{3.11}$$

Let us calculate the limit of G_{free} as $t \to t'$. In view of the Fourier inversion formula

$$\frac{1}{2\pi}\int_{-\infty}^{+\infty} e^{ikx} dk = \delta(x),$$

we have

$$\left(\frac{1}{2\pi\hbar}\right)^3 \int e^{\frac{i}{\hbar}\mathbf{p}\cdot(\mathbf{r}-\mathbf{r}')} d^3\mathbf{p} = \delta(\mathbf{r} - \mathbf{r}')$$

and hence

$$\lim_{t \to t'} G_{\text{free}}(\mathbf{r}, \mathbf{r}'; t, t') = \delta(\mathbf{r} - \mathbf{r}'). \tag{3.12}$$

The function G_{free} is a "propagator" for Schrödinger's equation.

Proposition 37. *Let $\psi' \in \mathcal{S}(\mathbb{R}^3_{\mathbf{r}})$. For $t \neq t'$ the function*

$$\psi(\mathbf{r}, t) = \int G_{\text{free}}(\mathbf{r}, \mathbf{r}'; t, t')\psi'(\mathbf{r}') d^3\mathbf{r}' \tag{3.13}$$

$(d^3\mathbf{r}' = dx'dy'dz')$ *is the solution of Schrödinger's Cauchy problem*

$$i\hbar\frac{\partial\psi}{\partial t} = -\frac{\hbar^2}{2m}\nabla_{\mathbf{r}}^2\psi, \quad \lim_{t\longrightarrow t'}\psi(\cdot,t) = \psi'. \tag{3.14}$$

Proof. The fact that ψ is a solution of Schrödinger's equation follows from (3.13) using (3.11). The equality $\lim_{t\to t'}\psi(\cdot,t') = \psi'$ immediately follows from (3.12) since we have

$$\lim_{t\longrightarrow t'}\psi(\mathbf{r},t) = \int G_{\text{free}}(\mathbf{r},\mathbf{r}';t,t')\psi'(\mathbf{r}')\,d^3\mathbf{r}'$$

$$= \int \delta(\mathbf{r}-\mathbf{r}')\psi'(\mathbf{r}')\,d^3\mathbf{r}'$$

$$= \psi'(\mathbf{r}). \qquad \square$$

3.2.3. An explicit expression for G_{free}

The free-particle propagator G_{free} is expressed in integral form by formula (3.10). It can actually be evaluated explicitly. For this purpose we will make use of the following well-known result from the theory of Fresnel integrals.

Lemma 38. *Let λ be a real number, $\lambda \neq 0$. Then*

$$\frac{1}{\sqrt{2\pi}}\int_{-\infty}^{+\infty} e^{-iuv}e^{i\lambda u^2/2}\,du = (e^{i\pi/4})^{\text{sign}(\lambda)}|\lambda|^{-1/2}e^{-iv^2/2\lambda}, \tag{3.15}$$

where $\text{sign}(\lambda) = +1$ *if* $\lambda > 0$ *and* $\text{sign}(\lambda) = -1$ *if* $\lambda < 0$.

For a proof, see any textbook on integral calculus.

Proposition 39. *The Green function G_{free} is given by the formula*

$$G_{\text{free}}(\mathbf{r},\mathbf{r}';t,t') = \left(\frac{m}{2\pi i\hbar(t-t')}\right)^{3/2}\exp\left(\frac{i}{\hbar}W_{\text{free}}(\mathbf{r},\mathbf{r}';t,t')\right), \tag{3.16}$$

where

$$\left(\frac{m}{2\pi i\hbar(t-t')}\right)^{3/2} = (e^{i\frac{3\pi}{4}})^{\text{sign}(t-t')}\left(\frac{m}{2\pi i\hbar|t-t'|}\right)^{3/2} \tag{3.17}$$

and W_{free} *is the free-particle generating function, i.e.,*

$$W_{\text{free}}(\mathbf{r}, \mathbf{r}'; t, t') = m \frac{(\mathbf{r} - \mathbf{r}')^2}{2(t - t')}. \tag{3.18}$$

Proof. Let $\mathbf{r} = (x, y, z)$ and $\mathbf{r}' = (x', y', z')$. We have $G_{\text{free}} = G_x \otimes G_y \otimes G_z$ where

$$G_x = \frac{1}{2\pi\hbar} \int_{-\infty}^{+\infty} \exp\left[\frac{i}{\hbar} \left(p_x(x - x') - \frac{p_x'^2}{2m}(t - t') \right) \right] dp_x'$$

and similar definitions for G_y and G_z. Setting $u = p_x$, $v = -(x - x')/\hbar$ and $\lambda = -(t - t')/m\hbar$ in (3.15) we get

$$\hbar\sqrt{2\pi} G_x = (e^{-i\frac{\pi}{4}})^{\text{sign}(t-t')} \sqrt{\frac{m\hbar}{|t - t'|}} \exp\left[\frac{i}{\hbar} m \frac{(x - x')^2}{2(t - t')} \right].$$

Performing similar calculations with G_y and G_z we get (3.16).   □

Remark 40. Formula (3.17) corresponds to the argument choices $\arg i = \pi/2$ and

$$\arg(t - t') = \begin{cases} 0 & \text{if } t - t' > 0, \\ \pi & \text{if } t - t' < 0. \end{cases} \tag{3.19}$$

The result above can be extended without difficulty to the case of systems with an arbitrary number n of degrees of freedom. In fact, let

$$i\hbar \frac{\partial \psi}{\partial t} = -\sum_{j=1}^{n} \frac{\hbar^2}{2m_j} \frac{\partial^2 \psi}{\partial x_j^2}$$

be Schrödinger's equation for such a system. The corresponding Green function is then given by

$$G_{\text{free}}(x, x'; t, t') = \left(\frac{m \cdots m_n}{2\pi i\hbar(t - t')} \right)^{n/2} \exp\left(\frac{i}{\hbar} W_{\text{free}}(x, x'; t, t') \right), \tag{3.20}$$

where the argument of $t - t'$ is given by (3.19); W_{free} is, as before, the free-particle generating function:

$$W_{\text{free}}(x, x'; t, t') = \sum_{j=1}^{n} m_j \frac{(x_j - x_j')^2}{2(t - t')}. \tag{3.21}$$

3.3. Schrödinger's Equation

Until now we have been discussing the matter wave associated with a single particle, moving in the physical three-dimensional space $\mathbb{R}_r^3$. We now ascribe to the n-dimensional configuration space $\mathbb{R}_x^n$ as much physical reality as we do to $\mathbb{R}_r^3$. As Holland [160, p. 278] emphasizes, this is indeed a radical step, but it is a necessary one if we wish to extend the theory of the Schrödinger equation to embrace many-particle systems.

3.3.1. A classical propagator

Let us now try to generalize the derivation of the Schrödinger equation for a system of free particles to the case where a potential is present; i.e., the Hamiltonian function is now

$$H(z, t) = \sum_{j=1}^{n} \frac{p_j^2}{2m_j} + V(x, t).$$

Recall that the free propagator is given by

$$G_{\text{free}}(x, x'; t, t') = \left(\frac{m_1 \cdots m_n}{2\pi i \hbar (t - t')} \right)^{n/2} \exp\left(\frac{i}{\hbar} W_{\text{free}}(x, x'; t, t') \right), \quad (3.22)$$

where

$$W_{\text{free}}(x, x'; t, t') = \sum_{j=1}^{n} m_j \frac{(x_j - x_j')^2}{2(t - t')} \quad (3.23)$$

is the free-particle generating function. In view of Proposition 33, the associated Van Vleck density is

$$\rho_{\text{free}}(x, x'; t, t') = \left(\frac{m_1 \cdots m_n}{t - t'} \right)^n \quad (3.24)$$

so that we may rewrite the propagator as

$$G_{\text{free}}(x, x'; t, t') = \left(\frac{1}{2\pi i \hbar} \right)^{n/2} \sqrt{\rho_{\text{free}}}(x, x'; t, t') \exp\left(\frac{i}{\hbar} W_{\text{free}}(x, x'; t, t') \right).$$

$$(3.25)$$

This suggests that in the general case the propagator might be given by the following "obvious" extension of formula (3.25),

$$G_{cl}(x, x'; t, t') = \left(\frac{1}{2\pi i\hbar} \right)^{n/2} \sqrt{\rho(x, x'; t, t')} \exp \left(\frac{i}{\hbar} W(x, x'; t, t') \right),$$

(3.26)

where W is the generating function determined by the Hamiltonian H and ρ the corresponding Van Vleck density (we are writing G_{cl} where the subscript cl stands for "classical": the function G_{cl} is, after all, constructed — as is G_{free} — using only objects from classical mechanics). That this Ansatz is not in general quite true is shown by Proposition 42 below, whose proof requires the following lemma.

Lemma 41. *The square root $a = \sqrt{\rho}$ of Van Vleck's density satisfies the equation*

$$\frac{\partial a}{\partial t} + \sum_{j=1}^{n} \frac{1}{m_j} \frac{\partial W}{\partial x_j} \frac{\partial a}{\partial x_j} + \frac{1}{2} a \sum_{j=1}^{n} \frac{1}{m_j} \frac{\partial^2 W}{\partial x_j^2} = 0.$$

(3.27)

Proof. Writing the continuity equation (2.83) in the form

$$\frac{\partial \rho}{\partial t} + \rho \operatorname{div} v + v \cdot \nabla_x \rho = 0$$

formula (3.27) follows, replacing ρ with a^2 and observing that the velocity vector v is given by

$$v = \left(\frac{1}{m_1} \frac{\partial W}{\partial x_1}, \ldots, \frac{1}{m_n} \frac{\partial W}{\partial x_n} \right). \qquad \Box$$

Proposition 42. *Assume that the generating function W determined by H exists for all t such that $0 < |t - t'| < \varepsilon$.*

(i) *For fixed x', t' the function G_{cl} satisfies the equation*

$$i\hbar \frac{\partial G_{cl}}{\partial t} = (\widehat{H} - Q) G_{cl},$$

(3.28)

where the function Q is given by

$$Q = -\sum_{j=1}^{n} \frac{\hbar^2}{2m_j} \frac{\partial^2 \sqrt{\rho}}{\partial x_j^2}.$$

(3.29)

(ii) *As $t \to t'$ the function G has the limit*

$$\lim_{t \to t'} G_{\mathrm{cl}}(x, x'; t, t') = \delta(x - x').$$ (3.30)

Proof. For notational simplicity we give the proof in the case $n = 1$.

(i) Setting $R = (2\pi i\hbar)^{-n/2}\sqrt{\rho}$ we have

$$i\hbar \frac{\partial G_{\mathrm{cl}}}{\partial t} = e^{\frac{i}{\hbar}W} \left(-\frac{\partial W}{\partial t} R + i\hbar \frac{\partial R}{\partial t} \right)$$ (3.31)

and, similarly,

$$\frac{\partial^2 G_{\mathrm{cl}}}{\partial x_j^2} = e^{\frac{i}{\hbar}W} \left[\frac{\partial^2 R}{\partial x_j^2} + \frac{2i}{\hbar} \frac{\partial R}{\partial x_j} \frac{\partial W}{\partial x_j} + \frac{i}{\hbar} R \frac{\partial^2 W}{\partial x_j^2} - \frac{1}{\hbar^2} \left(\frac{\partial W}{\partial x_j} \right)^2 R \right].$$

It follows that

$$i\hbar \frac{\partial G_{\mathrm{cl}}}{\partial t} - \widehat{H} G_{\mathrm{cl}} = -R e^{\frac{i}{\hbar}W} \left(\frac{\partial W}{\partial t} - H(x, \nabla_x W, t) \right)$$

$$+ e^{\frac{i}{\hbar}W} \left[i\hbar \frac{\partial R}{\partial t} + \sum_{j=1}^{n} \frac{\hbar^2}{2m_j} \frac{\partial^2 R}{\partial x_j^2} \right.$$

$$\left. + \frac{i\hbar}{m_j} \frac{\partial R}{\partial x_j} \frac{\partial W}{\partial x_j} + \frac{i\hbar}{2m_j} R \frac{\partial^2 W}{\partial x_j^2} \right].$$

Taking Hamilton–Jacobi's equation into account we have

$$\frac{\partial W}{\partial t} - H(x, \nabla_x W, t) = 0$$

and hence

$$i\hbar \frac{\partial G_{\mathrm{cl}}}{\partial t} - \widehat{H} G_{\mathrm{cl}}$$

$$= e^{\frac{i}{\hbar}W} \left[i\hbar \frac{\partial R}{\partial t} + \sum_{j=1}^{n} \frac{\hbar^2}{2m_j} \frac{\partial^2 R}{\partial x_j^2} + \frac{i\hbar}{m_j} \frac{\partial R}{\partial x_j} \frac{\partial W}{\partial x_j} + \frac{i\hbar}{2m_j} R \frac{\partial^2 W}{\partial x_j^2} \right]$$

$$= e^{\frac{i}{\hbar}W} \sum_{j=1}^{n} \frac{\hbar^2}{2m_j} \frac{\partial^2 R}{\partial x_j^2} + i\hbar e^{\frac{i}{\hbar}W} \left[\frac{\partial R}{\partial t} + \sum_{j=1}^{n} \frac{1}{m_j} \frac{\partial R}{\partial x_j} \frac{\partial W}{\partial x_j} + \frac{1}{2m_j} R \frac{\partial^2 W}{\partial x_j^2} \right].$$

In view of equation (3.27) the expression in the square bracket is zero, so we have

$$i\hbar\frac{\partial G_{\text{cl}}}{\partial t} - \widehat{H}G_{\text{cl}} = e^{\frac{i}{\hbar}W}\sum_{j=1}^{n}\frac{\hbar^2}{2m_j}\frac{\partial^2 R}{\partial x_j^2}$$

which is (3.28).

(ii) In view of formula (2.55) in Proposition 31 we have, for $t - t' \to 0$,

$$W(x, x'; t, t') = \sum_{j=1}^{n} m_j\frac{(x_j - x_j')^2}{2(t - t')} + O(t - t') \tag{3.32}$$

hence the Van Vleck determinant satisfies the estimate

$$\rho(x, x'; t, t') = \frac{m_1 \cdots m_n}{(t - t')^n}\det\left(I_{n\times n} + O((t - t')^2)\right).$$

Noting that

$$\det[I_{n\times n} + O((t - t')^2)] = 1 + O((t - t')^2) \tag{3.33}$$

this yields

$$\rho(x, x'; t, t') = \frac{m_1 \cdots m_n}{(t - t')^n}(1 + O((t - t')^2) \tag{3.34}$$

for $\Delta t \to 0$ so that

$$\sqrt{\rho}(x, x'; t, t') = \frac{(m_1 \cdots m_n)^{1/2}}{(t - t')^{n/2}}(1 + O((t - t')^2). \tag{3.35}$$

Combining formulas (3.32) and (3.35), we see that

$$G_{\text{cl}}(x, x'; t, t') = G_{\text{free}}(x, x'; t, t') + O\left((t - t')^2\right)$$

and hence

$$\lim_{t \to t'} G_{\text{cl}}(x, x'; t, t') = \lim_{t \to t'} G_{\text{free}}(x, x'; t, t') = \delta(x - x')$$

as claimed. $\square$

It follows from the result above that when we have an arbitrary potential, the function G_{cl} is the propagator of an integro-differential equation. In fact, for every $\psi' \in \mathcal{S}(\mathbb{R}^n_x)$ the function

$$\psi(x,t) = \int G_{cl}(x,x';t,t')\psi'(x')\,d^n x' \qquad (3.36)$$

satisfies the integro-differential Cauchy problem

$$i\hbar\frac{\partial\psi}{\partial t} = (\widehat{H} - \widehat{Q})\psi, \quad \psi(\cdot,t') = \psi'(\cdot), \qquad (3.37)$$

where $\widehat{Q}$ is the operator defined by

$$\widehat{Q}\psi(x,t) = \int Q(x,x';t,t')G_{cl}(x,x';t,t')\psi'(x')\,d^n x' \qquad (3.38)$$

the function Q being given by formula (3.29). In fact, differentiation under the integration sign on the right-hand side of the expression (3.36) leads to the equality

$$i\hbar\frac{\partial\psi}{\partial t} - \widehat{H}\psi = \int \left(i\hbar\frac{\partial G_{cl}}{\partial t} - \widehat{H}G_{cl}\right)\psi'd^n x'$$

$$= \frac{\hbar^2}{2m}\int \frac{\nabla^2_x\sqrt{|\rho|}}{\sqrt{|\rho|}}G_{cl}\psi'd^n x,$$

which is (3.37).

3.3.2. An exact propagator

There is however one very important case where G_{cl} is an exact propagator, namely when the potential function is a polynomial function of degree at most two in the position variables. Such a potential function can always be written in the form

$$V(x,t) = \tfrac{1}{2}M(t)x \cdot x + m(t) \cdot x,$$

where $M(t)$ is a symmetric matrix, possibly depending (in a smooth way) on time t and $m(t)$ a vector (possibly also depending smoothly on t).

Let us state and prove this essential result; we will see in the next chapter that it can be interpreted as a mathematical property of the metaplectic group.

Proposition 43. *The function G_{cl} is a true propagator, that is*

$$i\hbar\frac{\partial G_{\text{cl}}}{\partial t} = \widehat{H}G_{\text{cl}}, \quad G_{\text{cl}}(x, x'; t', t') = \delta(x - x') \tag{3.39}$$

if (and only if) the potential is a polynomial of maximal degree 2 in the position variables x_j. In this case the function

$$\psi(x, t) = \int G_{\text{cl}}(x, x'; t, t')\psi'(x')\, d^n x'$$

satisfies the Schrödinger equation with initial datum ψ' at time t':

$$i\hbar\frac{\partial \psi}{\partial t} = \widehat{H}\psi, \quad \psi(x, t') = \psi'(x).$$

Proof. It suffices to show that the term Q defined by (3.29) is equal to zero if and only if the potential is a polynomial of maximal degree 2 in the position variables x_j. Suppose that

$$V(x, t) = \tfrac{1}{2}M(t)x \cdot x + m(t) \cdot x$$

and assume first that $m(t) = 0$ so the Hamiltonian function is

$$H(z, t) = \sum_{j=1}^{n} \frac{p_j^2}{2m_j} + \frac{1}{2}M(t)x \cdot x. \tag{3.40}$$

This is a (homogeneous) quadratic function of the x, p variables, hence the flow it determines consists of symplectic matrices

$$S_{t,t'} = \begin{pmatrix} A(t, t') & B(t, t') \\ C(t, t') & D(t, t') \end{pmatrix}$$

and the generating function $W(x, x'; t, t')$ is thus a quadratic form

$$W(x, x') = \tfrac{1}{2}P(t, t')x \cdot x - L(t, t')x \cdot x' + \tfrac{1}{2}Q(t, t')x' \cdot x', \tag{3.41}$$

where

$$P(t, t') = D(t, t')B(t, t')^{-1},$$
$$L(t, t') = B^T(t, t')^{-1},$$
$$Q(t, t') = B(t, t')^{-1}A(t, t')$$

(see Proposition 3). The Van Vleck density is thus given by formula (2.76):

$$\rho(x, x'; t, t') = \det B(t, t')^{-1};$$

it is independent of the variables x, x' and hence

$$\frac{\partial^2}{\partial x_j^2} \sqrt{\rho(x, x'; t, t')} = 0 \quad \text{for } j = 1, \dots, n. \tag{3.42}$$

It follows that the term Q in formula (3.29) vanishes so that we have

$$i\hbar \frac{\partial G_{\text{cl}}}{\partial t} = \widehat{H} G_{\text{cl}}$$

and G_{cl} is thus in this case a true propagator. The case $m(t) \neq 0$ is treated in a similar way: the flow determined by the Hamiltonian (3.40) consists of affine symplectomorphisms; in view of Proposition 27 the generating function can thus be written

$$W'(x, x'; t, t') = W(x, x'; t, t') + \alpha(t, t') \cdot x + \alpha'(t, t') \cdot x',$$

where $W(x, x'; t, t')$ is of the type (3.41) above and $\alpha(t, t')$, $\alpha'(t, t')$ vectors; it follows that the relations (3.42) will still hold, and one concludes as in the homogeneous case. That the condition is also necessary is clear when $n = 1$, since in this case the vanishing of the second x-derivative of $\sqrt{\rho}$ immediately implies that ρ is a quadratic polynomial; the general case can then readily be proven by induction on the dimension n of configuration space. $\qquad \square$

3.3.3. The short-time propagator

Huygens' principle is well known from optics; it says that any point on a wavefront may be regarded as a source of spherical or circular wavelets. The sum of all such secondary wavelets emanating from the front with a speed equal to the speed of propagation of the waves is the same as the wavefront itself (it is implicitly assumed that the secondary waves travel only in the forward direction). Recall (equation (1.85)) that an infinitesimal change in phase of a light ray is given by $\Delta\Phi = 2\pi L/\lambda$.

Let us return to the classical propagator

$$G_{\text{cl}}(x, x'; t, t') = \left(\frac{1}{2\pi i\hbar}\right)^{n/2} \sqrt{\rho(x, x'; t, t')} \exp\left(\frac{i}{\hbar} W(x, x'; t, t')\right) \tag{3.43}$$

constructed above. As we have seen above the "classical wavefunction"

$$\psi(x, t) = \int G_{\text{cl}}(x, x'; t, t') \psi'(x') d^n x'$$

is not in general a solution of the Schrödinger equation since it satisfies the integro-differential equation

$$i\hbar\frac{\partial\psi}{\partial t} = \widehat{H}\psi - \widehat{Q}\psi$$

where the extra term $\widehat{Q}\psi$ is given by

$$\widehat{Q}\psi(x,t) = \int Q(x,x';t,t')G_{\mathrm{cl}}(x,x';t,t')\psi'(x')d^n x',$$

$$Q = -\sum_{j=1}^{n}\frac{\hbar^2}{2m_j}\left(\frac{1}{\sqrt{\rho}}\frac{\partial^2\sqrt{\rho}}{\partial x_j^2}\right)$$

(formula (3.37)). It easily follows from formula (3.35) that we have $\widehat{Q}\psi = O(\Delta t^2)$ for $\Delta t = t - t' \to 0$, and hence the classical wavefunction satisfies

$$i\hbar\frac{\partial\psi}{\partial t} = \widehat{H}\psi + O(\Delta t^2).$$

This strongly suggests that the classical propagator G_{cl} might be a good approximation to the true propagator G for small time intervals $t - t' = \Delta t$. That this guess is true follows from the following considerations, which ultimately lead to the construction of a "short-time propagator" whose expression is very simple, because it can be expressed directly in terms of the potential V in the Hamiltonian. Recall from Section 2.2.3 that the correct asymptotic expression for the generating function is given by

$$W(x,x';t,t') = \widetilde{W}(x,x';t,t') + O(\Delta t^2) \qquad (3.44)$$

(cf. equation (2.55)), where

$$\widetilde{W}(x,x';t,t') = \sum_{j=1}^{n} m_j\frac{(x_j - x'_j)^2}{2\Delta t} - \widetilde{V}(x,x',t')\Delta t, \qquad (3.45)$$

the function $\widetilde{V}$ being the average

$$\widetilde{V}(x,x',t') = \int_0^1 V(\lambda x + (1-\lambda)x',t')d\lambda. \qquad (3.46)$$

Let us define

$$\widetilde{G}(x,x';t,t') = \left(\frac{1}{2\pi i\hbar}\right)^{n/2}\sqrt{\widetilde{\rho}}(x,x';t,t')\exp\left(\frac{i}{\hbar}\widetilde{W}(x,x';t,t')\right),$$

$$(3.47)$$

where $\widetilde{\rho}$ is (up to the sign) the determinant of the Hessian of $\widetilde{W}$:

$$\widetilde{\rho} = \det(-\widetilde{W}''_{xx'}) \tag{3.48}$$

(cf. formula (2.74) in Proposition 33).

Proposition 44. *Let ψ be the exact solution of Schrödinger's equation*

$$i\hbar\frac{\partial\psi}{\partial t} = -\sum_{j=1}^{n}\frac{\hbar^2}{2m_j}\frac{\partial^2\psi}{\partial x_j^2} + V\psi, \quad \psi(x,t') = \psi'(x),$$

that is

$$\psi(x,t) = \int G(x,x';t,t')\psi'(x')d^n x'.$$

The function

$$\widetilde{\psi}(x,t) = \int \widetilde{G}(x,x';t,t')\psi'(x')d^n x'$$

satisfies

$$\psi(x,t) - \widetilde{\psi}(x,t) = O\left(\Delta t^2\right). \tag{3.49}$$

Proof. Let us set as usual $\Delta t = t - t'$; expanding the exact solution ψ to first order and using $\psi(x,t') = \psi'(x)$ yields

$$\psi(x,t) = \psi'(x) + \frac{\partial\psi}{\partial t}(x,t')\Delta t + O\left(\Delta t^2\right)$$

and hence, using Schrödinger's equation and the initial condition $\psi(x,t') = \psi'(x)$ to evaluate $\partial\psi/\partial t$,

$$\psi(x,t) = \left[1 + \frac{\Delta t}{i\hbar}\left(-\sum_{j=1}^{n}\frac{\hbar^2}{2m_j}\frac{\partial^2}{\partial x_j^2} + V(x,t')\right)\right]\psi'(x) + O(\Delta t^2). \tag{3.50}$$

We are going to prove that the function $\widetilde{\psi}(x,t)$ has the same expansion to order $O\left(\Delta t^2\right)$; the proposition will follow. We first notice that the Van Vleck-type determinant $\widetilde{\rho}$ is explicitly given, taking formula (3.45) into account, by

$$\widetilde{\rho}(x, x'; t, t') = \det\left(\frac{1}{\Delta t}M + \widetilde{V}''_{xx'}(x, x', t')\Delta t\right),$$

where M is the mass matrix (the diagonal matrix with positive entries the masses m_j) and

$$\widetilde{V}''_{xx'} = \left(\frac{\partial^2 \widetilde{V}}{\partial x_j \partial x'_k}\right)_{1 \leq j, k \leq n}$$

is the matrix of mixed second derivatives of $\widetilde{V}$. We have

$$\frac{1}{\Delta t}M + \widetilde{V}''_{xx'}\Delta t = \frac{1}{\Delta t}M(I_{n \times n} + M^{-1}\widetilde{V}''_{x,x'}\Delta t^2)$$

$$= \frac{1}{\Delta t}M(I_{n \times n} + O_{n \times n}(\Delta t^2)),$$

where $O_{n \times n}(\Delta t^2)$ is an $n \times n$ matrix whose entries are $O(\Delta t^2)$; taking the determinants of both sides and noting that $\det M = m_1 \cdots m_n$, we get

$$\widetilde{\rho}(x, x'; t, t') = \frac{m_1 \cdots m_n}{(\Delta t)^n}\det\left(I_{n \times n} + O_{n \times n}(\Delta t^2)\right)$$

$$= \frac{m_1 \cdots m_n}{(\Delta t)^n}(1 + O(\Delta t^2))$$

(cf. formula (3.33)). Notice that this shows that $\widetilde{\rho}$ is, up to a term $O(\Delta t^2)$, just the free particle Van Vleck density

$$\rho_{\text{free}}(t) = \frac{m_1 \cdots m_n}{(\Delta t)^n}$$

and hence

$$\sqrt{\widetilde{\rho}} = \sqrt{\rho_{\text{free}}}(1 + O(\Delta t^2))$$

(we are disregarding here any discussion of arguments of the square root) so we have shown that

$$\widetilde{\psi}(x, t) = \left(\frac{1}{2\pi i\hbar}\right)^{n/2}\sqrt{\rho_{\text{free}}(t)}\int e^{\frac{i}{\hbar}\widetilde{W}(x, x'; t, t')}\psi'(x')d^n x' + O(\Delta t^2).$$

Let us next estimate $\widetilde{\psi}(x,t)$ for $\Delta t \to 0$. Setting

$$W_{\text{free}}(x,x';t,t') = \sum_{j=1}^{n} m_j \frac{(x_j - x_j')^2}{2\Delta t}$$

(it is the free particle generating function) we have

$$e^{\frac{i}{\hbar}\widetilde{W}} = \exp\left(\frac{i}{\hbar}(W_{\text{free}} - \widetilde{V}\Delta t)\right) = e^{\frac{i}{\hbar}W_{\text{free}}}\left(1 - \frac{i}{\hbar}\widetilde{V}\Delta t\right) + O(\Delta t^2)$$

and hence

$$\widetilde{\psi}(x,t) = \left(\tfrac{1}{2\pi i \hbar}\right)^{n/2} \sqrt{\rho_{\text{free}}}(t) \int e^{\frac{i}{\hbar}W_{\text{free}}}\left(1 - \frac{i}{\hbar}\widetilde{V}\Delta t\right) \psi'(x')d^n x' + O(\Delta t^2).$$

We can write

$$\widetilde{\psi} = \widetilde{\psi}_1 - \frac{i\Delta t}{\hbar}\widetilde{\psi}_2 + O(\Delta t^2), \tag{3.51}$$

where

$$\widetilde{\psi}_1 = \left(\frac{1}{2\pi i \hbar}\right)^{n/2} \sqrt{\rho_{\text{free}}} \int e^{\frac{i}{\hbar}W_{\text{free}}}\psi'(x')d^n x' \tag{3.52}$$

and

$$\widetilde{\psi}_2 = \left(\frac{1}{2\pi i \hbar}\right)^{n/2} \sqrt{\rho_{\text{free}}} \int e^{\frac{i}{\hbar}W_{\text{free}}}\widetilde{V}\psi'(x')d^n x'. \tag{3.53}$$

The function $\widetilde{\psi}_1$ is just the solution of the free particle Schrödinger equation

$$i\hbar\frac{\partial \psi}{\partial t} = -\sum_{j=1}^{n}\frac{\hbar^2}{2m_j}\frac{\partial^2 \psi}{\partial x_j^2}, \quad \psi(x,t') = \psi'(x)$$

and thus satisfies (cf. (3.50))

$$\widetilde{\psi}_1(x,t) = \left[1 + \frac{\Delta t}{i\hbar}\left(-\sum_{j=1}^{n}\frac{\hbar^2}{2m_j}\frac{\partial^2}{\partial x_j^2}\right)\right]\psi(x,t') + O\left(\Delta t^2\right). \tag{3.54}$$

The integral in formula (3.53) for $\widetilde{\psi}_2$ can easily be evaluated using the method of stationary phase (see, e.g., [100, Lemma 239]), yielding

$$\int e^{\frac{i}{\hbar}W_{\text{free}}(x,x';t,t')}\widetilde{V}(x,x',t')\psi'(x')d^n x'$$

$$= \frac{(2\pi i\hbar\Delta t)^{n/2}}{m_1\cdots m_n}\widetilde{V}(x,x;t')\psi'(x) + O\left(\Delta t^2\right).$$

Since $\widetilde{V}(x,x;t') = V(x,t')$ we thus have

$$\int e^{\frac{i}{\hbar}W_{\text{free}}(x,x';t,t')}\widetilde{V}(x,x',t')\psi'(x')d^n x'$$

$$= \frac{(2\pi i\hbar\Delta t)^{n/2}}{m_1\cdots m_n}V(x,t')\psi'(x) + O\left(\Delta t^2\right)$$

and hence, taking (3.54) and (3.51) into account

$$\widetilde{\psi}(x,t) = \left[1 + \frac{\Delta t}{i\hbar}\left(-\sum_{j=1}^{n}\frac{\hbar^2}{2m_j}\frac{\partial^2}{\partial x_j^2}\right)\right]\psi'(x)$$

$$- \frac{i\Delta t}{\hbar}V(x,t')\psi'(x) + O(\Delta t^2);$$

the asymptotic formula (3.49) follows in view of (3.50). $\square$

Corollary 45. *Let ψ_{cl} be the classical wavefunction defined by*

$$\psi_{\text{cl}}(x,t) = \int G_{\text{cl}}(x,x';t,t')\psi'(x')d^n x'.$$

We have

$$\psi(x,t) - \psi_{\text{cl}}(x,t) = O\left(\Delta t^2\right) \tag{3.55}$$

for $\Delta t = t - t' \to 0$.

Proof. We have

$$G_{\mathrm{cl}}(x, x'; t, t') = \left(\frac{1}{2\pi i \hbar} \right)^{n/2} \sqrt{\rho(x, x'; t, t')} \exp \left(\frac{i}{\hbar} \widetilde{W}(x, x'; t, t') \right)$$

$$+ O(\Delta t^2), \tag{3.56}$$

where

$$\rho(x, x'; t, t') = \det \left(-\frac{\partial^2 \widetilde{W}(x, x'; t, t')}{\partial x_j \partial x'_k} \right)_{1 \leq j, k \leq n}. \tag{3.57}$$

It turns out that this formula can be somewhat improved. Noting that $\det(I_{n \times n} + O(\Delta t^2)) = 1 + O(\Delta t^2)$, we thus have

$$\rho(x, x'; t, t') = \frac{m_1 \cdots m_n}{(\Delta t)^n} (1 + O(\Delta t^2)). \tag{3.58}$$

Write

$$\widetilde{\rho}(\Delta t) = \frac{m_1 \cdots m_n}{(\Delta t)^n}, \tag{3.59}$$

which is just the Van Vleck density for the free particle Hamiltonian. We thus have

$$\rho(x, x'; t, t') = \widetilde{\rho}(\Delta t)(1 + O(\Delta t^2)) \tag{3.60}$$

and hence we can rewrite formula (3.56) as

$$G(x, x'; t, t') = \left(\frac{1}{2\pi i \hbar} \right)^{n/2} \sqrt{\widetilde{\rho}(\Delta t)} \tag{3.61}$$

$$\times \exp \left(\frac{i}{\hbar} \widetilde{W}(x, x'; t, t') \right) + O(\Delta t^2). \qquad \square$$

3.4. Bohm's Quantum Theory of Motion

In this section we review the basic ideas which led David Bohm in 1952 to give an alternative interpretation of quantum theory, nowadays known as "Bohmian mechanics", or "the theory of quantum motion" (see [29, 30, 32]).

3.4.1. Quantum trajectories and potential

Consider a (possibly time-dependent) Hamiltonian function

$$H(x, p, t) = \sum_{j=1}^{n} \frac{p_j^2}{2m_j} + V(x, t) \qquad (3.62)$$

and the corresponding quantum operator

$$\widehat{H} = \sum_{j=1}^{n} \frac{-\hbar^2}{2m_j} \frac{\partial^2}{\partial x_j^2} + V(x, t)$$

deduced from H by the formal substitution $(x, p) \mapsto (x, -i\hbar\nabla_x)$. Let $\psi = \psi(x, t)$ be a solution of the time-dependent Schrödinger equation

$$i\hbar \frac{\partial \psi}{\partial t}(x, t) = \widehat{H}\psi(x, t) \qquad (3.63)$$

with initial condition

$$\psi(x, 0) = \psi_0(x) \qquad (3.64)$$

(the initial time is chosen as $t' = 0$ for simplicity, but any other value is as good for what follows).

In their original formulation Bohm's postulates for a quantum theory of motion were enounced for a single particle evolving in ordinary space. We are stating them directly for many-body systems:

- *A physical system consists of a wave propagating in configuration space $\mathbb{R}_x^n$ together with a system of particles moving under the guidance of that wave.*
- *This wave is represented by a complex function $\psi = \psi(x, t)$ whose time-evolution is governed by the Schrödinger equation with Hamiltonian obtained by quantization of the classical Hamiltonian function of the system of particles.*
- *The motion of the system in configuration space is given as the solution x^ψ to the first-order differential equation*

$$\dot{x}(t) = \sum_{j=1}^{n} \frac{1}{m_j} \frac{\partial \Phi}{\partial x_j}(x(t), t) \qquad (3.65)$$

where Φ is the phase of ψ.

These postulates — which constitute by themselves sufficient conditions for a theory of motion — are usually complemented by a subsidiary condition (the Born statistical interpretation of quantum mechanics):

- *The probability of finding the system of particles at time t in a measurable subset Ω of the configuration space $\mathbb{R}^n_x$ is*

$$P(\Omega, t) = \frac{\int_\Omega |\psi(x,t)|^2 d^n x}{\int |\psi(x,t)|^2 d^n x}.$$

Writing as above ψ in polar form $Re^{i\Phi/\hbar}$ and assuming that $R > 0$, the functions $R = R(x,t)$ and $\Phi = \Phi(x,t)$ satisfy the following coupled system of partial differential equations

$$\frac{\partial R^2}{\partial t} + \sum_{j=1}^n \frac{1}{m_j} \frac{\partial}{\partial x_j} \left(R^2 \frac{\partial \Phi}{\partial x_j} \right) = 0, \tag{3.66a}$$

$$\frac{\partial \Phi}{\partial t} + \sum_{j=1}^n \frac{1}{2m_j} \left(\frac{\partial \Phi}{\partial x_j} \right)^2 + V + Q^\psi = 0, \tag{3.66b}$$

where the function $Q^\psi = Q^\psi(x,t)$ is given by

$$Q^\psi = -\sum_{j=1}^n \frac{\hbar^2}{2m_j R} \frac{\partial^2 R}{\partial x_j^2}. \tag{3.67}$$

In physics, this function is called the *quantum potential*. Equation (3.66b) satisfied by the phase Φ can therefore be written

$$\frac{\partial \Phi}{\partial t}(x,t) + H(x, \nabla_x \Phi, t) + Q^\psi(x,t) = 0. \tag{3.68}$$

Mathematically speaking, this is just the Hamilton–Jacobi equation satisfied by the new Hamiltonian function

$$H^\psi(z,t) = H(z,t) + Q^\psi(x,t). \tag{3.69}$$

Observe that even if the potential V does not depend on time, the Hamiltonian function H^ψ will generically be time-dependent. Also, the quantum potential Q^ψ does not depend on the size of the wavefunction ψ, but it depends on its *form*: the quantum potential can thus be large even when the wavefunction is spread out by propagation across large distances; it is thus highly *nonlocal* in nature. We mention that a common misconception is that (3.68) reduces to the classical Hamilton–Jacobi equation in the limit $\hbar \to 0$. This is generally not true, because the quotients $(\partial^2 R/\partial x_j^2)/R$ usually depend on $\hbar$ and might very well grow faster than $\hbar^{-2}$ when $\hbar \to 0$, causing the quantum potential to blow up in the limit.

The quantum potential acting on a system of particles is not, as the ordinary potential is, a preassigned function of the position coordinates (and time). With each system is potentially associated an infinity of functions Q^ψ, each corresponding to a solution of the Schrödinger equation of the system. As soon as we have more than one particle this property has significant consequences. As Bohm [31] puts it, "...*the quantum potential is capable of constituting a nonlocal connection, depending directly on the state of the whole in a way that is not reducible to a preassigned relationship among the parts...*" (also see [18]). This property becomes apparent if one considers the Hamiltonian function for a system of N particles in pairwise interaction without external potential:

$$H = \sum_{j=1}^{N} \frac{\mathbf{p}_j^2}{2m} + \sum_{k<j} V(\mathbf{r}_j - \mathbf{r}_k)$$

($\mathbf{r}_j = (x_j, y_j, z_j)$ is the coordinate vector of the jth particle and $\mathbf{p}_j = (p_{x,j}, p_{y,j}, p_{z,j})$ its momentum vector). The potential energy is here uniquely specified by the nature of the particles (for instance, $V(\mathbf{r}_j - \mathbf{r}_k)$ might be the gravitational potential exerted by the jth particle on the kth particle, and only depends on the masses of these particles). However the interparticle quantum potential $Q^\psi(\mathbf{r}_1, \ldots, \mathbf{r}_N, t)$ is determined by the wavefunction ψ which is *external* to the system. This is really a novel feature of quantum mechanics; as Holland (see [160, §7.1.3]) puts it, whereas in classical mechanics the whole is the sum of its parts, in quantum mechanics the whole is prior to the parts.

We observe that equation (3.66a) satisfied by the amplitude R can be viewed as continuity equation; in fact, setting $\rho = R^2$ it is equivalent to the familiar equation

$$\frac{\partial \rho}{\partial t} + \mathrm{div}(\rho v) = 0 \qquad (3.70)$$

where the velocity field $v^\psi = (v_1, \ldots, v_n)$ is defined by

$$v^\psi(x,t) = \left(\frac{1}{m_1} \frac{\partial \Phi}{\partial x_1}(x,t), \ldots, \frac{1}{m_n} \frac{\partial \Phi}{\partial x_n}(x,t) \right); \qquad (3.71)$$

equivalently, the *momentum field* is defined by the innocent looking equation

$$p^\psi(x,t) = \nabla_x \Phi(x,t). \tag{3.72}$$

Using Born's statistical interpretation of the wavefunction, this momentum field has a very simple meaning, which has been observed by many mathematicians (see, e.g., [197, pp. 95–96]). Recall that the mean values of the position and momentum vectors $x = (x_1, \ldots, x_n)$ and $p = (p_1, \ldots, p_n)$ in the state ψ are given by the formulas

$$\langle x \rangle = \frac{\langle x\psi | \psi \rangle}{\langle \psi | \psi \rangle}, \quad \langle p \rangle = \frac{\langle -i\hbar \nabla_x \psi | \psi \rangle}{\langle \psi | \psi \rangle}. \tag{3.73}$$

Proposition 46. *Let* $\psi = Re^{i\Phi/\hbar}$ *be a normalized wavefunction:* $\|\psi\|_{L^2} = 1$. *Assume that* ψ *vanishes at infinity in all the variables* $x_1, \ldots, x_n$. *Then the mean value (in the state ψ) of the momentum at time* t *is given by*

$$\langle p(t) \rangle = \langle p^\psi(\cdot, t)\psi | \psi \rangle = \int p^\psi(x,t)|\psi(x,t)|^2 d^n x \tag{3.74}$$

and is thus the expected value of the momentum field with respect to the probability density $|\psi(x,t)|^2$.

Proof. Omitting the variable t, we have, in view of the second formula (3.73),

$$\langle p \rangle = i\hbar \int (\nabla_x \psi^*(x))\psi(x)d^n x.$$

By the product rule

$$\nabla_x \psi^*(x) = \nabla_x(Re^{-\frac{i}{\hbar}\Phi}) = \left(\nabla_x R - \frac{i}{\hbar} R\nabla_x \Phi \right) e^{-\frac{i}{\hbar}\Phi}$$

hence

$$\langle p \rangle = i\hbar \int \left(\nabla_x R(x) - \frac{i}{\hbar} R(x)\nabla_x \Phi(x) \right) R(x)d^n x$$

$$= i\hbar \int R(x)\nabla_x R(x)d^n x + \int R(x)^2 \nabla_x \Phi(x)d^n x$$

$$= i\hbar \int R(x)\nabla_x R(x)d^n x + \langle p^\psi \psi | \psi \rangle.$$

There remains to show that the integral on the right-hand side vanishes; for this it suffices to note that for $j = 1, \ldots, n$ we have

$$\int_{-\infty}^{\infty} R(x) \frac{\partial R}{\partial x_j}(x) dx_j = \frac{1}{2} \int_{-\infty}^{\infty} \frac{\partial}{\partial x_j}(R(x))^2 dx_j$$

$$= \frac{1}{2} \left[(R(x))^2 \right]_{x_j = -\infty}^{x_j = \infty} = 0$$

since ψ (and hence R) vanishes at infinity in x_j. ☐

While it may be difficult in practice to solve explicitly the continuity equation (3.70) there is an important relation between the initial value $\rho(x, 0)$ and $\rho(x(t), t)$: we have

$$\rho(x(t), t) \left| \det \frac{\partial x(t)}{\partial x(t')} \right| = \rho(x(t'), t'). \tag{3.75}$$

This relation is easily verified [100, 208] by differentiating its left-hand side with respect to t and using (3.70): one finds the value 0; hence the left-hand side is constant and equal to the right-hand side taking $t = 0$. In the case where $\rho = R^2$, the relation shows that the amplitude R changes following the rule

$$R(x^\psi(t), t) \left| \det \frac{\partial x^\psi(t)}{\partial x} \right|^{1/2} = R(x, 0). \tag{3.76}$$

We can rewrite this equality formally as

$$R(x^\psi(t), t) |dx^\psi(t)|^{1/2} = R(x, 0) |dx|^{1/2}, \tag{3.77}$$

where the "square roots" $dx^\psi(t)^{1/2}$ and $dx^{1/2}$ are interpreted as half-forms as we have shown in [97, 98, 100]. Notice that these relations immediately imply, almost without any calculation, the well-known property that the norm of the wavefunction is conserved in time. In fact, for fixed t we have, using any of the equivalent formulas (3.76) or (3.77),

$$\int |\psi(x, t)|^2 d^n x = \int |R(x^\psi(t), t)|^2 d^n x^\psi(t)$$

$$= \int |R(x, 0)|^2 d^n x$$

$$= \int |\psi(x, 0)|^2 d^n x;$$

this sequence of equalities is valid as soon as $\psi_0 = \psi(\cdot, 0) \in L^2(\mathbb{R}^n)$.

When a vector potential is present the treatment is quite similar. Assume that the Hamiltonian function has the form

$$H = \frac{1}{2m}\left(p - A(x,t)\right)^2 + V(x,t), \tag{3.78}$$

where $p = (p_1, \ldots, p_n)$, $x = (x_1, \ldots, x_n)$ (we assume for notational simplicity that all the masses are equal to m). A tedious but straightforward calculation shows that if $\psi = Re^{i\Phi/\hbar}$ is a solution of the corresponding Schrödinger equation, then R and Φ satisfy the coupled system of differential equations

$$\frac{\partial \Phi}{\partial t} + \frac{1}{2m}(\nabla_x S - A)^2 + V + Q^\psi = 0, \tag{3.79}$$

$$\frac{\partial R^2}{\partial t} + \frac{1}{m}\nabla_x \cdot [R^2(\nabla_x \Phi - A)] = 0; \tag{3.80}$$

here the quantum potential Q is given by the same expression (3.67) as in the scalar case, which reads here

$$Q^\psi = -\frac{\hbar^2}{2mR}\sum_{j=1}^{n}\frac{\partial^2 R}{\partial x_j^2}.$$

One then postulates (see the discussion in [160, §3.11.2]) that the momentum field is given, as in classical mechanics, by

$$p(x,t) = \nabla_x \Phi(x,t) - A(x,t)$$

and the quantum trajectories are then found by integrating the equations

$$m\dot{x} = \nabla_x \Phi(x,t) - A(x,t). \tag{3.81}$$

Let us briefly discuss the effect of a change of gauge

$$(A,V) \longmapsto \left(A + \nabla_x \chi, V - \frac{\partial \chi}{\partial t}\right) \tag{3.82}$$

on the quantum motion. Recall from Chapter 1 (subsection 1.3.2) that such a transformation does not affect the motion of a system in configuration space. In the quantum case the situation is similar. For this we need the following classical result.

Lemma 47. *Assume that ψ is a solution of Schrödinger's equation*

$$i\hbar\frac{\partial\psi}{\partial t} = \left[\frac{1}{2m}\left(-i\hbar\nabla_x - A(x,t)\right)^2 + V(x,t)\right]\psi.$$

Then $\psi^\chi = e^{i\chi/\hbar}$ is a solution of

$$i\hbar\frac{\partial\psi}{\partial t} = \left[\frac{1}{2m}\left(-i\hbar\nabla_x - A(x,t) - \nabla_x\chi\right)^2 + V(x,t) - \frac{\partial\chi}{\partial t}\right]\psi.$$

The proof is purely computational, and left to the reader. It follows that if we perform the gauge transformation (3.82) the phase Φ of ψ becomes $\Phi + \chi$ while the amplitude R is unchanged; the equation of motion (3.81) thus becomes

$$m\dot{x} = \nabla_x(\Phi(x,t) + \chi(x,t)) - (A(x,t) + \nabla_x\chi(x,t))$$
$$= \nabla_x\Phi(x,t) - A(x,t)$$

and is thus unchanged.

3.4.2. The quantum phase space flow

Let us return to Bohm's equation of motion (3.65); recall that we defined in (3.69) the Hamiltonian function

$$H^\psi(z,t) = H(z,t) + Q^\psi(x,t) \tag{3.83}$$

obtained by replacing the potential V with $V + Q$ where Q is the quantum potential (3.67):

$$Q^\psi = -\sum_{j=1}^n \frac{\hbar^2}{2m_j R}\frac{\partial^2 R}{\partial x_j^2}; \tag{3.84}$$

the latter is defined at all points (x,t) where $R(x,t) \neq 0$.

We are going to see that quantum motion can be interpreted in terms of Hamilton's equations for the function H^ψ. The proof is not immediately obvious, and requires some calculations involving the modified Hamilton–Jacobi equation (3.68).

Proposition 48. *The quantum motion of a system is the projection on configuration space of the solution of Hamilton's equation for H^ψ:*

$$\dot{x}_j^\psi(t) = \frac{1}{m_j}p_j^\psi(t), \ \ \dot{p}_j^\psi(t) = -\frac{\partial}{\partial x_j}(V(x^\psi(t),t) + Q^\psi(x^\psi(t),t)). \tag{3.85}$$

Proof. (For an alternative proof see [125].) We will choose all masses m_j equal to m for notational simplicity. Equation (3.65) allows us to define the momentum by

$$p(t) = \nabla_x \Phi(x(t), t);$$

taking the time-derivative we get

$$\dot{p}(t) = \frac{\partial(\nabla_x \Phi)}{\partial t}(x(t), t) + \nabla_x \cdot \nabla_x \Phi(x(t), t) \dot{x}(t),$$

that is, taking the Bohm equation of motion (3.65) into account

$$\dot{p}(t) = \frac{\partial(\nabla_x \Phi)}{\partial t}(x(t), t) + \nabla_x \cdot \nabla_x \Phi(x(t), t) \frac{1}{m} \nabla_x \Phi(x(t), t)$$

$$= \left(\frac{\partial}{\partial t} + \frac{1}{m} \nabla_x \Phi(x, t) \nabla_x \right) \nabla_x \Phi(x, t).$$

On the other hand, differentiating Hamilton–Jacobi's equation

$$\frac{\partial \Phi}{\partial t}(x, t) + \frac{1}{2m}(\nabla_x \Phi(x, t))^2 + V(x, t) + Q^\psi(x, t) = 0$$

with respect to the position variables yields

$$\frac{\partial(\nabla_x \Phi)}{\partial t}(x, t) + \frac{1}{m} \nabla_x \Phi(x, t) \nabla_x \cdot \nabla_x \Phi(x, t)$$

$$+ \nabla_x(V(x, t) + Q^\psi(x, t)) = 0$$

which we can rewrite as

$$\left(\frac{\partial}{\partial t} + \frac{1}{m} \nabla_x \Phi(x, t) \nabla_x \right) \nabla_x \Phi(x, t) = -\nabla_x(V(x, t) + Q^\psi(x, t)).$$

Since x is arbitrary in this equality, we may replace it with $x(t)$; comparison with the expression for $\dot{p}(t)$ above shows that we have

$$\dot{p}(t) = -\nabla_x(V(x(t), t) + Q^\psi(x(t), t))$$

which is the second equation (3.85); the first equation is just the definition of the Bohm momentum. □

In classical mechanics, the value of the Hamiltonian function along a curve of solutions of the Hamilton equations is the energy of the system. Let us discuss the notion of energy from the Bohmian point of view, where energy is usually defined by

$$E(x^\psi(t), t) = -\frac{\partial \Phi}{\partial t}(x^\psi(t), t). \tag{3.86}$$

In view of the quantum Hamilton–Jacobi equation (3.68) and taking the relation $p^\psi(t) = \nabla_x \Phi(x^\psi(t), t)$ into account, we thus have

$$E(x^\psi(t), t) = H^\psi(x^\psi(t), p^\psi(t), t). \tag{3.87}$$

Notice that since the quantum potential is generically time-dependent, we do not have conservation of energy in time.

In the Bohm interpretation, the quantum trajectory followed by the system of particles is given by the system of first-order differential equations together with the boundary (initial value) conditions

$$x_j^\psi(0) = x_j, \quad p_j^\psi(0) = \frac{\partial \Phi}{\partial x_j}(x, 0). \tag{3.88}$$

These equations are obviously Hamiltonian, as they are rigorously equivalent to

$$\dot{z}^\psi(t) = J\nabla_z H^\psi(z^\psi(t), t), \quad H^\psi = H + Q^\psi. \tag{3.89}$$

Let us denote (f_t^ψ) the phase space flow determined by the Hamiltonian function $H^\psi = H + Q^\psi$. Regardless of any physical or ontological interpretation this flow may be studied using the standard methods described in the previous sections (also see Holland's approach [160, 161]). We will call (f_t^ψ) the *Bohmian flow*. Recall now that equations (3.89) must be complemented by the initial conditions (3.88), which we rewrite as

$$x^\psi(0) = x, \quad p^\psi(0) = \nabla_x \Phi(x, 0). \tag{3.90}$$

The geometrical meaning of these conditions are — or should be! — quite obvious: the initial phase space point $z = (x, p)$ at time $t = 0$ lies on the n-dimensional manifold

$$\mathcal{V}_0 : p = \nabla_x \Phi(x, 0). \tag{3.91}$$

We notice that this manifold is Lagrangian: every pair of vectors (T_z, T_z') tangent to $\mathcal{V}_0$ at some point z are skew-orthogonal, that is $\sigma(T_z, T_z') = 0$ (σ the symplectic form). As time evolves, the initial phase space point z

is transformed into the point $z^{\psi}(t) = f_t^{\psi}(z)$, and this point lies on the manifold

$$\mathcal{V}_t^{\psi} : p = \nabla_x \Phi(x, t) \tag{3.92}$$

which is Lagrangian as well (the Lagrangian character of a manifold is conserved by symplectomorphisms and hence by Hamiltonian flows [5, 100, 108]). We refer again to Holland's paper [161] for a detailed and careful study of the Hamiltonian properties of the Bohmian flow.

3.4.3. Short-time quantum trajectories

Let us determine the quantum potential Q corresponding to the propagator $G = G(x, x_0; t, t_0)$ for small values of $\Delta t = t - t_0$ using the asymptotic formulas we proved in Chapter 2 (we are following here de Gosson and Hiley [124] and de Gosson *et al.* [126]). Recall that in view of Huygens' principle, the propagator describes an isotropic source of point-like particles emanating from the point x_0 at initial time t_0. We have, by definition,

$$Q = -\sum_{j=1}^{n} \frac{\hbar^2}{2m_j \sqrt{\rho}} \frac{\partial^2 \sqrt{\rho}}{\partial x_j^2}.$$

We have, using (3.60),

$$\sqrt{\rho} = \sqrt{\widetilde{\rho}(\Delta t)}(1 + O(\Delta t^2))$$

and hence

$$\frac{\partial^2 \sqrt{\rho}}{\partial x_j^2} = O((\Delta t)^2)).$$

From this it follows that the quantum potential associated with the propagator satisfies

$$Q(x, x_0; \Delta t) = O(\Delta t^2). \tag{3.93}$$

The discussion above suggests that the quantum trajectory of a sharply located particle should be identical with the classical (Hamiltonian) trajectory for short times. Let us show this is indeed the case. If we want to monitor the motion of such a particle, we have of course to specify its initial momentum which gives its direction of propagation at time $t_0 = 0$;

we set

$$p(0) = p_0. \tag{3.94}$$

The trajectory in position space is obtained by solving the system of differential equations

$$\dot{x} = \hbar \, \mathrm{Im} \left(\sum_{j=1}^{n} \frac{1}{m_j G} \frac{\partial G}{\partial x_j} \right), \quad x(0) = x_0; \tag{3.95}$$

replacing G with its approximation

$$\widetilde{G}(x, x_0; \Delta t) = \left(\frac{1}{2\pi i \hbar} \right)^{n/2} \sqrt{\widetilde{\rho}(\Delta t)} \exp \left(\frac{i}{\hbar} \widetilde{W}(x, x_0; \Delta t) \right)$$

we have, since $G - \widetilde{G} = O(\Delta t^2)$ in view of (3.61),

$$\dot{x} = \hbar \, \mathrm{Im} \left(\sum_{j=1}^{n} \frac{1}{m_j \widetilde{G}} \frac{\partial \widetilde{G}}{\partial x_j} \right) + O(\Delta t^2).$$

A straightforward calculation, using the expression (2.55) for the approximate action $\widetilde{W}(x, x_0; \Delta t)$, leads to the relation

$$\dot{x}(\Delta t) = \frac{x(\Delta t) - x_0}{\Delta t} - M^{-1} \nabla_x \widetilde{V}(x(\Delta t), x_0) \Delta t + O(\Delta t^2) \tag{3.96}$$

(cf. the proof of [100, Lemma 248]); here $M = \mathrm{diag}(m_1, \ldots, m_n)$ is the "mass-matrix". This equation is singular for $\Delta t = 0$ hence the initial condition $x(0) = x_0$ is not sufficient for finding a unique solution. This is of course consistent with the fact that (3.96) describes an arbitrary particle emanating from x_0; to single out one quantum trajectory we have to use the additional condition (3.94) giving the direction of the particle at time $t = 0$. We thus have

$$x(\Delta t) = x_0 + M^{-1} p_0 \Delta t + O(\Delta t^2);$$

in particular $x(\Delta t) = x_0 + O(\Delta t)$ and hence, by continuity,

$$\nabla_x \widetilde{V}(x(\Delta t), x_0) = \nabla_x \widetilde{V}(x_0, x_0) + O(\Delta t).$$

Let us calculate $\nabla_x \widetilde{V}(x_0, x_0)$. We have, taking definition (3.46) into account,

$$\nabla_x \widetilde{V}(x, x_0) = \int_0^1 \lambda \nabla_x V(\lambda x + (1 - \lambda) x_0, 0) d\lambda$$

and hence

$$\nabla_x \widetilde{V}(x_0, x_0) = \int_0^1 \lambda \nabla_x V(x_0, 0) d\lambda = \frac{1}{2} \nabla_x V(x_0, 0).$$

We can thus rewrite equation (3.96) as

$$\dot{x}(\Delta t) = \frac{x(\Delta t) - x_0}{\Delta t} - \frac{1}{2} M^{-1} \nabla_x V(x_0, 0) \Delta t + O(\Delta t^2).$$

Let us now differentiate both sides of this equation with respect to Δt:

$$\ddot{x}(t) = \frac{x(t) - x_0}{(\Delta t)^2} + \frac{\dot{x}(t)}{\Delta t} - \frac{1}{2} M^{-1} \nabla_x V(x_0, 0) + O(\Delta t), \tag{3.97}$$

that is, replacing $\dot{x}(\Delta t)$ by the value given by (3.96),

$$\dot{p}(t) = M\ddot{x}(t) = -\nabla_x V(x_0, 0) + O(\Delta t). \tag{3.98}$$

Solving this equation we get

$$p(t) = p_0 - \nabla_x V(x_0, 0) \Delta t + O(\Delta t^2). \tag{3.99}$$

Summarizing, the solutions of the Hamilton equations are given by

$$x(\Delta t) = x_0 + \frac{p_0}{m} \Delta t + O(\Delta t^2), \tag{3.100}$$

$$p(\Delta t) = p_0 - \nabla_x V(x_0, 0) \Delta t + O(\Delta t^2). \tag{3.101}$$

These equations are, up to the error terms $O(\Delta t^2)$, the equations of motion of a classical particle moving under the influence of the potential V; there is no trace of the quantum potential, which is being absorbed by the terms $O(\Delta t^2)$. The motion is thus identical with the classical motion on time scales of order $O(\Delta t^2)$.

This result puts Bohm's original perception, which led him to the causal interpretation, on a firm mathematical footing. He writes [31]

Indeed it had long been known that when one makes a certain approximation (WKB) Schrödinger's equation becomes equivalent to the classical Hamilton–Jacobi theory. At a certain point I asked myself: What would happen, in the demonstration of this equivalence, if we did not make this approximation? I saw immediately that there would be an additional potential, representing a kind of force, that would be acting on the particle.

The source of this "force" was the quantum potential. In our approach we see that while any classical potential acts immediately, the quantum potential does not. From this fact two consequences follow.

Firstly, it gives a rigorous treatment of the "watched pot" effect. If we keep observing a particle that, if unwatched, would make a transition from one quantum state to another, then it will no longer make that transition. The unwatched transition occurs when the quantum potential grows to produce the transition. Continuously observing the particle does not allow the quantum potential to develop so the transition does not take place; we are going to study the quantum Zeno effect below. Secondly, in the situation when the quantum potential decreases continuously with time, the quantum trajectory continuously deforms into a classical trajectory. This means that there is no need to appeal to decoherence to reach the classical domain.

3.4.4. The quantum Zeno effect

By studying the short time propagator, we have found that any classical potential kicks in immediately to $O((\Delta t)^2)$. On the other hand the quantum potential does not contribute to the equation of motion to this order. In other words the short-time propagator is totally classical in *all* quantum processes in the limit $\Delta t \to 0$. This has a direct consequence for the Zeno effect; our exposition follows de Gosson and Hiley [125], which contains a detailed discussion of the relation between implicate order and the quantum Zeno effect.

The Bohm approach, as discussed in [32], is based on the assumption that the world has a continuous existence with each particle having simultaneously a definite position and a definite momentum. This momentum is not an eigenvalue of the momentum operator for the state under consideration, but the real part of the weak value of the momentum operator at a point in space. This was shown in detail by Leavens [182] and in [153]. Furthermore these values, and this is a very significant point, can now be determined by experiment as shown in [82]. Let us begin by introducing some notation. We have seen that the datum of the propagator $G_0 = G(x, x_0; t, t_0)$ determines a quantum potential Q^Ψ and thus Hamilton equations associated with $H^\Psi = H + Q^\Psi$. We now choose as initial time $t_0 = 0$ and denote the corresponding quantum potential by Q^0 and set $H^0 = H + Q^0$. After time Δt we make a position measurement and find that the particle is located at a point x_1. The future evolution of the particle is now governed by the *new* propagator $G_1 = G(x, x_1; t, t_0)$, leading to a new quantum potential Q^1 and to a new Hamiltonian H^1; repeating this until time t we thus have a sequence of points $x_0, x_1, \ldots, x_N = x$ and a corresponding sequence of Hamiltonian functions $H^0, H^1, \ldots, H^N$ determined by the quantum potentials $Q^0, Q^1, \ldots, Q^N$. We denote by (f_{t,t_0}^0), $(f_{t,t_1}^1), \ldots, (f_{t,t_{N-1}}^{N-1})$ the time-dependent flows determined by the Hamiltonian functions $H^0, H^1, \ldots, H^N$; we have set here $t_1 = t_0 + \Delta t$, $t_2 = t_1 + \Delta t$ and so on. Repeating the observation procedure, we get a sequence of successive equalities

$$(x_1, p_1) = f_{t_1, t_0}^0(x_0, p_0)$$

$$(x_2, p_2) = f_{t_2, t_1}^1(x_1, p_1)$$

$$\vdots$$

$$(x, p) = f_{t, t_{N-1}}^{N-1}(x_{N-1}, p_{N-1})$$

which implies that the final point $x = x_N$ observed at time t is expressed in terms of the initial point x_0 by the formula

$$(x, p) = f_{t, t_{N-1}}^{N-1} \cdots f_{t_2, t_1}^1 f_{t_1, t_0}^0(x_0, p_0). \tag{3.102}$$

Denote now by (g_{t,t_0}^0), $(g_{t,t_1}^1), \ldots, (g_{t,t_{N-1}}^{N-1})$ the approximate flows determined by the equations

$$(x_1, p_1) = \left(x_0 + \frac{p_0}{m}\Delta t, p_0 - \nabla_x V(x_0)\Delta t\right)$$

$$(x_2, p_2) = \left(x_1 + \frac{p_1}{m}\Delta t, p_1 - \nabla_x V(x_1)\Delta t\right)$$

$$\vdots$$

$$(x, p) = \left(x_{N-1} + \frac{p_{N-1}}{m}\Delta t, p_{N-1} - \nabla_x V(x_{N-1})\Delta t\right).$$

The sequence of estimates

$$f_{t_k, t_{k-1}}^0(x_{k-1}, p_{k-1}) - g_{t_k, t_{k-1}}^0(x_{k-1}, p_{k-1}) = O(\Delta t^2) \tag{3.103}$$

implies (using the Lie–Trotter formula, see, e.g., [48], or [100, Appendix]) that we have

$$\lim_{N \to \infty} g_{t,t_{N-1}}^{N-1} \cdots g_{t_2,t_1}^1 g_{t_1,0}^0 (x_0, p_0) = \lim_{N \to \infty} f_{t,t_{N-1}}^{N-1} \cdots f_{t_2,t_1}^1 f_{t_1,0}^0 (x_0, p_0).$$

(3.104)

The argument goes as follows (for a detailed proof see [100]): since we have $g_{t_k,t_{k-1}}^k = f_{t_k,t_{k-1}}^k + O(\Delta t^2)$ the product is approximated by

$$g_{t,t_{N-1}}^{N-1} \cdots g_{t_2,t_1}^1 g_{t_1,t_0}^0 = f_{t,t_{N-1}}^{N-1} \cdots f_{t_2,t_1}^1 f_{t_1,t_0}^0 + NO(\Delta t^2) \qquad (3.105)$$

and since $\Delta t = t/N$ we have $NO(\Delta t^2) = O(\Delta t)$ which goes to zero when $N \to \infty$.

Now, recall our remark that the quantum potential is absent from the approximate flows $g_{t_k,t_{k-1}}^k$; using again the Lie–Trotter formula together with short-time approximations to the Hamiltonian flow (f_t) determined by the classical Hamiltonian H, we get

$$\lim_{N \to \infty} g_{t,t_{N-1}}^{N-1} \cdots g_{t_2,t_1}^1 g_{t_1,0}^0 (x_0, p_0) = f_t \qquad (3.106)$$

and hence

$$\lim_{N \to \infty} f_{t,t_{N-1}}^{N-1} \cdots f_{t_2,t_1}^1 f_{t_1,0}^0 (x_0, p_0) = f_t, \qquad (3.107)$$

which shows that the observed trajectory is the classical one. We have thus shown how a detailed mathematical examination of the deeper symplectic structure that underlies the Bohm approach predicts that if a quantum particle is watched continuously, it will follow a classical trajectory. Another illustration of how continuous observations gives rise to a quantum Zeno effect, which, in this case, suppresses an atomic transition, has already been given in [32]. They considered the transition in an Auger-like particle and showed that the perturbed wavefunction, which is proportional to Δt for times less than $1/\Delta E$ (ΔE is the energy released in the transition), will never become large and therefore cannot make a significant contribution to the quantum potential. Again for this reason no transition will take place.

Chapter 4

The Metatron

For Bohm there was no classical point-like particle following quantum trajectories. But instead, at the fundamental level, there was a basic process or activity which left a "track" in, say, a bubble chamber. Thus the track could be explained poetically by the *enfolding* and *unfolding* of an invariant form in the overall underlying process. We will see in this chapter that it is the metaplectic structure that gives rise to the quantum properties. Since the classical and quantum motions are related but different, it was proposed in [100] to call the object that obeys the Bohmian law of motion a *Metatron*. We choose this term rather than the usual term "particle", because we are talking about an excitation induced by the metaplectic representation of the underlying Hamiltonian evolution, rather than a classical object. To fully appreciate this notion we have to be aware of some subtle mathematical details. While ray mechanics is obtained from symplectic geometry (as discussed in chapter 1), the wave properties emerge from the properties of the two-fold covering group of the symplectic group: the metaplectic group. It is in this covering group that Schrödinger's equation appears *mathematically*, without any reference to some physical process: it is already built inside the symplectic geometry. In fact, to every Hamiltonian flow we can associate a "metaplectic flow". However, this transformation is more general than a simple geometric change within a given explicate order. It should rather be considered, in the terminology of Bohm and Hiley, as a *metamorphosis* in which everything alters in a thorough-going manner while some subtle and highly implicit features remain invariant. This metamorphosis can be regarded as an enfolding into the whole followed by an unfolding.

Indeed a deeper investigation suggests that the Metatron is more like an invariant feature of an underlying extended process, which elsewhere (de Gosson [115], de Gosson and Hiley [125]) we have argued that the term *quantum blob* may be more suggestive. We will study the notion of quantum blob, which is closely related to the uncertainty principle, in Chapter 5.

4.1. Introduction

In Chapter 3 we briefly discussed Stone's theorem following which a strongly continuous group (U_t) of unitary operators on $L^2(\mathbb{R}^n)$ can be formally written $U_t = e^{-i\widehat{H}t/\hbar}$ where the infinitesimal generator $\widehat{H}$ is a (generally unbounded) self-adjoint operator on $L^2(\mathbb{R}^n)$. Setting $\psi = e^{-i\widehat{H}t/\hbar}\psi_0$ this leads to the abstract "Schrödinger equation"

$$i\hbar\frac{\partial\psi}{\partial t} = \widehat{H}\psi, \quad \psi(x,0) = \psi_0(x).$$

In this chapter we will see that the operator $\widehat{H}$ (the "quantum Hamiltonian") can be explicitly determined in a *canonical way* when the group (U_t) is associated (in a way we will explain) with the Hamiltonian flow determined by an arbitrary quadratic Hamiltonian function H, i.e., a function of the type

$$H(z) = \tfrac{1}{2}Mz \cdot z.$$

The fact that we really have a *canonical* correspondence shows, *a posteriori*, that our constructs in the last chapter where we produced (Proposition 43) exact solutions of the Schrödinger equation for Hamiltonian functions of the type "kinetic energy plus quadratic potential" were not just accidental: there really is something going on! The method we will use is, in a sense, more important than the result itself (which is just a generalization of Proposition 43). It namely highlights the role played by the metaplectic representation of the symplectic group; it shows that the symplectic group (and hence Hamiltonian mechanics) already implicitly contains quantum mechanics, which unfolds when one passes to the metaplectic representation.

This fact, which seems to have been known by mathematicians working in the field of harmonic analysis for quite some time, usually provokes mixed feelings among most physicists; we are actually going to make things even worse: we will show later in this chapter that Schrödinger's equation can be derived from Hamilton's equations for all Hamiltonians.

We do this by using an extension of the metaplectic representation, relying on the symplectic covariance properties of quantization. More precisely, we will show that with every Hamiltonian flow (f_t^H) we are able to associate (again in a canonical way) a strongly continuous group (U_t) of unitary operators whose infinitesimal generator $\widehat{H}$ is precisely the Weyl quantization of the Hamiltonian function H determining the flow (f_t^H). Thus, conceptually, Schrödinger's equation and Hamilton's equations of motion are mathematically *equivalent*!

Now, we have been taught by the Founding Fathers that there is no way to *deduce* quantum mechanics from classical mechanics. So our claim cannot be true, and we have probably used some dubious *légerdemain* to derive Schrödinger's equation, the master equation of non-relativistic quantum mechanics. However, what we have done is to complete the task that Schrödinger himself set out to do. In his original article [256], Schrödinger noted that classical mechanics was a complete analogue of geometric optics and what he was looking for was "an *undulatory mechanics*", the most obvious route being the Hamiltonian picture based on wave theory. Unfortunately, the necessary mathematics did not exist at the time this work was being undertaken. Nevertheless Schrödinger did arrive at his equation, even though he, himself, realized the steps leading to it were "not unambiguous"! Our work shows how Hamilton's equations of motion studied in the first chapter can be lifted onto the covering group of the symplectic group to produce Schrödinger's equation, thus rigorously completing the task Schrödinger had set himself. This result enables us to see the relationship between classical and quantum mechanics in a new way.

4.2. The Metaplectic Group

The metaplectic group $\mathrm{Mp}(2n, \mathbb{R})$ has, as such, a rather recent history although its implicit appearance can probably be traced back to Fresnel's and Gouy's work in optics around 1820; see the historical accounts in [139, 140]. The first rigorous constructions of $\mathrm{Mp}(2n, \mathbb{R})$ as a group seem to have been initiated by the work of [259] and van Hove [280]. Shale [261] remarked about a decade later that the metaplectic representation of the symplectic group is to bosons what the spin representation of the rotation group $\mathrm{SO}(2n, \mathbb{R})$ is to fermions. The study of the metaplectic group was generalized by Weil [290] to arbitrary fields, following earlier work of Siegel on number theory. It seems that Maslov was the first to observe in 1965, following the work of Buslaev [43], the role played by the metaplectic group in semiclassical mechanics; we are actually more or less following Buslaev's presentation here. Maslov's theory was clarified and improved by Leray [184], who used the properties of $\mathrm{Mp}(2n, \mathbb{R})$ to define a new mathematical structure, Lagrangian Analysis, which was subsequently extended by one of the present authors. Also see [67] who address the metaplectic group from a somewhat different point of view.

There are essentially two (closely related, and of course equivalent) ways to define the metaplectic group; both use sets of generators consisting of particular "elementary" unitary operators. We mention that there exists a third, more elusive, way to define the metaplectic group. It consists in using the irreducibility of the displacement operators (corresponding to the so-called Schrödinger representation of the Heisenberg group). This approach, which at first sight seems simpler, however has a major drawback: it only defines metaplectic operators up to an unknown (and arbitrary) phase; it thus only allows us strictly speaking to construct a projective representation of the symplectic group. Using general principles one can then adjust the phases in order to obtain a "true" representation of the double cover of the symplectic group, i.e., the metaplectic representation. We will *not* use this approach in this book.

4.2.1. Quadratic Fourier transforms

We are going to mimic the construction of free symplectic matrices of Chapter 1 at the operator level; a rule of thumb is that the standard symplectic matrix $J = \begin{pmatrix} 0 & I \\ -I & 0 \end{pmatrix}$ should become a Fourier transform, more precisely the operator $\widehat{J}$ defined by

$$\widehat{J}\psi(x) = \left(\frac{1}{2\pi i \hbar}\right)^{n/2} \int e^{-\frac{i}{\hbar} x \cdot x'} \psi(x') d^n x'. \tag{4.1}$$

The presence of the imaginary unit i in the factor preceding the integral might be a little bit intriguing. Its presence is due to the fact that we want to get correct phases when taking products of operators (it is in fact related to the "Gouy phase"; see the discussion in [139]). The argument of i is chosen to be $\pi/2$ so that

$$\left(\frac{1}{2\pi i \hbar}\right)^{n/2} = e^{-in\pi/4} \left(\frac{1}{2\pi \hbar}\right)^{n/2}.$$

Let now S_W be a free symplectic matrix with generating function W (Proposition 2):

$$S_W = \begin{pmatrix} A & B \\ C & D \end{pmatrix}, \quad \det B \neq 0$$

and

$$W(x, x') = \tfrac{1}{2} DB^{-1} x \cdot x - (B^T)^{-1} x \cdot x' + \tfrac{1}{2} B^{-1} A x' \cdot x'. \tag{4.2}$$

To S_W we associate the *two* operators $\widehat{S}_{W,m}$ and $\widehat{S}_{W,m+2} = -\widehat{S}_{W,m}$ defined by

$$\widehat{S}_{W,m} \psi(x) = \left(\frac{1}{2\pi i \hbar} \right)^{n/2} \Delta_m(W) \int e^{\frac{i}{\hbar} W(x,x')} \psi(x') d^n x', \tag{4.3}$$

where

$$\Delta_m(W) = i^m \sqrt{|\det(B^T)^{-1}|}. \tag{4.4}$$

The number m is an integer (sometimes called "Maslov index" [100]) determined by the choice of square root of the determinant $\det(B^T)^{-1}$:

$$m = \frac{1}{\pi} \arg \det(B^T)^{-1} = \frac{1}{\pi} \arg \det B. \tag{4.5}$$

For instance, if $\det B > 0$ then we can choose either $m = 0$ or $m = 2$ depending on which square root we choose; similarly, if $\det B < 0$ then $m = 1$ or $m = 2$. Note that since m only appears in the exponent of i^m it is defined modulo 4: once a choice of m is made, we can replace it, if we wish, by $m + 4k$ (k an arbitrary integer) without affecting the definition of $\Delta_m(W)$ and hence of the operator $\widehat{S}_{W,m}$. It turns out that the operators $\widehat{S}_{W,m}$ (which are *a priori* only defined for "nice" functions ψ, say $\psi \in \mathcal{S}(\mathbb{R}^n)$) can be extended into *unitary* operators acting on $L^2(\mathbb{R}^n)$. This is not quite obvious from the definition (4.3), and we will prove it later. Let us say that the idea is that $\widehat{S}_{W,m}$ is a kind of generalized Fourier transform (this is why we call it a *quadratic Fourier transform*). In fact, assume that $W(x, x') = -x \cdot x'$. This is just the generating function of the standard symplectic matrix J, in which case the operator $\widehat{S}_{W,m}$ reduces to

$$\widehat{J}\psi(x) = \left(\frac{1}{2\pi i \hbar} \right)^{n/2} \int e^{-\frac{i}{\hbar} x \cdot x'} \psi(x') d^n x'.$$

Taking for granted that the operators $\widehat{S}_{W,m}$ are invertible, and that their inverses are operators of the same type, they generate a group (of unitary operators). We call that group the metaplectic group, and denote it by $\text{Mp}(2n, \mathbb{R})$. To justify this definition we now need to work a little bit. The main step consists in showing that the operators $\widehat{S}_{W,m}$ can be written as products of simpler operators; we will also need to define a natural projection of $\text{Mp}(2n, \mathbb{R})$ onto the symplectic group $\text{Sp}(2n, \mathbb{R})$ making the metaplectic group into a covering group (of order two) of the symplectic group.

Recall from Chapter 1, Section 1.2.3, that the symplectic matrices

$$M_L = \begin{pmatrix} L^{-1} & 0 \\ 0 & L^T \end{pmatrix}, \quad V_P = \begin{pmatrix} I & 0 \\ -P & I \end{pmatrix}$$

together with J, generate $\text{Sp}(2n, \mathbb{R})$. We are going to see that, similarly, $\text{Mp}(2n, \mathbb{R})$ is generated by the operators $\widehat{M}_{L,m}$ and $\widehat{V}_P$ together with the Fourier transform $\widehat{J}$; this will follow from the fact that each quadratic Fourier transform $\widehat{S}_{W,m}$ can be written as a product

$$\widehat{S}_{W,m} = \widehat{V}_{-P}\widehat{M}_{L,m}\widehat{J}\widehat{V}_{-Q}$$

exactly in the same way as any free symplectic matrix can be written as

$$S_W = V_{-P}M_L J V_{-Q}$$

(formula (1.31) of Chapter 1). The operators $\widehat{M}_{L,m}$ and $\widehat{V}_{-P}$ are defined by

$$\widehat{M}_{L,m}\psi(x) = i^m \sqrt{|\det L|}\,\psi(Lx), \qquad (4.6)$$

where L is an invertible $n \times n$ matrix, the integer m being defined by

$$m = \frac{1}{\pi}\arg\det L \qquad (4.7)$$

(as in the definition of the integer m in (4.5) we have two choices for each L), and

$$\widehat{V}_P\psi(x) = e^{-\frac{i}{2\hbar}Px^2}\psi(x), \qquad (4.8)$$

where P is a symmetric $n \times n$ matrix; we use the notation $Px^2 = Px \cdot x$. The operators $\widehat{M}_{L,m}$ and $\widehat{V}_P$ have the obvious group properties

$$\widehat{M}_{L,m}\widehat{M}_{L',m'} = \widehat{M}_{L'L,m+m'}, \quad \widehat{M}_{L,m}^{-1} = \widehat{M}_{L^{-1},-m} \qquad (4.9)$$

(beware the ordering in the first formula: it is $L'L$, and *not* LL', that appears on the right-hand side), and

$$\widehat{V}_P\widehat{V}_{P'} = \widehat{V}_{P+P'}, \quad (\widehat{V}_P)^{-1} = \widehat{V}_{-P}. \tag{4.10}$$

In addition, these operators are unitary:

$$\langle \widehat{M}_{L,m}\psi | \widehat{M}_{L,m}\psi \rangle = \langle \psi | \psi \rangle,$$

$$\langle \widehat{V}_P\psi | \widehat{V}_P\psi \rangle = \langle \psi | \psi \rangle$$

(we leave the elementary proof of these properties to the reader).

It follows from formulas (4.9) and (4.10) that the operators $\widehat{M}_{L,m}$ and $\widehat{V}_P$ themselves generate a group. It is a group of "local" operators; we will explain below what this means.

In addition to (4.9), (4.10) the operators $\widehat{M}_{L,m}$, $\widehat{V}_P$ satisfy the following intertwining relations (the proofs are omitted, because they are straightforward to check):

$$\widehat{M}_{L,m}\widehat{V}_P = \widehat{V}_{L^T P L}\widehat{M}_{L,m}, \quad \widehat{J}\widehat{M}_{L,m} = \widehat{M}_{(L^T)^{-1},m}\widehat{J},$$

$$\widehat{V}_P\widehat{M}_{L,m} = \widehat{M}_{L,m}\widehat{V}_{(L^T)^{-1}PL^{-1}}, \quad \widehat{J}^{-1}\widehat{M}_{L,m} = \widehat{M}_{(L^T)^{-1},m}\widehat{J}^{-1}. \tag{4.11}$$

These formulas are very useful when one has to perform products of quadratic Fourier transforms. They will, for instance, allow us to find very easily the inverse of $\widehat{S}_{W,m}$. Let us first prove the following factorization result (cf. Lemma 7).

Lemma 49. *Let* $W = (P, L, Q)$ *be a quadratic form, and* m *defined by* (4.7). *We have the factorization*

$$\widehat{S}_{W,m} = \widehat{V}_{-P}\widehat{M}_{L,m}\widehat{J}\widehat{V}_{-Q}. \tag{4.12}$$

In particular, $\widehat{S}_{W,m}$ *is a unitary operator (as a product of unitary operators).*

Proof. Since $\widehat{V}_P\widehat{V}_{-P} = I$ we have, by definition of $\widehat{S}_{W,m}$:

$$\widehat{V}_P\widehat{S}_{W,m}\widehat{V}_Q\psi(x) = \left(\frac{1}{2\pi i}\right)^{n/2}\sqrt{\det L}\int e^{-iLx\cdot x'}\psi(x')\,d^n x',$$

that is

$$\widehat{V}_P \widehat{S}_{W,m} \widehat{V}_Q \psi = \widehat{M}_{L,m} \widehat{J} \psi.$$

It follows that we have $\widehat{V}_P \widehat{S}_{W,m} \widehat{V}_Q = \widehat{M}_{L,m} \widehat{J}$, hence (4.12) since $\widehat{V}_P^{-1} = \widehat{V}_{-P}$ and $\widehat{V}_Q^{-1} = \widehat{V}_{-Q}$. $\qquad\square$

Lemma 49 allows us to prove very easily that the inverse of a quadratic Fourier transform is also a quadratic Fourier transform.

Corollary 50. *The inverse of the quadratic Fourier transform* $\widehat{S}_{W,m}$ *is given by:*

$$(\widehat{S}_{W,m})^{-1} = \widehat{S}_{W^*,m^*} \quad \text{with} \quad \begin{cases} W^*(x,x') = -W(x',x) \\ m^* = n - m \bmod 4. \end{cases} \tag{4.13}$$

(*That is, if* $W = (P, L, Q)$, *then* $W^* = (-Q, -L^T, -P)$.)

Proof. In view of (4.12) we have

$$(\widehat{S}_{W,m})^{-1} = \widehat{V}_Q \widehat{J}^{-1} (\widehat{M}_{L,m})^{-1} \widehat{V}_P$$

and a straightforward calculation, using the formulas in (4.12), shows that

$$\widehat{J}^{-1} (\widehat{M}_{L,m})^{-1} = \widehat{J}^{-1} \widehat{M}_{L^{-1},m} = \widehat{M}_{L^T,m} \widehat{J}^{-1} = \widehat{M}_{-L^T, n-m} \widehat{J}$$

and hence

$$(\widehat{S}_{W,m})^{-1} = \widehat{V}_Q \widehat{M}_{-L^T, n-m} \widehat{J} \widehat{V}_P$$

from which (4.13) immediately follows. $\qquad\square$

It follows from this result that $\mathrm{Mp}(2n, \mathbb{R})$ is a group of unitary operators: as we have seen, $\mathrm{Mp}(2n, \mathbb{R})$ contains the identity (in fact $\widehat{S}_{W,m} \widehat{S}_{W^*,m^*} = \widehat{S}_{W,m} (\widehat{S}_{W,m})^{-1} = I$), and the inverse of a product $\widehat{S} = \widehat{S}_{W_1,m_1} \cdots \widehat{S}_{W_k,m_k}$ is

$$\widehat{S}^{-1} = \widehat{S}_{W_k^*,m_k^*} \cdots \widehat{S}_{W_1^*,m_1^*}$$

and is hence also in $\mathrm{Mp}(2n, \mathbb{R})$. It also follows from the factorization formula (4.12) that the elements of $\mathrm{Mp}(2n, \mathbb{R})$ are unitary operators acting on the space $L^2(\mathbb{R}_x^n)$ of square integrable functions: the operators $\widehat{V}_{-P}$, $\widehat{M}_{L,m}$ and $\widehat{J}$ are defined on the Schwartz space $\mathcal{S}(\mathbb{R}_x^n)$ and are obviously unitary for the L^2-norm, hence so is $\widehat{S}_{W,m}$. Every $\widehat{S} \in \mathrm{Mp}(2n, \mathbb{R})$ being, by definition, a product of the $\widehat{S}_{W,m}$ is therefore also a unitary operator on $L^2(\mathbb{R}_x^n)$.

It follows from the discussion above that the metaplectic group $\mathrm{Mp}(2n, \mathbb{R})$ is generated by the set of all operators $\widehat{M}_{L,m}$, $\widehat{V}_P$ together with the Fourier transform $\widehat{J}$. We could in fact have defined $\mathrm{Mp}(2n, \mathbb{R})$ as being the group generated by these "simpler" operators, but this would have led to more inconveniences than advantages, because we would then have lost the canonical relationship between free symplectic matrices and quadratic Fourier transforms.

It follows from Lemma 6 (Chapter 1) regarding products of free symplectic matrices that we have the following criterion for deciding whether a product of quadratic Fourier transforms is itself a quadratic Fourier transform.

Lemma 51. *We have $\widehat{S}_{W,m}\widehat{S}_{W',m'} = \widehat{S}_{W'',m''}$ for some pair (W'', m'') if and only if*

$$\det(P' + Q) \neq 0 \tag{4.14}$$

and in this case $W'' = (P'', L'', Q'')$ is given by formula (1.29).

(The proof of this result is somewhat technical, and will not be given here; see [184] or [94, 96].)

4.2.2. The covering projection Π

Let us now investigate more precisely the relation between $\mathrm{Mp}(2n, \mathbb{R})$ and its "classical shadow" $\mathrm{Sp}(2n, \mathbb{R})$. It will result in the construction of a projection $\Pi : \mathrm{Mp}(2n, \mathbb{R}) \to \mathrm{Sp}(2n, \mathbb{R})$ which is a surjective group homomorphism with kernel $\{-I, +I\}$.

Recall that the operators $\pm\widehat{S}_{W,m}$ are associated with the free symplectic matrices S_W by the integral formula (B.6). The relation between the set $\mathrm{Sp}_0(2n, \mathbb{R})$ of all free symplectic matrices, and the set $\mathrm{Mp}_0(2n, \mathbb{R})$ of all quadratic Fourier transforms is one-to-two: with every S_W is associated exactly two pairs (W, m) corresponding to the two possible choices (modulo 4π) of the argument of $\det(-W''_{x,x'})$.

We denote by Π_0 the mapping $\mathrm{Mp}_0(2n, \mathbb{R}) \to \mathrm{Sp}_0(2n, \mathbb{R})$ which to each quadratic Fourier transform $\widehat{S}_{W,m}$ associates the free symplectic matrix S_W.

The mapping Π_0 has the following properties, which makes it a perfect candidate for being a "partial" projection:

Lemma 52. *The mapping* $\Pi_0 : \mathrm{Mp}_0(2n, \mathbb{R}) \to \mathrm{Sp}_0(2n, \mathbb{R})$ *satisfies*

$$\begin{cases} \Pi_0((\widehat{S}_{W,m})^{-1}) = (S_W)^{-1} \\ \Pi_0(\widehat{S}_{W,m}\widehat{S}_{W',m'}) = S_W S_{W'} \end{cases} \tag{4.15}$$

when $S_W S_{W'}$ *is itself a free symplectic matrix.*

Proof. Recall that $\widehat{S}_{W,m}^{-1} = \widehat{S}_{W^*,m^*}$ where $W^*(x,x') = -W(x',x)$ and $m^* = n - m$ (see Corollary 50). If $S_{W^*} = \Pi_0(\widehat{S}_{W^*,m^*})$ then

$$(x,p) = S_{W^*}(x',p') \Longleftrightarrow \begin{cases} p = \nabla_x W^*(x,x') \\ p' = -\nabla_{x'} W^*(x,x') \end{cases}$$

that is

$$(x,p) = S_{W^*}(x',p') \Longleftrightarrow \begin{cases} p = -\nabla_{x'} W(x,x') \\ p' = \nabla_x W(x,x') \end{cases}$$

and hence $(x,p) = S_{W^*}(x',p')$ is equivalent to $(x',p') = S_W(x',p')$, which proves the first formula in (4.15). The proof of the second formula (4.15) follows from Lemma 51, using Proposition 6; see [96, pp. 86–88]. $\qquad\square$

The formulae (4.15) suggest that it might be possible to extend the projection Π_0 to a globally defined homomorphism

$$\Pi : \mathrm{Mp}(2n, \mathbb{R}) \longrightarrow \mathrm{Sp}(2n, \mathbb{R})$$

which is at the same time a group homomorphism.

Theorem 53. (i) *The mapping* Π_0 *which to* $\widehat{S}_{W,m} \in \mathrm{Mp}(2n, \mathbb{R})$ *associates* $S_W \in \mathrm{Sp}(2n, \mathbb{R})$ *can be extended into a mapping* $\Pi : \mathrm{Mp}(2n, \mathbb{R}) \to \mathrm{Sp}(2n, \mathbb{R})$ *such that*

$$\Pi(\widehat{S}\widehat{S}') = \Pi(\widehat{S})\Pi(\widehat{S}'). \tag{4.16}$$

(ii) *That mapping* Π *is determined by the condition*

$$(x,p) = \Pi(\widehat{S}_{W,m})(x',p') \Longleftrightarrow \begin{cases} p = \nabla_x W(x,x') \\ p' = -\nabla_{x'} W(x,x'). \end{cases}$$

(iii) Π *is surjective* ($=onto$) *and two-to-one, hence* Π *is a covering mapping, and* $\mathrm{Mp}(2n, \mathbb{R})$ *is a double cover of* $\mathrm{Sp}(2n, \mathbb{R})$.

We refer to Appendix B for a discussion of this important result.

4.3. The Weyl Quantization of Observables

4.3.1. The problem of quantization

Quantum mechanics in its modern form has its historical origins in the work of Bohr, Born, Heisenberg, Jordan, Pauli, von Neumann, Schrödinger, Weyl and Wigner in the mid-1920s. Already in the early years (1925–1926) quantum physicists were confronted with the problem of ordering, which consisted of finding an unambiguous procedure for associating a self-adjoint operator with a given "classical observable" (mathematicians would call it a *real symbol*). The oldest quantization procedure was actually suggested by Schrödinger himself who associated (in the case $n = 1$) the partial differential operator

$$\widehat{H} = -\frac{\hbar^2}{2m} \frac{\partial^2}{\partial x^2} + V(x)$$

with the Hamiltonian function

$$H(x, p) = \frac{p^2}{2m} + V(x).$$

Schrödinger's empirical prescription thus consisted of the formal substitution $p \to -i\hbar \partial/\partial x$ in the Hamiltonian function. It was quickly agreed that the correspondence rule

$$x_j \to x_j, \quad p_j \to -i\hbar \partial/\partial x_j$$

could be successfully applied to any set of position and momentum variables, thus turning the Hamiltonian function

$$H = \sum_{j=1}^{n} \frac{1}{2m_j} p_j^2 + V(x_1, \ldots, x_n) \tag{4.17}$$

into the partial differential operator

$$\widehat{H} = \sum_{j=1}^{n} \frac{-\hbar^2}{2m_j} \frac{\partial^2}{\partial x_j^2} + V(x_1, \ldots, x_n). \tag{4.18}$$

However, it soon became apparent that these rules lead to fundamental ambiguities when applied to more general observables involving products of functions of x_j and p_j. For instance, what should the operator corresponding to the magnetic Hamiltonian

$$H = \sum_{j=1}^{n} \frac{1}{2m_j} \left(p_j - A_j(x_1, \ldots, x_n)\right)^2 + V(x_1, \ldots, x_n) \qquad (4.19)$$

be? Even in the simple case of the product $x_j p_j = p_j x_j$ the correspondence rule led to the *a priori* equally good answers $-i\hbar x_j \partial/\partial x_j$ and $-i\hbar(\partial/\partial x_j)x_j$ which differ by the quantity $i\hbar$; things became even more complicated when one came (empirically) to the conclusion that the right answer is the "symmetric rule"

$$x_j p_j \rightarrow -\frac{1}{2}i\hbar \left(x_j \frac{\partial}{\partial x_j} + \frac{\partial}{\partial x_j} x_j \right) \qquad (4.20)$$

corresponding to the splitting $x_j p_j = \frac{1}{2}(x_j p_j + p_j x_j)$. In 1926, Born and Jordan [34] proposed the more general rule

$$\text{(BJ)} \quad x_j^m p_j^\ell \rightarrow \frac{1}{\ell+1} \sum_{k=0}^{\ell} \widehat{p}_j^{\,\ell-k} \widehat{x}_j^{\,m} \widehat{p}_j^{\,k}. \qquad (4.21)$$

for the quantization of monomials (see Appendix C for details). We refer to [77] for a readable analysis cast in a "modern" language of Born and Jordan's argument; the older papers [46] by Castellani and [55] by Crehan also contain valuable information. Born and Jordan's rules (4.21) were actually soon superseded by Weyl's quantization procedure: in his mathematical study of the foundations of quantum mechanics, Weyl proposed in [295] a rule which leads, for monomials, to the replacement of the Born–Jordan prescription with

$$\text{(Weyl)} \quad x_j^m p_j^\ell \rightarrow \frac{1}{2^\ell} \sum_{k=0}^{\ell} \binom{\ell}{k} \widehat{p}_j^{\,\ell-k} \widehat{x}_j^{\,m} \widehat{p}_j^{\,k}. \qquad (4.22)$$

Weyl's rule (which coincides with the Born–Jordan rule when $m + n = 2$) nowadays plays an important role in mathematical analysis (especially, the theory of pseudo-differential operators), and in physics it has become the preferred quantization scheme. This is mainly due to two reasons: first to real observables correspond (formally) self-adjoint operators; this is a very desirable property since a rule of thumb in quantum mechanics is that to a real observable should correspond an operator with real eigenvalues (which are, in quantum mechanics, the values that the observable can actually take). Another advantage of the Weyl correspondence is of a more subtle nature: it is the symplectic covariance property. This property which is actually *characteristic* of the Weyl correspondence among all other pseudo-differential calculi says that if we perform a linear *symplectic* change of variables in the symbol, then the resulting operator is conjugated to the original by any of the two metaplectic operators corresponding to the change of variable. A third property, which is actually less welcome for physical purposes, is that the Weyl correspondence is invertible. More precisely, one shows, using the kernel theorem from distribution theory, that there is a one-to-one correspondence between continuous operators $\mathcal{S}(\mathbb{R}^n) \to \mathcal{S}'(\mathbb{R}^n)$ and (distributional) symbols (= observables). While this property is very agreeable from a mathematical perspective, invertibility poses severe epistemological problems, because it is not physically founded (see the discussion in [173]): there might be quantum systems without classical analogue, and these cannot be detected if one uses Weyl quantization. It seems (but this is still more or less an open question, and a topic of current ongoing research) that the Born–Jordan scheme (or rather a pseudo-differential extension thereof) might be in the end the "correct quantization scheme", because it is *not* invertible. See [114, 116, 132] for recent quantum-mechanically oriented developments in the topic of Born–Jordan quantization. We have summarized the main properties of Born–Jordan quantization in Appendix C.

4.3.2. The displacement and reflection operators

There are several ways of introducing Weyl quantization for general observables. None of them is totally trivial (except, of course, for the monomial rule (4.22)). We will use here harmonic analysis, following the lines in [108, 112]; in the Appendix B of Littlejohn's seminal paper [189] a very readable review of Weyl quantization is given.

The phase space translation operators $T(z_0) : z \mapsto z + z_0$ are symplectomorphisms (because the Jacobian matrix of a translation is the identity, and is hence symplectic). Consider now the simple Hamiltonian function (called the "displacement Hamiltonian" for a reason that will become obvious)

$$H_{z_0}(z) = p \cdot x_0 - p_0 \cdot x = \sigma(z, z_0).$$

The associated Hamilton equations are $\dot{x} = x_0$, $\dot{p} = p_0$ hence the flow is given by the formula

$$f_t^{H_{z_0}}(z) = z + tz_0$$

and we have $T(z_0) = f_1^{H_{z_0}}$. Consider next the associated Schrödinger equation

$$i\hbar \frac{\partial \psi}{\partial t} = (-i\hbar x_0 \cdot \partial_x - p_0 \cdot x)\psi, \quad \psi(x, 0) = \psi_0(x) \qquad (4.23)$$

obtained from H_{z_0} by using the standard prescription (4.18). The solution of this equation can be formally written as

$$\psi(x, t) = \widehat{D}(z_0, t)\psi_0(x) = e^{-\frac{i}{\hbar} t\sigma(\hat{z}, z_0)}\psi_0(x). \qquad (4.24)$$

Using for instance the method of characteristics, or a direct calculation, one sees that an explicit formula for this solution is given by

$$\widehat{D}(z_0, t)\psi_0(x) = e^{\frac{i}{\hbar}(tp_0 \cdot x - \frac{1}{2}t^2 p_0 \cdot x_0)}\psi_0(x - tx_0). \qquad (4.25)$$

By definition, the displacement (or Heisenberg, or Heisenberg–Weyl) operator $\widehat{D}(z_0)$ is the operator obtained by setting the value of t equal to one:

$$\widehat{D}(z_0) = \widehat{D}(z_0, 1) = e^{-\frac{i}{\hbar}\sigma(\hat{z}, z_0)}. \qquad (4.26)$$

In view of formula (4.25) above its action on functions is thus given by

$$\widehat{D}(z_0)\psi(x) = e^{\frac{i}{\hbar}(p_0 \cdot x - \frac{1}{2}p_0 \cdot x_0)}\psi(x - x_0). \qquad (4.27)$$

It is clear that $\widehat{D}(z_0)$ is an unitary operator on $L^2(\mathbb{R}^n)$: we have

$$||\widehat{D}(z_0)\psi||_{L^2} = ||\psi||_{L^2} \qquad (4.28)$$

for every $\psi \in L^2(\mathbb{R}^n)$. Clearly the displacement operators are automorphisms of $\mathcal{S}(\mathbb{R}^n)$ which extend, by duality, into operators

$$\widehat{D}(z_0) : \mathcal{S}'(\mathbb{R}^n) \to \mathcal{S}'(\mathbb{R}^n).$$

The following relations satisfied by the Heisenberg operators are considered by many mathematicians and physicists almost as "mythic", in the sense that they are supposed to contain the essence of quantum mechanics. This view is however questionable, because these operators (and thus their commutation relations) can be defined using only classical arguments (the Hamilton–Jacobi theory together with the notion of phase of a Lagrangian manifold: see [108]).

Proposition 54. *The Heisenberg–Weyl operators satisfy the relations*

$$\widehat{D}(z_0)\widehat{D}(z_1) = e^{\frac{i}{\hbar}\sigma(z_0,z_1)}\widehat{D}(z_1)\widehat{D}(z_0) \tag{4.29}$$

and

$$\widehat{D}(z_0 + z_1) = e^{-\frac{i}{2\hbar}\sigma(z_0,z_1)}\widehat{D}(z_0)\widehat{D}(z_1) \tag{4.30}$$

for all $z_0, z_1 \in \mathbb{R}^{2n}$.

We omit the proof of these relations here since they are of a pure algebraic nature; see for instance [108, 112] or [83].

Closely related to the displacement operators are the reflection (or Grossmann–Royer) operators $\widehat{\Pi}(z_0)$: by definition, the reflection operator $\widehat{\Pi}(z_0)$ is the operator $\widehat{\Pi}(z_0) : \mathcal{S}(\mathbb{R}^n) \to \mathcal{S}(\mathbb{R}^n)$ defined by the formulas

$$\widehat{\Pi}(0)\psi(x) = \psi(-x) \tag{4.31}$$

and

$$\widehat{\Pi}(z_0) = \widehat{D}(z_0)\widehat{\Pi}(0)\widehat{D}(z_0)^{-1}. \tag{4.32}$$

Clearly, as for the displacement operators, the reflection operators extend into mappings

$$\widehat{\Pi}(z_0) : \mathcal{S}'(\mathbb{R}^n) \to \mathcal{S}'(\mathbb{R}^n).$$

Using the expression (4.27) of the displacement operator it is straightforward to check that we have the explicit definition

$$\widehat{\Pi}(z_0)\psi(x) = e^{\frac{2i}{\hbar}p_0 \cdot (x-x_0)}\psi(2x_0 - x). \tag{4.33}$$

The reflection operators are linear unitary involutions of $\mathcal{S}(\mathbb{R}^n)$:

$$||\widehat{\Pi}(z_0)\psi||_{L^2} = ||\psi||_{L^2}, \ \widehat{\Pi}(z_0)\widehat{\Pi}(z_0) = I_d. \tag{4.34}$$

The nice thing about the displacement and reflection operators is that they allow us to define without effort two mathematical objects playing a central role in phase space quantum mechanics: the Wigner function (or transform) [158, 299] and the ambiguity function.

Definition 55. Let $\psi \in \mathcal{S}(\mathbb{R}^n)$ and $z \in \mathbb{R}^{2n}$. The Wigner function (or transform, or distribution) of ψ is given by

$$W\psi(z) = \left(\frac{1}{\pi\hbar}\right)^n \langle \psi | \widehat{\Pi}(z)\psi \rangle. \tag{4.35}$$

The ambiguity function of ψ is defined by

$$A\psi(z) = \left(\frac{1}{2\pi\hbar}\right)^n \langle \psi | \widehat{D}(z)\psi \rangle. \tag{4.36}$$

Using the explicit expression (4.33) for $\widehat{\Pi}(z)\psi$ we have

$$\langle \psi | \widehat{\Pi}(z)\psi \rangle = \int e^{\frac{2i}{\hbar} p_0 \cdot (x - x_0)} \psi(2x_0 - x)\psi^*(x) d^n x;$$

setting $2x_0 - x = x_0 + \frac{1}{2}y$ and replacing thereafter (x_0, p_0) with (x, p) this yields, after multiplication by $(1/\pi\hbar)^n$:

$$W\psi(z) = \left(\frac{1}{2\pi\hbar}\right)^n \int e^{-\frac{i}{\hbar} p \cdot y} \psi\left(x + \frac{1}{2}y\right) \psi^*\left(x - \frac{1}{2}y\right) d^n y \tag{4.37}$$

which is the familiar textbook expression of the Wigner function. A similar calculation shows that the ambiguity function is given by

$$A\psi(z) = \left(\frac{1}{2\pi\hbar}\right)^n \int e^{-\frac{i}{\hbar} p \cdot y} \psi\left(y + \frac{1}{2}x\right) \psi^*\left(y - \frac{1}{2}x\right) d^n y. \tag{4.38}$$

We will study in detail the properties of Wigner and ambiguity functions later on in this book.

4.3.3. Weyl correspondence: The preferred quantization rule

We now have the tools we need to define and study the Weyl quantization procedure (or "Weyl correspondence"); this is essentially Weyl's original definition, recast in modern notation.

Definition 56. Let $a \in \mathcal{S}(\mathbb{R}^{2n})$ (or, more generally, $a \in \mathcal{S}'(\mathbb{R}^{2n})$). The Weyl operator with symbol a is the operator $\widehat{A}$ defined by

$$\widehat{A}\psi(x) = \left(\frac{1}{\pi\hbar}\right)^n \int a(z_0)\widehat{\Pi}(z_0)\psi(x) d^{2n} z_0. \tag{4.39}$$

We will write $\widehat{A} \overset{\text{Weyl}}{\longleftrightarrow} a$ ("Weyl correspondence") or $\widehat{A} = \text{Op}^{\text{W}}(a)$.

The interest of the definition above — which might seem at first sight somewhat unusual to experts in pseudodifferential operator theory — is that it can be written in terms of the distributional brackets $\langle\langle \cdot, \cdot \rangle\rangle$ on $\mathbb{R}^{2n}$ as

$$\widehat{A}\psi = \left(\frac{1}{\pi\hbar}\right)^n \langle\langle a(\cdot), \widehat{\Pi}(\cdot)\psi \rangle\rangle. \tag{4.40}$$

Since $\widehat{\Pi}(\cdot)\psi$ is in the Schwartz space $\mathcal{S}(\mathbb{R}^n)$ if and only if ψ is, this bracket makes *de facto* sense for any symbol $a \in \mathcal{S}'(\mathbb{R}^{2n})$, and we can thus avoid any discussion of convergence problems, as opposed to the standard definition of Weyl operators which goes as follows: let us replace in the definition (4.39) $\widehat{\Pi}(z_0)\psi(x)$ by its explicit value (4.33):

$$\widehat{A}\psi(x) = \left(\frac{1}{\pi\hbar}\right)^n \iint a(z_0)e^{\frac{2i}{\hbar}p_0(x-x_0)}\psi(2x_0 - x))d^n p_0 d^n x_0;$$

setting $y = 2x_0 - x$ and $p = p_0$ we get

$$\widehat{A}\psi(x) = \left(\frac{1}{2\pi\hbar}\right)^n \iint e^{\frac{i}{\hbar}p(x-y)}a(\tfrac{1}{2}(x+y), p)\psi(y)d^n p d^n y. \tag{4.41}$$

This formula is often taken in the mathematical literature as the (formal) definition of the Weyl operator with symbol a (see [108, 112, 262], and the references therein). One should however be careful when using this formula because of convergence problems in the integral. In fact if ψ is in the Schwartz space $\mathcal{S}(\mathbb{R}^n)$ and if a decreases at infinity sufficiently fast, the integral is absolutely convergent; however in more realistic cases, where $a(x, p)$ is bounded by, say a function with polynomial growth, the integral on the right-hand side will not be convergent in the variable p.

Weyl operators can be defined using the notion of *kernel*. By definition, the (operator) kernel of an operator $\widehat{A}$ is the function (or distribution) $K_{\widehat{A}}$ such that

$$\widehat{A}\psi(x) = \int K(x, y)\psi(y)d^n y. \tag{4.42}$$

Comparison with definition (4.41) shows that the kernel of $\widehat{A}$ is given by the expression

$$K(x, y) = \left(\frac{1}{2\pi\hbar}\right)^n \int e^{\frac{i}{\hbar}p(x-y)}a\left(\frac{1}{2}(x+y), p\right)d^n p. \tag{4.43}$$

Notice that the right-hand side can be viewed as a Fourier transform; the inversion formula then allows us to express the Weyl symbol in terms of the kernel:

$$a(x,p) = \int e^{-\frac{i}{\hbar}p \cdot y} K\left(x + \frac{1}{2}y, x - \frac{1}{2}y\right) d^n y. \qquad (4.44)$$

In addition, K can be given a sense when a is a tempered distribution, and this allows us to define $\widehat{A}$ for arbitrary $a \in \mathcal{S}'(\mathbb{R}^{2n})$ by (4.42) viewing the integral on the right-hand side as the distributional bracket $\langle K, \psi \rangle$ which is defined for every $\psi \in \mathcal{S}(\mathbb{R}^n)$.

A very important result from functional analysis, which we will discuss later is the so-called Schwartz kernel theorem. It says that if an operator A (*a priori* only defined on $\mathcal{S}(\mathbb{R}^n)$) is continuous from $\mathcal{S}(\mathbb{R}^n)$ to $\mathcal{S}'(\mathbb{R}^n)$ then it is represented by a kernel, i.e., there exists a tempered distribution K such that

$$A\psi(x) = \int K(x,y)\psi(y)d^n y$$

(the integral being viewed as the distributional bracket $\langle \cdot, \cdot \rangle$). It follows, in particular, that every continuous operator $\mathcal{S}(\mathbb{R}^n) \to \mathcal{S}'(\mathbb{R}^n)$ can be viewed as a Weyl operator: it suffices to define its Weyl symbol by formula (4.44)! This fact is actually somewhat of an annoyance in quantum mechanics, because it shows that if Weyl quantization is the correct rule, then every quantum observable has a classical counterpart, namely the corresponding Weyl symbol. In short, every operator can be "dequantized", and this is perhaps not a very realistic assumption: see the discussion in Kauffmann's paper [173].

Proposition 57. *The Weyl correspondence* $a \overset{Weyl}{\longleftrightarrow} \widehat{A}$ *is linear and one-to-one.*

(i) *If* $\widehat{A} = \mathrm{Op}^{\mathrm{W}}(a)$ *and* $\widehat{A} = \mathrm{Op}^{\mathrm{W}}(a')$ *then* $a = a'$.
(ii) *In particular* $\mathrm{Op}^{\mathrm{W}}(1) = I_{\mathrm{d}}$ *(the identity operator on* $\mathcal{S}'(\mathbb{R}^n)$*).*

Proof. (Cf. [112, Chapter 10]). The linearity of the correspondence $a \overset{Weyl}{\longleftrightarrow} \widehat{A}$ is obvious; to show that it is one-to-one, it thus suffices to demonstrate that if $\widehat{A}\psi = 0$ for all $\psi \in \mathcal{S}(\mathbb{R}^n)$ then $a = 0$. But $\widehat{A}\psi = 0$ is equivalent (formula (4.40)) to $\langle a(\cdot), \widehat{\Pi}(\cdot)\psi \rangle = 0$, that is to $a = 0$ since $\psi \in \mathcal{S}(\mathbb{R}^n)$ is arbitrary. To show that if $a = 1$ then $\widehat{A}$ is the identity, it suffices to note that by the second formula (4.40) we have

$$\widehat{A}\psi(x) = \left(\frac{1}{\pi\hbar}\right)^n \langle\langle 1, \widehat{\Pi}(\cdot)\psi(x)\rangle\rangle$$

$$= \int \left(\int e^{\frac{2i}{\hbar}p_0\cdot(x-x_0)}d^n p_0\right) \psi(2x_0 - x)d^n x_0$$

$$= \int \delta(2(x - x_0))\psi(2x_0 - x)d^n x_0$$

$$= \psi(x)$$

so that $\widehat{A}\psi = \psi$ for all $\psi \in \mathcal{S}(\mathbb{R}^n)$. By continuity, using the density of $\mathcal{S}(\mathbb{R}^n)$, we also have $\widehat{A}\psi = \psi$ for all $\psi \in \mathcal{S}'(\mathbb{R}^n)$ hence $\widehat{A} = I_{\mathrm{d}}$ on $\mathcal{S}'(\mathbb{R}^n)$.
□

The symplectic Fourier transform F_σ is defined, for $a \in \mathcal{S}(\mathbb{R}^{2n})$, by

$$F_\sigma a(z) = \left(\frac{1}{2\pi\hbar}\right)^n \int e^{-\frac{i}{\hbar}\sigma(z,z')}a(z')d^{2n}z'. \tag{4.45}$$

We will often use the shorthand notation $a_\sigma = F_\sigma a$. We list below the main properties of the symplectic Fourier transform; they are easily deduced from the corresponding properties of the ordinary Fourier transform on $\mathbb{R}^{2n}_z$,

$$Fa(z) = \left(\frac{1}{2\pi\hbar}\right)^n \int e^{-\frac{i}{\hbar}z\cdot z'}a(z')d^{2n}z'. \tag{4.46}$$

• The Fourier transforms F_σ and F are related by the formula

$$F_\sigma a(z) = Fa(Jz) = F(a \circ J)(z). \tag{4.47}$$

In particular F_σ is a linear automorphism $F_\sigma : \mathcal{S}(\mathbb{R}^{2n}) \to \mathcal{S}(\mathbb{R}^{2n})$ which extends by duality into an automorphism $F_\sigma : \mathcal{S}'(\mathbb{R}^{2n}) \to \mathcal{S}'(\mathbb{R}^{2n})$.
• The symplectic Fourier transform is involutive and unitary:

$$F_\sigma \circ F_\sigma = I_{\mathrm{d}}, \quad ||F_\sigma a||_{L^2} = ||a||_{L^2}. \tag{4.48}$$

• For $a \in \mathcal{S}'(\mathbb{R}^{2n})$ and $S \in \mathrm{Sp}(2n,\mathbb{R})$ we have

$$F_\sigma a(Sz) = F_\sigma(a \circ S)(z). \tag{4.49}$$

• The symplectic Fourier transform satisfies the Plancherel formula

$$\langle F_\sigma a | b\rangle = \langle a | F_\sigma b\rangle. \tag{4.50}$$

Let a be function, or a distribution, defined on phase space $\mathbb{R}^{2n} \equiv \mathbb{R}^n_x \times \mathbb{R}^n_p$. We assume that a is "well-behaved" in the sense that its Fourier transform exists (for instance, a can be a tempered distribution):

$$\widehat{A}\psi(x) = \left(\frac{1}{2\pi\hbar}\right)^n \int a_\sigma(z_0)\widehat{D}(z_0)\psi(x)d^{2n}z_0; \qquad (4.51)$$

equivalently,

$$\widehat{A}\psi(x) = \left(\frac{1}{\pi\hbar}\right)^n \int a(z_0)\widehat{\Pi}(z_0)\psi(x)d^{2n}z_0. \qquad (4.52)$$

Recall that we write $\widehat{A} \overset{\text{Weyl}}{\longleftrightarrow} a$ or $a \overset{\text{Weyl}}{\longleftrightarrow} \widehat{A}$ or also $\widehat{A} = \text{Op}^{\text{W}}(a)$ to denote the "Weyl correspondence". The function a_σ is often called the "twisted" symbol of $\widehat{A}$ (it is sometimes also called the "covariant symbol"; see [189]).

We are now going to study an essential property of Weyl quantization, that of *symplectic covariance*. It turns out that this property is actually characteristic of Weyl quantization, in the sense that there is no other quantization scheme satisfying this covariance property (see [114, 119, 120]).

Proposition 58. *Let $S \in \text{Sp}(2n, \mathbb{R})$ and $\widehat{S} \in \text{Mp}(2n, \mathbb{R})$ be any one of the two metaplectic operators with $S = \pi^{\text{Mp}}(\widehat{S})$.*
  (i) *We have*

$$\widehat{S}\widehat{D}(z_0)\widehat{S}^{-1} = \widehat{D}(Sz_0) \quad and \quad \widehat{S}\widehat{\Pi}(z_0)\widehat{S}^{-1} = \widehat{\Pi}(Sz_0). \qquad (4.53)$$

  (ii) *For every Weyl operator $\widehat{A} \overset{\text{Weyl}}{\longleftrightarrow} a$ we have the correspondence*

$$a \circ S \overset{\text{Weyl}}{\longleftrightarrow} \widehat{S}^{-1}\widehat{A}\widehat{S}. \qquad (4.54)$$

  (iii) *For every $\psi \in L^2(\mathbb{R}^{2n})$ we have*

$$W(\widehat{S}\psi)(z) = W\psi(S^{-1}z) \quad and \quad A(\widehat{S}\psi)(z) = A\psi(S^{-1}z). \qquad (4.55)$$

Proof. (i) A detailed proof is given in [108, 112]; it consists in verifying separately the first relation (4.53) on the generators $\widehat{V}_{-P}$, $\widehat{M}_{L,m}$ and $\widehat{J}$ of the metaplectic group. The second formula (4.53) follows using the relation (4.33) expressing the reflection operator $\widehat{\Pi}(z_0)$ in terms of the displacement operator $\widehat{D}(z_0)$. (ii) Let us denote $\widehat{B}$ the Weyl operator with symbol $a \circ S$.

We have

$$\widehat{B}\psi(x) = \left(\frac{1}{2\pi\hbar}\right)^n \int a_\sigma(Sz_0)\widehat{D}(z_0)\psi(x)d^{2n}z_0$$

that is, performing the change of variables $Sz \mapsto z$ and taking into account the fact that $\det S = 1$ since S is symplectic,

$$\widehat{B}\psi(x) = \left(\frac{1}{2\pi\hbar}\right)^n \int a_\sigma(z_0)\widehat{D}(S^{-1}z_0)\psi(x)d^{2n}z_0.$$

By formula (4.53) above we have $\widehat{S}^{-1}\widehat{D}(z_0)\widehat{S} = \widehat{D}(S^{-1}z_0)$ and hence

$$\widehat{B}\psi(x) = \left(\frac{1}{2\pi\hbar}\right)^n \int a_\sigma(z_0)\widehat{S}^{-1}\widehat{D}(z_0)\widehat{S}\psi(x)d^{2n}z_0$$

$$= \left(\frac{1}{2\pi\hbar}\right)^n \widehat{S}^{-1}\left(\int a_\sigma(z_0)\widehat{D}(z_0)d^{2n}z_0\right)\widehat{S}\psi(x)$$

$$= \widehat{S}^{-1}\widehat{A}(\widehat{S}\psi)(x)$$

which is formula (4.54). (iv) The formulas (4.55) immediately follow from the symplectic covariance relations (4.53) in view of the definitions (5.4) and (4.36). $\qquad\square$

Note that the phase space translation operators $T(z_0) : z \mapsto z + z_0$ satisfy the intertwining formula $ST(z_0)S^{-1} = T(Sz_0)$ for every $S \in \mathrm{Sp}(2n, \mathbb{R})$; formula (4.53) can thus be viewed as the metaplectic analogue of this relation. Also note that for a given symplectic matrix S the choice of the corresponding metaplectic operator $\widehat{S}$ is irrelevant, since both choices only differ by a sign, which does not affect the right-hand sides of (4.53) and (4.54).

The Weyl correspondence also satisfies a covariance property with respect to the displacement operators: denoting by $T(z_0)$ the translation operator $z \mapsto z + z_0$ and defining $T(z_0)a(z) = a(z - z_0)$ we have

$$T(z_0)a \overset{\text{Weyl}}{\longleftrightarrow} \widehat{D}(z_0)\widehat{A}\widehat{D}(z_0)^{-1}. \tag{4.56}$$

Combining this property with the symplectic covariance relation (4.54) one can prove that Weyl operators are covariant under the action of the inhomogeneous metaplectic group, which is the group of unitary operators generated by $\text{Mp}(2n, \mathbb{R})$ and the displacement operators (see [112, §10.3.2]).

Here are three examples.

Orthogonal projections

Projection operators play a very important role in quantum mechanics; they are idempotents. Let $\psi \in L^2(\mathbb{R}^n)$ be a normalized vector and consider the orthogonal projection

$$\widehat{\rho}_\psi : L^2(\mathbb{R}^n) \to \{\alpha\psi : \alpha \in \mathbb{C}\}.$$

For each $\phi \in L^2(\mathbb{R}^n)$ we have the explicit expression

$$\widehat{\rho}_\psi \phi = \langle \psi | \phi \rangle \psi. \tag{4.57}$$

The Weyl symbol π_ψ of $\widehat{\rho}_\psi$ and the Wigner transform $W\psi$ of ψ are related by the formula

$$\pi_\psi(z) = (2\pi\hbar)^n W\psi(z). \tag{4.58}$$

In fact, using the integral representation (4.41),

$$\widehat{\rho}_\psi \phi(x) = \left(\frac{1}{2\pi\hbar}\right)^n \iint e^{\frac{i}{\hbar} p_0 \cdot (x-y)} \pi_\psi \left(\frac{1}{2}(x+y), p_0\right) \phi(y) d^n p_0 d^n y$$

$$= \left(\frac{1}{2\pi\hbar}\right)^n \int \left[\int e^{\frac{i}{\hbar} p_0 \cdot (x-y)} \pi_\psi \left(\frac{1}{2}(x+y), p_0\right) d^n p_0\right] \phi(y) d^n y$$

and since (4.57) can be rewritten

$$\widehat{\rho}_\psi \phi(x) = \int \psi^*(y)\phi(y)\psi(x) d^n y$$

we have the identity

$$\left(\frac{1}{2\pi\hbar}\right)^n \int e^{\frac{i}{\hbar} p_0 \cdot (x-y)} \pi_\psi \left(\frac{1}{2}(x+y), p_0\right) d^n p_0 = \psi^*(y)\psi(x)$$

since ϕ is arbitrary. Replacing x with $x + \frac{1}{2}y$ and y with $x - \frac{1}{2}y$ this is the same thing as

$$\left(\frac{1}{2\pi\hbar}\right)^n \int e^{\frac{i}{\hbar}p_0 \cdot y} \pi_\psi(x, p_0) d^n p_0 = \psi^* \left(x - \frac{1}{2}y\right) \psi \left(x + \frac{1}{2}y\right);$$

noting that the integral on the left-hand side is the Fourier transform of π_ψ multiplied by $(2\pi\hbar)^{n/2}$ we have

$$\pi_\psi(x, p_0) = \int e^{-\frac{i}{\hbar}p_0 \cdot y} \psi^* \left(x - \frac{1}{2}y\right) \psi \left(x + \frac{1}{2}y\right) d^n y$$

hence formula (4.58) in view of definition (6.29) of the Wigner transform.

Monomials

Any good quantization theory should satisfy the prescription

$$x_j p_j \longleftrightarrow \tfrac{1}{2}(\widehat{X_j P_j} + \widehat{P_j X_j}). \tag{4.59}$$

Let us check that the Weyl correspondence $\widehat{A} \overset{\text{Weyl}}{\longleftrightarrow} a$ satisfies this requirement. It suffices to consider the case $n = 1$; we have, taking $a(z) = xp$,

$$2\pi\hbar\widehat{A}\psi(x) = \frac{1}{2} \iint e^{\frac{i}{\hbar}p(x-y)} px\psi(y)dydp + \frac{1}{2} \iint e^{\frac{i}{\hbar}p(x-y)} py\psi(y)dydp.$$

Formula (4.59) follows in view of the obvious equalities

$$\iint e^{\frac{i}{\hbar}p(x-y)} xp\psi(y)dydp = 2\pi\hbar x \widehat{P}\psi(x),$$

$$\iint e^{\frac{i}{\hbar} \cdot (x-y)} py\psi(y)dydp = 2\pi\hbar\widehat{P}(x\psi)(x).$$

A similar argument, involving some more calculations, shows that the Weyl correspondence satisfies the general monomial rule (4.22), that is

$$x_j^m p_j^\ell \overset{\text{Weyl}}{\longleftrightarrow} \frac{1}{2^\ell} \sum_{k=0}^{\ell} \binom{\ell}{k} \widehat{p}_j^{\ell-k} \widehat{x}_j^m \widehat{p}_j^k. \tag{4.60}$$

The displacement operator

Since we have formally $\widehat{D}(z_0) = e^{-\frac{i}{\hbar}\sigma(\hat{z}, z_0)}$ (formula (4.26)) an educated guess is that the Weyl symbol of $\widehat{D}(z_0)$ should be the function $e^{-\frac{i}{\hbar}\sigma(z, z_0)}$.

Let us show this is indeed the case, that is we have $e^{-\frac{i}{\hbar}\sigma(\hat{z},z_0)} \overset{\text{Weyl}}{\longleftrightarrow}$ $e^{-\frac{i}{\hbar}\sigma(z,z_0)}$. Let $\widehat{A}_{z_0}$ be the operator with Weyl symbol $e^{-\frac{i}{\hbar}\sigma(z,z_0)}$. We have, using definition (4.52) of Weyl operators

$$
\begin{aligned}
\widehat{A}_{z_0}\psi(x) &= \iint e^{-\frac{i}{\hbar}\sigma(z',z_0)}\widehat{\Pi}(z')\psi(x)d^np'd^nx' \\
&= \left(\frac{1}{\pi\hbar}\right)^n \iint e^{-\frac{i}{\hbar}(p'x_0 - p_0\cdot x')}e^{\frac{2i}{\hbar}p'(x-x')}\psi(2x'-x)d^np'd^nx' \\
&= \left(\frac{1}{\pi\hbar}\right)^n \iint e^{\frac{i}{\hbar}p_0x'}e^{\frac{i}{\hbar}p'(-x_0+2x-2x')}\psi(2x'-x)d^np'd^nx' \\
&= \left(\frac{1}{\pi\hbar}\right)^n \int e^{\frac{i}{\hbar}p_0\cdot x'}\left[\int e^{\frac{i}{\hbar}p'(-x_0+2x-2x')}d^np'\right]\psi(2x_0-x)d^nx' \\
&= 2^n \int e^{\frac{i}{\hbar}p_0x'}\delta(-x_0+2x-2x')\psi(2x_0-x)d^nx'.
\end{aligned}
$$

Setting $y = -2x'+2x-x_0$ we get, taking into account the explicit expression (4.27) of the displacement operator,

$$
\begin{aligned}
\widehat{A}_{z_0}\psi(x) &= \int e^{\frac{i}{2\hbar}p_0(2x-y-x_0)}\delta(y)\psi(x-y-x_0)d^ny \\
&= e^{\frac{i}{2\hbar}p_0(2x-x_0)}\psi(x-x_0) \\
&= \widehat{D}(z_0)\psi(x).
\end{aligned}
$$

4.3.4. Adjoints and products

The formal adjoint $\widehat{A}^\dagger$ of an operator $\widehat{A} : \mathcal{S}(\mathbb{R}^n) \to \mathcal{S}(\mathbb{R}^n)$ is defined by the formula

$$
\langle \widehat{A}\psi|\phi\rangle = \langle\psi|\widehat{A}^\dagger\phi\rangle
$$

for all ψ and ϕ in $\mathcal{S}(\mathbb{R}^n)$.

Proposition 59. *The formal adjoint $\widehat{A}^\dagger$ of the Weyl operator $\widehat{A} = \text{Op}^{\text{W}}(a)$ is the Weyl operator $\widehat{A}^\dagger = \text{Op}^{\text{W}}(a^*)$. In particular, $\widehat{A}$ is formally self-adjoint if and only if the symbol a is real.*

The proof of this result is given in Appendix D. Weyl quantization thus satisfies one of the major requirements of quantum mechanics, namely that to a real observable should correspond a self-adjoint operator (whose eigenvalues are thus *real numbers*).

We now assume that the Weyl operators

$$\widehat{A} = \left(\frac{1}{2\pi\hbar}\right)^n \int a_\sigma(z)\widehat{D}(z)d^{2n}z,$$

$$\widehat{B} = \left(\frac{1}{2\pi\hbar}\right)^n \int b_\sigma(z)\widehat{D}(z)d^{2n}z$$

can be composed and set $\widehat{C} = \widehat{A}\widehat{B}$. Writing

$$\widehat{C} = \left(\frac{1}{2\pi\hbar}\right)^n \int c_\sigma(z)\widehat{D}(z)d^{2n}z,$$

we want to express the symbol c in terms of the symbols a and b. This will lead us to the notions of twisted product and twisted convolution.

Proposition 60. *Let* $\widehat{A} = \mathrm{Op}^{\mathrm{W}}(a)$ *and* $\widehat{B} = \mathrm{Op}^{\mathrm{W}}(b)$ *be Weyl operators.*
(i) *When the operator* $\widehat{C} = \widehat{A}\widehat{B}$ *is defined we have* $\widehat{C} = \mathrm{Op}^{\mathrm{W}}(c)$ *with*

$$c(z) = \left(\frac{1}{4\pi\hbar}\right)^{2n} \iint e^{\frac{i}{2\hbar}\sigma(z',z'')} a\left(z + \frac{1}{2}z'\right) b\left(z - \frac{1}{2}z''\right) d^{2n}z'd^{2n}z''. \tag{4.61}$$

(ii) *The symplectic Fourier transform* $c_\sigma = F_\sigma c$ *of* c *is given by*

$$c_\sigma(z) = \left(\frac{1}{2\pi\hbar}\right)^n \int e^{\frac{i}{2\hbar}\sigma(z,z')} a_\sigma(z - z')b_\sigma(z')d^{2n}z'. \tag{4.62}$$

The proof of this result is somewhat technical; we give it in Appendix D. Notice that performing a change of variable (4.62) is equivalent to

$$c_\sigma(z) = \left(\frac{1}{2\pi\hbar}\right)^n \int e^{-\frac{i}{2\hbar}\sigma(z,z')} a_\sigma(z')b_\sigma(z - z')d^{2n}z'. \tag{4.63}$$

The composition formulas (4.62), (4.62) always make sense when the symbols a and b are in the Schwartz space $\mathcal{S}(\mathbb{R}^{2n})$ of rapidly decreasing functions, but they can be extended to more general situations. They are very important in several contexts. We will use them in our study of the density matrix in Chapter 6. The importance of these formulas justifies the following definition.

Definition 61. Let a and b be two symbols.

(i) The function $a \times_\hbar b$ given by

$$(a \times_\hbar b)(z) = \left(\frac{1}{4\pi\hbar}\right)^{2n} \iint e^{\frac{i}{2\hbar}\sigma(z',z'')} a\left(z + \frac{1}{2}z'\right) b$$
$$\times \left(z - \frac{1}{2}z''\right) d^{2n}z' d^{2n}z''$$

is called the twisted product (or Moyal product) of a and b.

(ii) The function $a \star_\hbar b$ given by

$$a \star_\hbar b(z) = \left(\frac{1}{2\pi\hbar}\right)^n \int_{\mathbb{R}^{2n}} e^{\frac{i}{2\hbar}\sigma(z,z')} a_\sigma(z - z') b_\sigma(z') dz'$$

is called the twisted convolution (or star product) of a and b.

Notice that in general $a \times_\hbar b \neq b \times_\hbar a$ and $a \star_\hbar b \neq b \star_\hbar a$ (because in general $\widehat{A}\widehat{B} \neq \widehat{B}\widehat{A}$). In view of the result above, we have, since the symplectic Fourier transform is its own inverse,

$$a \star_\hbar b = F_\sigma(a \times_\hbar b) \quad \text{and} \quad a \times_\hbar b = F_\sigma(a \star_\hbar b). \tag{4.64}$$

We are going to see that these notions can be interpreted in terms of a pseudodifferential calculus on phase space.

Bopp pseudo-differential operators

Recall that the displacement operator $\widehat{D}(z_0)$ acts on functions (or tempered distributions) defined on $\mathbb{R}^n$ via the formula

$$\widehat{D}(z_0)\psi(x) = e^{\frac{i}{\hbar}(p_0 \cdot x - \frac{1}{2}p_0 \cdot x_0)}\psi(x - x_0);$$

a natural step is to extend the domain of $\widehat{D}(z_0)$ by letting it act on functions (or distributions) defined on phase space $\mathbb{R}^{2n}$. This approach was actually initiated in [108] and detailed in [112] in connection with the study of the phase space Schrödinger equation.

For $z_0 \in \mathbb{R}^{2n}$ we define an operator $\widetilde{T}(z_0)$ acting on functions $\Psi \in \mathcal{S}(\mathbb{R}^{2n})$ by

$$\widetilde{T}(z_0)\Psi(z) = e^{-\frac{i}{\hbar}\sigma(z,z_0)}\Psi\left(z - \frac{1}{2}z_0\right). \tag{4.65}$$

Recall that the usual displacement operator $\widehat{D}(z_0) = e^{-\frac{i}{\hbar}\sigma(\widehat{z},z_0)}$ can be viewed as the time-one propagator for the Schrödinger equation associated with the translation Hamiltonian $H_{z_0}(z) = \sigma(z,z_0)$; the operator with this

Weyl symbol is $\widehat{H}_{z_0}(z) = \sigma(\widehat{z}, z_0)$ with $\widehat{z} = (x, -i\hbar\partial_x)$ and the solution of the corresponding Schrödinger equation

$$i\hbar\frac{\partial}{\partial t}\psi = \widehat{H}_{z_0}\psi \ , \ \ \psi(x,0) = \psi_0(x)$$

is formally given by $\psi(x,t) = e^{it\sigma(\widehat{z}, z_0)/\hbar}\psi_0(x)$. To define the operators $\widetilde{T}(z_0)$ one proceeds exactly in the same way, except that we replace the Hamiltonian operator $\widehat{H}_{z_0}(z)$ with $\widetilde{H}_{z_0}(z) = \sigma(\widetilde{z}, z_0)$ where

$$\widetilde{z} = z + \tfrac{1}{2}i\hbar J\partial_z; \tag{4.66}$$

this corresponds to the so-called "Bopp shifts" $(x, p) \mapsto (\ , \widetilde{p})$ defined by

$$x_j \mapsto \widetilde{x}_j = x_j + \frac{1}{2}i\hbar\frac{\partial}{\partial p_j}, \quad p_j \mapsto \widetilde{u}_j = p_j - \frac{1}{2}i\hbar\frac{\partial}{\partial x_j} \tag{4.67}$$

and which were first introduced by Bopp [33] and later exploited by Kubo [177]. This new correspondence principle leads us to the "phase space Schrödinger equation"

$$i\hbar\frac{\partial}{\partial t}\Psi(z,t) = \sigma(\widetilde{z}, z_0)\Psi(z,t), \quad \Psi(z,0) = \Psi_0(z)$$

whose solution is

$$\Psi(z,t) = e^{\frac{i}{\hbar}t\sigma(\widetilde{z}, z_0)}\Psi_0(z) = e^{-\frac{i}{\hbar}t\sigma(z, z_0)}\Psi_0\left(z - \frac{1}{2}tz_0\right).$$

We thus have

$$\Psi(z,t) = \widetilde{T}(z_0)\Psi(z) = e^{\frac{i}{\hbar}\sigma(\widetilde{z}, z_0)}\Psi_0(z).$$

It turns out that the operators $\widetilde{T}(z_0)$ satisfy commutation relations which are similar to those (4.29) and (4.30) satisfied by the displacement operators: we have

$$\widetilde{T}(z_0 + z_1) = e^{-\frac{i}{2\hbar}\sigma(z_0, z_1)}\widetilde{T}(z_0)\widetilde{T}(z_1), \tag{4.68}$$

$$\widetilde{T}(z_1)\widetilde{T}(z_0) = e^{-\frac{i}{\hbar}\sigma(z_0, z_1)}\widetilde{T}(z_0)\widetilde{T}(z_1) \tag{4.69}$$

for all $z_0, z_1 \in \mathbb{R}^{2n}$. In fact, formula (4.69) follows from the following chain of equalities:

$$\widetilde{T}(z_1)\widetilde{T}(z_0) = e^{\frac{i}{2\hbar}\sigma(z_1, z_0)}\widetilde{T}(z_1 + z_0)$$

$$= e^{-\frac{i}{2\hbar}\sigma(z_0, z_1)}\widetilde{T}(z_0 + z_1)$$

$$= e^{-\frac{i}{\hbar}\sigma(z_0, z_1)}\widetilde{T}(z_0)\widetilde{T}(z_1).$$

The proof of (4.68) is similar and left to the reader.

In analogy with the formula

$$\widehat{A}\psi(x) = \left(\frac{1}{2\pi\hbar}\right)^n \int a_\sigma(z_0)\widehat{D}(z_0)\psi(x)d^{2n}z_0$$

defining a Weyl operator, we define the *Bopp operator* $\widetilde{A}$ with symbol $a \in \mathcal{S}'(\mathbb{R}^{2n})$ by the formula

$$\widetilde{A}\Psi(z) = \left(\frac{1}{2\pi\hbar}\right)^n \int a_\sigma(z_0)\widetilde{T}(z_0)\Psi(z)d^{2n}z_0;$$

we will write $\widetilde{A} \overset{\text{Bopp}}{\longleftrightarrow} a$ ("Bopp correspondence") or simply $\widetilde{A} = \widetilde{\text{Op}}(a)$.

Let us view the linear operator $\widetilde{A}$ as a Weyl operator $\widetilde{A} : \mathcal{S}(\mathbb{R}^{2n}) \to \mathcal{S}'(\mathbb{R}^{2n})$. One proves (see [112, Chapter 18; 110]) the following.

Proposition 62. *Viewing $\widetilde{A}$ as a Weyl operator $\mathcal{S}(\mathbb{R}^{2n}) \to \mathcal{S}(\mathbb{R}^{2n})$ its Weyl symbol $\widetilde{a} \overset{\text{Weyl}}{\longleftrightarrow} \widetilde{A}$ is obtained from the usual symbol $a \overset{\text{Weyl}}{\longleftrightarrow} \widehat{A}$ by*

$$\widetilde{a}(z,\zeta) = a(z - \tfrac{1}{2}J\zeta) \tag{4.70}$$

that is, setting $\zeta = (\zeta_x, \zeta_p)$,

$$\widetilde{a}(z,\zeta) = a(x - \tfrac{1}{2}\zeta_p, p + \tfrac{1}{2}\zeta_x). \tag{4.71}$$

Note that (z, ζ) is the generic point of the $4n$-dimensional phase space $\mathbb{R}^{2n} \times \mathbb{R}^{2n}$; the variable $\zeta = (\zeta_x, \zeta_p) \in \mathbb{R}^n \times \mathbb{R}^n$ is viewed as the dual variable of $z = (x, p)$.

The result above justifies the interpretation of $\widetilde{A}$ as the operator obtained from the usual Weyl symbol a by the Bopp quantization rule; it motivates the notation

$$\widetilde{A} = \widetilde{\text{Op}}(a) = a(x + \tfrac{1}{2}i\hbar\partial_p, p - \tfrac{1}{2}i\hbar\partial_x) \tag{4.72}$$

we introduced above.

The relation with the Moyal product is the following. Recall that the twisted convolution $c = a \star_\hbar b$ is defined by

$$c(z) = \left(\frac{1}{4\pi\hbar}\right)^{2n} \iint e^{\frac{i}{2\hbar}\sigma(z',z'')}a\left(z + \frac{1}{2}z'\right)b\left(z - \frac{1}{2}z''\right)d^{2n}z'd^{2n}z'';$$

equivalently, the symplectic Fourier transform of c is given by the formula

$$c_\sigma(z) = \left(\frac{1}{2\pi\hbar}\right)^n \int e^{\frac{i}{2\hbar}\sigma(z,z')} a_\sigma(z-z') b_\sigma(z') d^{2n} z'.$$

In particular, it is easy to see that (in the case $n = 1$)

$$x \star_\hbar a = \left(x + \frac{1}{2} i\hbar \frac{\partial}{\partial p}\right) a, \quad p \star_\hbar a = \left(p - \frac{1}{2} i\hbar \frac{\partial}{\partial x}\right) a. \tag{4.73}$$

It turns out that, more generally, the twisted convolution $a \star_\hbar b$ of two functions a, b can always be written as

$$a \star_\hbar b = a(x + \tfrac{1}{2} i\hbar \partial_p, p - \tfrac{1}{2} i\hbar \partial_x) b, \tag{4.74}$$

where $\widetilde{A}$ is the Bopp pseudodifferential operator (4.72). For a detailed treatment of the Bopp approach to the twisted product and convolution, see [112, Chapter 18; 110]. We note that some authors in the physical literature use a formalism similar to (4.74) without any mathematical motivation; for instance given a Hamiltonian function

$$H(x, p) = \frac{p^2}{2m} + V(x)$$

they "Bopp quantize" it by considering the operator

$$H\left(x + \frac{1}{2} i\hbar \frac{\partial}{\partial p}, p - \frac{1}{2} i\hbar \frac{\partial}{\partial x}\right) = \frac{1}{2m}\left(p - \frac{1}{2} i\hbar \frac{\partial}{\partial x}\right)^2 + V\left(x + \frac{1}{2} i\hbar \frac{\partial}{\partial p}\right).$$

This expression is however of course totally meaningless, because these authors fail to give a definition of the term $V(x + \tfrac{1}{2} i\hbar \frac{\partial}{\partial p})$ for arbitrary potentials.

4.4. From Classical to Quantum

In this section we intend to show that the quantum world formally emerges from classical mechanics in its Hamiltonian formulation. We first have to give a precise analysis of Hamiltonian flows from a modern perspective, where the group of Hamiltonian symplectomorphisms plays a central role. This will allow us, in passing, to show that Bohmian phase space trajectories are deformations of the usual Hamiltonian trajectories.

4.4.1. The group of Hamiltonian symplectomorphisms

The group $\mathrm{Ham}(2n, \mathbb{R})$ is the connected component of the group $\mathrm{Symp}(2n, \mathbb{R})$ of all symplectomorphisms of the symplectic phase space $(\mathbb{R}^{2n}, \sigma)$. Each of its points is the value of a Hamiltonian flow at some time t. The study of the various algebraic and topological properties of the group $\mathrm{Ham}(2n, \mathbb{R})$ is a very active area of current research; see [159] or [240]. We are following here with some modifications and complements the exposition in [108].

We will say that a symplectomorphism f of the standard symplectic space $(\mathbb{R}^{2n}, \sigma)$ is *Hamiltonian* (*function*) if there exists a (real) function $H \in C^\infty(\mathbb{R}^{2n+1}_{z,t})$ and a number $a \in \mathbb{R}$ such that $f = f_a^H$. Taking $a = 0$ it is clear that the identity is a Hamiltonian symplectomorphism. The set of all Hamiltonian symplectomorphisms is denoted by $\mathrm{Ham}(2n, \mathbb{R})$. We are going to see that it is a connected and normal subgroup of $\mathrm{Symp}(2n, \mathbb{R})$; but let us first state the following result which is a precise way of formulating the "canonical invariance of Hamilton's equations under canonical transformations" as it is called in the older literature (e.g., [92]).

Proposition 63. *Let g be a symplectomorphism. The flows (f_t^H) and $(f_t^{H \circ g})$ are conjugate; more precisely we have*

$$f_t^{H \circ g} = g^{-1} \circ f_t^H \circ g. \tag{4.75}$$

For a proof we refer to [108, Proposition 2.50].

Proposition 64. *Let (f_t^H) and (f_t^K) be Hamiltonian flows. Then:*

$$f_t^H f_t^K = f_t^{H \# K} \quad \text{with } H \# K(z,t) = H(z,t) + K((f_t^H)^{-1}(z),t), \tag{4.76}$$

$$(f_t^H)^{-1} = f_t^{\bar{H}} \quad \text{with } \bar{H}(z,t) = -H(f_t^H(z),t). \tag{4.77}$$

We omit the proof of this result; see [108, Proposition 2.50; 159, 240]. Let us now show that $\mathrm{Ham}(2n, \mathbb{R})$ is a group, as claimed.

Proposition 65. $\mathrm{Ham}(2n, \mathbb{R})$ *is a normal and connected subgroup of the group* $\mathrm{Symp}(2n, \mathbb{R})$ *of all symplectomorphisms of* $(\mathbb{R}^{2n}_z, \sigma)$.

Proof. Let us show that if $f, g \in \mathrm{Ham}(2n, \mathbb{R})$ then $fg^{-1} \in \mathrm{Ham}(2n, \mathbb{R})$. We begin by remarking that if $f = f^H$ for some $a \neq 0$ then we also have $f = f_1^{H^a}$ where $H^a(z, t) = aH(z, at)$. In fact, setting $t^a = at$ we have

$$\frac{dz^a}{dt} = J\partial_z H^a(z^a, t) \Longleftrightarrow \frac{dz^a}{dt^a} = J\partial_z H(z^a, t^a)$$

and hence $f_t^{H^a} = f_{at}^H$. We may thus assume that $f = f_1^H$ and $g = f_1^K$ for some Hamiltonians H and K. Now, using successively (4.76) and (4.77) we have

$$fg^{-1} = f_1^H (f_1^K)^{-1} = f_1^{H \# \bar{K}}$$

hence $fg^{-1} \in \mathrm{Ham}(2n, \mathbb{R})$. That $\mathrm{Ham}(2n, \mathbb{R})$ is a normal subgroup of $\mathrm{Symp}(2n, \mathbb{R})$ immediately follows from formula (4.75) above: if g is a symplectomorphism and $f \in \mathrm{Ham}(2n, \mathbb{R})$ then

$$f_1^{H \circ g} = g^{-1} f_1^H g \in \mathrm{Ham}(2n, \mathbb{R}) \tag{4.78}$$

so we are done. $\qquad\square$

The result above motivates the following definition.

Definition 66. The set $\mathrm{Ham}(2n, \mathbb{R})$ of all Hamiltonian symplecto-morphisms equipped with the law $fg = f \circ g$ is called the group of Hamiltonian symplectomorphisms of the standard symplectic space $(\mathbb{R}_z^{2n}, \sigma)$.

The topology of $\mathrm{Symp}(2n, \mathbb{R})$ is defined by specifying the convergent sequences: we will say that $\lim_{j \to \infty} f_j = f$ in $\mathrm{Symp}(2n, \mathbb{R})$ if and only if for every compact set $\mathcal{K}$ in $\mathbb{R}_z^{2n}$ the sequences $(f_{j|\mathcal{K}})$ and $(D(f_{j|\mathcal{K}}))$ converge uniformly towards $f_{|\mathcal{K}}$ and $D(f_{|\mathcal{K}})$, respectively. The topology of $\mathrm{Ham}(2n, \mathbb{R})$ is the topology induced by $\mathrm{Symp}(2n, \mathbb{R})$.

We are now going to prove a deep and beautiful result due to Banyaga [11]. It essentially says that a path of time-one Hamiltonian symplectomorphisms passing through the identity at time zero is itself Hamiltonian. It will follow that $\mathrm{Ham}(2n, \mathbb{R})$ is a connected group.

Let $t \mapsto f_t$ be a path in $\mathrm{Ham}(2n, \mathbb{R})$, defined for $0 \leq t \leq 1$ and starting at the identity: $f_0 = I$. We will call such a path a *one-parameter family of Hamiltonian symplectomorphisms*. Thus, each f_t is equal to some symplectomorphism $f_1^{H_t}$. A striking — and not immediately obvious! — fact is that each path $t \mapsto f_t$ is itself the flow of a Hamiltonian function. We are following here the somewhat simpler presentation due to Wang [284].

Proposition 67 (Banyaga). *Let (f_t) be a one-parameter family of Hamiltonian transformations such that $f_0 = I$. Then $f_t = f_t^H$ where*

$$H(z,t) = -\int_0^1 z^T J\left(\frac{d}{dt}f_t \circ f_t^{-1}\right)(\lambda z)d\lambda. \qquad (4.79)$$

Proof. The f_t being symplectomorphisms satisfy $Df_t(z)^T JDf_t(z) = J$. Dropping the variable z and differentiating both sides of this equality with respect to t we get, setting $\dot{f}_t = df_t/dt$,

$$(D\dot{f}_t)^T JDf_t + Df_t^T JD(\dot{f}_t) = 0;$$

equivalently:

$$J(D\dot{f}_t)(Df_t)^{-1} = -\left[D(f_t)^T\right]^{-1} D(\dot{f}_t)^T J$$

$$= [J(D\dot{f}_t)(D(f_t)^{-1})]^T.$$

Using the chain rule and the identity $D(f_t)^{-1} = D(f_t^{-1})$ this equality can be rewritten

$$D(J\dot{f}_t \circ f_t^{-1}) = [DJ(\dot{f}_t \circ f_t^{-1})]^T$$

showing that the Jacobian of the function $J(\dot{f}_t \circ f_t^{-1})$ is symmetric. It follows from Poincaré's lemma that there exists for each $t \in I$ a function H_t such that $J(\dot{f}_t \circ f_t^{-1}) = \partial_z H_t$ and one verifies that this function is explicitly given by the formula

$$H_t(z) = \int_0^1 z^T J(\dot{f}_t \circ f_t^{-1})(\lambda z)d\lambda.$$

Setting $H(z,t) = -H_t(z)$ this is precisely formula (4.79). $\qquad \square$

In the linear case one can construct the Hamiltonian explicitly.

Proposition 68. *Let $t \mapsto S_t$, $0 \le t \le 1$, be a path in $\mathrm{Sp}(2n,\mathbb{R})$ such that $S_0 = I$ and $S_1 = S$. There exists a Hamiltonian function $H = H(z,t)$ such that S_t is the phase flow determined by the Hamilton equations $\dot{z} = J\partial_z H$. Writing*

$$S_t = \begin{pmatrix} A_t & B_t \\ C_t & D_t \end{pmatrix}, \qquad (4.80)$$

the Hamiltonian function is the quadratic form

$$H(z,t) = \tfrac{1}{2}[(\dot{D}_t C_t^T - \dot{C}_t D_t^T)x^2 + 2(\dot{D}_t A_t^T - \dot{C}_t B_t^T)p \cdot x$$
$$+ (\dot{B}_t A_t^T - \dot{A}_t B_t^T)p^2], \tag{4.81}$$

where $\dot{A}_t = dA_t/dt$, etc.

Proof. The matrices S_t being symplectic satisfy $S_t^T J S_t = J$. Differentiating both sides of this equality with respect to t and setting $\dot{S}_t = \frac{d}{dt} S_t$ we get $\dot{S}_t^T J S_t + S_t^T J \dot{S}_t = 0$, equivalently,

$$J \dot{S}_t S_t^{-1} = -(S_t^T)^{-1} \dot{S}_t^T J = (J \dot{S}_t S_t^{-1})^T.$$

This equality can be rewritten $J \dot{S}_t S_t^{-1} = (J \dot{S}_t S_t^{-1})^T$ hence the matrix $J \dot{S}_t S_t^{-1}$ is symmetric. Set $J \dot{S}_t S_t^{-1} = M_t = M_t^T$; then

$$\dot{S}_t = X_t S_t, \; X_t = -J M_t. \tag{4.82}$$

Define now

$$H(z,t) = -\tfrac{1}{2}(J X_t)z \cdot z;$$

using (4.82) one verifies that the phase flow determined by H consists precisely of the symplectic matrices S_t. $\qquad\square$

4.4.2. Hamiltonian path homotopies and quantum motion

Now comes an innocent-looking result, which seems to be very technical. But it will play a crucial role in our interpretation of quantum motion.

Proposition 69. *Let $(f_t^H)_{0 \le t \le 1}$ and $(f_t^K)_{0 \le t \le 1}$ be two arbitrary paths in* $\mathrm{Ham}(2n, \mathbb{R})$. *The paths $(f_t^H f_t^K)_{0 \le t \le 1}$ and $(f_t)_{1 \le t \le 1}$, where*

$$f_t = \begin{cases} f_{2t}^K & \text{when } 0 \le t \le \tfrac{1}{2}, \\[2mm] f_{2t-1}^H f_1^K & \text{when } \tfrac{1}{2} \le t \le 1, \end{cases}$$

are homotopic with fixed extremities (i.e., they can be continuously deformed into one another while keeping the origin I and the endpoint $f_1^H f_1^K$ invariant).

Proof. Let us construct explicitly a homotopy of the first path on the second, that is, a continuous mapping

$$h : [0,1] \times [0,1] \to \mathrm{Ham}(2n, \mathbb{R})$$

such that $h(t,0) = f_t^H f_t^K$ and $h(t,1) = f_t$. Define h by $h(t,s) = a(t,s)b(t,s)$ where a and b are functions

$$a(t,s) = \begin{cases} I & \text{for } 0 \le t \le \dfrac{s}{2}, \\[2mm] f_{(2t-s)/(2-s)}^H & \text{for } \dfrac{s}{2} \le t \le 1, \end{cases}$$

$$b(t,s) = \begin{cases} f_{2t/(2-s)}^K & \text{for } 0 \le t \le 1 - \dfrac{s}{2}, \\[2mm] f_1^K & \text{for } \dfrac{s}{2} \le t \le 1. \end{cases}$$

We have $a(t,0) = f_t^H$, $b(t,0) = f_t^K$ hence $h(t,0) = f_t^H f_t^K$; similarly

$$h(t,1) = \begin{cases} f_{2t}^K & \text{for } 0 \le t \le \dfrac{1}{2}, \\[2mm] f_{2t-1}^H f_1^K & \text{for } \dfrac{1}{2} \le t \le 1, \end{cases}$$

that is $h(t,1) = f_t$. $\qquad\square$

We are now going to exploit the Hamiltonian character of Bohm's equations of motion using the methods reviewed in the previous sections. The phase flow determined by the new Hamiltonian function H^ψ is denoted for brevity by $f_t^{H^\psi} = f_t^\psi$ and the corresponding time-dependent flow $(f_{t,t'}^\psi)$ is given by $f_{t,t'}^\psi = f_t^\psi \circ (f_{t'}^\psi)^{-1}$.

We now establish an essential property, namely that the Bohmian flow (f_t^ψ) is the product of two Hamiltonian phase space flows, one of which being the classical phase space flow (f_t^H) determined by H, while the other is associated with the choice of the initial wavefunction.

Proposition 70. *Let (f_t^ψ) be the quantum flow defined above. The symplectomorphisms f_t^ψ can be factored as*

$$f_t^\psi = f_t^H q_t^\psi, \tag{4.83}$$

where (f_t^H) is the phase flow determined by the Hamiltonian function H and (q_t^ψ) is the phase flow determined by the Hamiltonian function

$$K(z,t) = Q^\psi(x(t),t) \quad \text{with } (x(t),p(t)) = f_t^H(x,p). \tag{4.84}$$

Proof. We first observe that $K(z,t)$ indeed depends on both variables x and p since $x(t)$ does. In view of formula (4.76) in Proposition 64 we have $f_t^\psi = f_t^H q_t^\psi = f_t^{H\#K}$ where

$$H\#K(z,t) = H(z,t) + K((f_t^H)^{-1}(z),t)$$
$$= H(z,t) + Q^\psi(f_t^H(f_t^H)^{-1}(z),t)$$
$$= H(z,t) + Q^\psi(z,t),$$

which proves the identity (4.83). $\qquad\square$

Formula (4.83) shows that the Bohmian flow (f_t^ψ) can be viewed as a deformation of the Hamiltonian flow (f_t^H): introduce a "deformation parameter" τ and define $h_t^\tau = f_t^H q_\tau^\psi$. We have $h_t^0 = f_t^H$ and $h_t^t = f_t^\psi$ so when τ varies from 0 to t the mapping f_t^H is smoothly deformed into f_t^ψ.

4.4.3. Schrödinger's and Hamilton's equations

We now have the machinery we need to derive Schrödinger's equation from Hamilton's equation of motion by lifting the flows f_t^H of $\mathrm{Ham}(2n,\mathbb{R})$ onto a unitary representation of a covering structure. We are following here our exposition in [123].

In order to motivate our approach let us now briefly return to the property of the metaplectic representation of $\mathrm{Sp}(2n,\mathbb{R})$ which shows that with every family (S_t) of symplectic matrices depending smoothly on t and such that $S_0 = I$, we can associate, in a unique way, a family $(\widehat{S}_t)$ of unitary operators on $L^2(\mathbb{R}^n)$ belonging to the metaplectic group $\mathrm{Mp}(2n,\mathbb{R})$ and such that $\widehat{S}_0 = 1$ and $\widehat{S}_t\widehat{S}_{t'} = \widehat{S}_{t+t'}$ (it is a standard property of covering spaces).

Banyaga's theorem tells us that for every flow (S_t) there exists some Hamiltonian H, so that we can write $(S_t) = (f_t^H)$. Now if we use the correspondence $H \overset{\mathrm{Weyl}}{\longleftrightarrow} \widehat{H}$ we have $(S_t) = (F_t^H)$ where

$$\widehat{S}_t = e^{-i\widehat{H}t/\hbar}$$

which is the solution of the Schrödinger equation

$$i\hbar\frac{d}{dt}\widehat{S}_t = \widehat{H}\widehat{S}_t$$

Here we have again introduced a scaling parameter having the dimensions of action which, for convenience, we have denoted by $\hbar$. (Usually $\widehat{H}$ is not a bounded operator on $L^2(\mathbb{R}^n)$ so that the exponential has to be defined using some functional calculus.)

We now ask whether this property has an analogue for paths in the group $\text{Ham}(2n, \mathbb{R})$ of Hamiltonian canonical transformations. That is can we find a family (F_t^H) corresponding to (f_t^H) for every physically relevant Hamiltonian. To sharpen up the discussion, let us introduce the following notation:

- Denote by $\mathcal{P}\text{Ham}(2n, \mathbb{R})$ the set of all one-parameter families (f_t) in $\text{Ham}(2n, \mathbb{R})$ depending smoothly on t and passing through the identity at time $t = 0$. As we have remarked above, such a family of canonical transformations is always the flow (f_t^H) of some (usually time-dependent) Hamiltonian H (Banyaga's theorem).
- Denote by $\mathcal{P}U(L^2(\mathbb{R}^n))$ the set of all strongly continuous one-parameter families (F_t) of unitary operators on $L^2(\mathbb{R}^n)$ depending smoothly on t and such that F_0 is the identity operator, and having the following property: the domain of the infinitesimal generator $\widehat{H}$ of (F_t) contains the Schwartz space $\mathcal{S}(\mathbb{R}^n)$.

Recall that strong continuity for a one-parameter group (F_t) means that we have

$$\lim_{t \to t_0} F_t \psi = F_{t_0}^H \psi \tag{4.85}$$

and the infinitesimal generator of (F_t) is the operator defined by

$$A = i\hbar \frac{d}{dt} F_t \psi \bigg|_{t=0} = i\hbar \lim_{\Delta t \to 0} \frac{F_{\Delta t}\psi - \psi}{\Delta t} \tag{4.86}$$

for every $\psi \in L^2(\mathbb{R}^n)$ and every real number t_0; formally $F_t = e^{-i\hbar A/t}$. Formally we have Stone's theorem (for a proof see [2, Supplement 7.4B, pp. 529–535]).

Proposition 71 (Stone). *For every strongly continuous one-parameter group (F_t) of unitary operators on a Hilbert space $\mathcal{H}$ there exists a self-adjoint operator A on $L^2(\mathbb{R}^n)$ such that $F_t = e^{itA/\hbar}$; in particular A is closed and densely defined in $\mathcal{H}$. Conversely, if A is a self-adjoint operator on $\mathcal{H}$ then there exists a unique one-parameter unitary group (F_t) whose infinitesimal generator is A, that is $F_t = e^{itA/\hbar}$.*

Proposition 72. *There exists a bijective correspondence*

$$\mathcal{C} : \mathcal{P}U(L^2(\mathbb{R}^n)) \longleftrightarrow \mathcal{P}Ham(2n, \mathbb{R})$$

whose restriction to families (S_t) *of symplectic matrices reduces to the metaplectic representation, and which has the following symplectic covariance property: for every* (f_t) *in* $\mathcal{P}Ham(2n, \mathbb{R})$ *and for every* S *in* $Sp(2n, \mathbb{R})$ *we have*

$$\mathcal{C}(Sf_tS^{-1}) = (\widehat{S}F_t\widehat{S}^{-1}), \tag{4.87}$$

where $\widehat{S}$ *is any of the two operators in* $Mp(2n, \mathbb{R})$ *such that* $\pi^{Mp}(\widehat{S}) = S$. *This correspondence* $\mathcal{C}$ *is bijective and we have*

$$i\hbar\frac{d}{dt}F_t = \widehat{H}F_t \tag{4.88}$$

where $\widehat{H} \overset{Weyl}{\longleftrightarrow} H$, *the Hamiltonian function* H *being determined by* (f_t).

It is perhaps worth observing that it is always preferable to take the family (F_t) as the fundamental object, rather than $\widehat{H}$ (and hence Schrödinger's equation). This remark has already been made by Weyl who noticed that (F_t) is everywhere defined and consists of bounded operators, while $\widehat{H}$ is generically unbounded and only densely defined (see the discussion in [197] for a discussion of related questions).

We assume that (f_t) is the Hamiltonian flow determined by a time-independent Hamiltonian function $H = H(x, p)$, in which case $(f_t) = (f_t^H)$ is a one-parameter group, that is $f_t^H f_{t'}^H = f_{t+t'}^H$. For the time-dependent case we refer to [123]. We thus want to associate with (f_t^H) a strongly continuous one-parameter group $(F_t) = (F_t^H)$ of unitary operators on $L^2(\mathbb{R}^n)$ satisfying some additional conditions. We proceed as follows: let $\widehat{H}$ be the operator associated with H by the Weyl correspondence: $\widehat{H} \overset{Weyl}{\longleftrightarrow} H$ and define $\mathcal{C}(f_t^H) = (F_t)$ by $F_t = e^{-it\widehat{H}/\hbar}$. The Weyl operator $\widehat{H}$ is self-adjoint and its domain obviously contains $\mathcal{S}(\mathbb{R}^n)$. Let us show that the covariance property (4.87) holds. We have

$$\mathcal{C}(Sf_t^HS^{-1}) = \mathcal{C}(f_t^{H\circ S^{-1}})$$

that is, by definition of $\mathcal{C}$,

$$\mathcal{C}(Sf_t^HS^{-1}) = (e^{-it\widehat{H\circ S^{-1}}/\hbar}).$$

In view of the symplectic covariance property $a \circ S^{-1} \overset{\text{Weyl}}{\longleftrightarrow} \widehat{S}\widehat{A}\widehat{S}^{-1}$ of Weyl operators we have $\widehat{H \circ S^{-1}} = \widehat{S}\widehat{H}\widehat{S}^{-1}$, and hence

$$\mathcal{C}(Sf_t^H S^{-1}) = (e^{-it\widehat{S}\widehat{H}\widehat{S}^{-1}/\hbar}) = (\widehat{S}e^{-it\widehat{H}/\hbar}\widehat{S}^{-1})$$

which is property (4.87).

Let conversely (F_t) be in $\mathcal{P}U(L^2(\mathbb{R}^n))$; we must show that we can find a unique (f_t) in $\mathcal{P}\text{Ham}(2n, \mathbb{R})$ such that $\mathcal{C}(f_t) = (F_t)$. By Stone's theorem and our definition of $\mathcal{P}U(L^2(\mathbb{R}^n))$ there exists a unique self-adjoint operator A, densely defined, and whose domain contains $\mathcal{S}(\mathbb{R}^n)$. Thus A is continuous on $\mathcal{S}(\mathbb{R}^n)$ and one can prove (see [123]) that for each value of the parameter τ there exists an observable a such that $A \overset{\text{Weyl}}{\longleftrightarrow} a$; then $A = \widehat{H} \overset{\text{Weyl}}{\longleftrightarrow} H$ for some function $H = H(x, p)$ and we have $\mathcal{C}(f_t^H) = (F_t)$.

There remains to show that the correspondence $\mathcal{C}$ restricts to the metaplectic representation for semigroups $(f_t) = (S_t)$ in $\text{Sp}(2n, \mathbb{R})$; but this is clear since (S_t) is generated, as a flow, by a quadratic Hamiltonian, and the unitary one-parameter group of operators determined by such a function precisely consists of metaplectic operators.

Chapter 5

Uncertainties and Quantum Blobs

In the present chapter, we study the uncertainty principles of quantum mechanics from a rather new and unconventional point of view, based on previous work of ours. We show that not only the standard uncertainty inequalities (in their strong form, due to Robertson and Schrödinger) but also that Hardy's uncertainty principle (which is undeservedly less know in quantum mechanics) can be reformulated in a topological and hence *intrinsic* way by using the notion of symplectic capacity whose existence follows from a deep theorem of M. Gromov, the "symplectic non-squeezing theorem". Symplectic capacities are actually *classical* objects, closely related to the notion of *action*.

5.1. Quantum Indeterminacy: First Steps

Contrarily to what is often believed, the Heisenberg uncertainty inequalities $\Delta P_j \Delta X_j \geq \frac{1}{2}\hbar$, which we write

$$\Delta X_j \Delta P_j \geq \frac{1}{2}\hbar \qquad (5.1)$$

and their stronger form, the Robertson–Schrödinger inequalities

$$(\Delta X_j)^2(\Delta P_j)^2 \geq \Delta(X_j, P_j)^2 + \frac{1}{4}\hbar^2 \qquad (5.2)$$

are not statements about the accuracy of our measurements; their derivation assumes on the contrary perfect instruments! The correct interpretation of the inequalities (5.1) and (5.2) is the following (see the discussion in [238, p. 93]): if the same preparation procedure is repeated a large number of times on an ensemble of systems, and is followed by either a measurement of x_j, or by a measurement of p_j, the results obtained have standard deviations ΔX_j and ΔP_j satisfying (5.1). In his original heuristic "derivation" of the uncertainty relation $\Delta P \Delta X \approx h$ Heisenberg did not give a precise definition of the "uncertainties" ΔP and ΔX; in his paper [147] he established this expression as the minimum unavoidable disturbance of p caused by a measurement of x; the quantities ΔP and ΔX were not precisely defined, they were just viewed as experimental "errors". The interpretation of the quantities ΔP and ΔX as standard deviations, and the rigorous proof of (5.1) is due to Kennard [174]. For a modern discussion of disturbances caused by the instrumental setup in measurements see the work [41, 42] by Busch and his collaborators; for a different and somewhat conflicting point of view see [227].

5.1.1. The Wigner function revisited

Recall that we defined in Chapter 4 the Wigner or function of $\psi \in \mathbb{R}^n$ in terms of the reflection operator, from which we derived the familiar expression

$$ W\psi(z) = \left(\frac{1}{2\pi\hbar}\right)^n \int e^{-\frac{i}{\hbar}p \cdot y} \psi\left(x + \frac{1}{2}y\right) \psi^*\left(x - \frac{1}{2}y\right) d^n y. \quad (5.3) $$

Let us slightly generalize this definition. For ψ and ϕ in $\mathcal{S}(\mathbb{R}^n)$ we define the cross-Wigner function

$$ W(\psi, \phi)(z) = \left(\frac{1}{\pi\hbar}\right)^n \langle \phi | \widehat{\Pi}(z) \psi \rangle \quad (5.4) $$

and a straightforward calculation shows that we have the explicit formula

$$ W(\psi, \phi)(z) = \left(\frac{1}{2\pi\hbar}\right)^n \int e^{-\frac{i}{\hbar}p \cdot y} \psi\left(x + \frac{1}{2}y\right) \phi^*\left(x - \frac{1}{2}y\right) d^n y. \quad (5.5) $$

Of course $W(\psi, \psi) = W\psi$; moreover $W(\psi, \phi) = W(\phi, \psi)^*$ and hence, in particular $W\psi$ *is always a real function*. This has important consequences in quantum mechanics, because it allows to view $W\psi$ as the substitute for a probability density, having the "right" marginal properties.

Proposition 73. *Let ψ be both integrable and square integrable: $\psi \in L^1(\mathbb{R}^n) \cap L^2(\mathbb{R}^n)$. We have*

$$\int W\psi(z)d^n p = |\psi(x)|^2, \quad \int W\psi(z)d^n x = |F\psi(p)|^2. \tag{5.6}$$

In particular

$$\int W\psi(z)d^{2n}z = ||\psi||^2_{L^2} = ||F\psi||^2_{L^2} \tag{5.7}$$

and hence

$$\int W\psi(z)d^{2n}z = 1 \text{ if and only if } ||\psi||_{L^2} = 1. \tag{5.8}$$

Proof. In view of the inverse Fourier transform formula

$$\int e^{-\frac{i}{\hbar}p\cdot y}d^n p = (2\pi\hbar)^n \delta(y)$$

we have

$$\int W\psi(z)d^n p = \int \delta(y)\psi\left(x + \frac{1}{2}y\right)\psi^*\left(x - \frac{1}{2}y\right)d^n y$$

$$= \int \delta(y)|\psi(x)|^2 d^n y$$

$$= |\psi(x)|^2$$

which is the first formula (5.6). To prove the second formula, set $x' = x + \frac{1}{2}y$ and $x'' = x - \frac{1}{2}y$ in the right-hand side of the equality

$$\int W\psi(z)d^n x = \left(\frac{1}{2\pi\hbar}\right)^n \iint e^{-\frac{i}{\hbar}p\cdot y}\psi\left(x + \frac{1}{2}y\right)\psi^*\left(x - \frac{1}{2}y\right)d^n y d^n x$$

we get

$$\int W\psi(z)d^n x = \left(\frac{1}{2\pi\hbar}\right)^n \iint e^{-\frac{i}{\hbar}p\cdot x'}\psi(x')\left(e^{-\frac{i}{\hbar}p\cdot x''}\psi(x'')\right)^* d^n x' d^n x''$$

$$= \left(\frac{1}{2\pi\hbar}\right)^n \int e^{-\frac{i}{\hbar}p\cdot x'}\psi(x')d^n x' \left(\int e^{-\frac{i}{\hbar}p\cdot x''}\psi(x'')d^n x''\right)^*$$

$$= |F\psi(p)|^2.$$

$\square$

A similar argument shows that, more generally, the cross-Wigner transform satisfies

$$\int W(\psi, \phi)(z) d^n p = \psi(x) \phi^*(x), \tag{5.9}$$

$$\int W(\psi, \phi)(z) d^n x = F\psi(p)(F\phi(p))^*. \tag{5.10}$$

However, the Wigner function is not in general positive. In fact, a classical result of Hudson [164] ("Hudson's theorem") tells us that $W\psi$ is non-negative if and only if ψ is a Gaussian $Ce^{M(x-x_0)^2}$ where M is a complex matrix with negative real eigenvalues and $x_0 \in \mathbb{R}^n$. This generic non-positivity of the Wigner function has been (and still is) at the origin of fierce debates between proponents and opponents to "negative probabilities". Needless to say, these debates are more of a philosophical than of a mathematical nature, and will not be further discussed here. In fact, the Wigner transform is a member of a much more general class of quasi-probability distributions (the "Cohen class"), whose properties are well understood (see [119, Chapter 7]). Here is a very important result which shows that averages can be computed in quantum mechanics by weighting the observable with the Wigner function, which plays the role of a (nonpositive) quasi-probability density.

Proposition 74. *Let ψ be in $\mathcal{S}(\mathbb{R}^n)$ and assume that $\widehat{A} \overset{Weyl}{\longleftrightarrow} a$ is self-adjoint and maps $\mathcal{S}(\mathbb{R}^n)$ into $L^2(\mathbb{R}^n)$. We have*

$$\langle \widehat{A}\psi | \psi \rangle = \langle \psi | \widehat{A} | \psi \rangle = \int a(z) W\psi(z) d^{2n} z. \tag{5.11}$$

Proof. Using definition (5.4) of $W\psi(z)$ in terms of the Grossmann–Royer operators, we have, using Fubini's rule and definition (4.39) of the Weyl operator $\widehat{A} \overset{Weyl}{\longleftrightarrow} a$,

$$\int a(z) W\psi(z) d^{2n} z = \left(\frac{1}{\pi\hbar}\right)^n \int a(z) \langle \psi | \widehat{T}_{GR}(z) \psi \rangle d^{2n} z$$

$$= \left(\frac{1}{\pi\hbar}\right)^n \int a(z) \left[\int \psi^*(u) \widehat{T}_{GR}(z) \psi(u) d^{2n} u\right] d^{2n} z$$

$$= \left(\frac{1}{\pi\hbar}\right)^n \int \psi^*(u) \left[\int a(z) \widehat{T}_{GR}(z) \psi(u) d^{2n} z\right] d^{2n} u$$

$$= \langle \psi | \widehat{A}\psi \rangle. \qquad \square$$

It is customary in quantum mechanics to write

$$\text{Tr}(\widehat{A}\widehat{\rho}_\psi) = \int a(z)W\psi(z)d^{2n}z, \tag{5.12}$$

where $\text{Tr}(\widehat{A}\widehat{\rho}_\psi)$ is the "trace" of the composition of the operator $\widehat{A}$ with the projection $\widehat{\rho}_\psi$ on the ray $\{\lambda\psi : \lambda \in \mathbb{C}\}$. We will occasionally use this notation, but one should be careful because, strictly speaking, the product $\widehat{A}\widehat{\rho}_\psi$ need not always be a trace-class operator! (We will discuss this delicate issue in Section 5.2.1.)

Let $\psi \neq 0$ be a "pure quantum state", and let a be a real symbol, viewed as a "classical observable". The number

$$\langle\widehat{A}\rangle_\psi = \frac{\langle\psi|\widehat{A}|\psi\rangle}{\langle\psi|\psi\rangle} \tag{5.13}$$

is called the average value of the corresponding "quantum observable" in the quantum state ψ. If ψ is normalized, we recover, using the notation (5.12), the textbook formula

$$\langle\widehat{A}\rangle_\psi = \text{Tr}(\widehat{A}\widehat{\rho}_\psi) = \text{Tr}(\widehat{\rho}_\psi\widehat{A}) \tag{5.14}$$

(cf. formula (5.30) below), taking into account the *caveat* above.

More generally, the same argument shows that the product $\langle\widehat{A}\psi|\phi\rangle$ can be expressed in terms of the cross-Wigner transform (5.5):

$$\langle\widehat{A}\psi|\phi\rangle = \int a(z)W(\psi,\phi)(z)d^{2n}z \tag{5.15}$$

for ψ and ϕ in $\mathcal{S}(\mathbb{R}^n)$. This formula has applications in the theory of weak measurements and values (see [121, 122]). It is also an important formula from a mathematical point of view, since it can be taken as a redefinition of a Weyl operator: it shows quite explicitly how the Wigner phase space formalism and Weyl pseudodifferential calculus are intimately linked (see [108, 112] for a detailed discussion of this relation).

5.1.2. The Robertson–Schrödinger uncertainty relations

The textbook Heisenberg inequalities $\Delta P_j \Delta X_j \geq \frac{1}{2}\hbar$ are a weak form (in fact, a consequence) of the Schrödinger–Robertson uncertainty relations, which have the property of being invariant under the action of symplectic transformations (i.e., linear canonical transformations). It turns out that these relations have a precise meaning even in the classical world.

The covariance matrix of a quasi-distribution

In what follows ρ is a real-valued function defined on $\mathbb{R}^{2n}$ and such that

$$\int \rho(z)d^{2n}z = 1 \tag{5.16}$$

and

$$\int (1 + |z|^2)|\rho(z)|d^{2n}z < \infty. \tag{5.17}$$

We do not assume that $\rho \geq 0$ so ρ is not in general a true probability density; typically ρ is the Wigner function of a quantum state (or, more generally, the Wigner transform of a mixed quantum state, see Proposition 80 in Section 5.2), so we will call ρ a "quasi-distribution". Notice that condition (A.3) ensures us that the Fourier transform $F\rho$ is twice continuously differentiable.

Introducing the notation $z_\alpha = x_\alpha$ if $1 \leq \alpha \leq n$ and $z_\alpha = p_{\alpha-n}$ if $n + 1 \leq \alpha \leq 2n$ we define the covariances

$$\Delta(Z_\alpha, Z_\beta) = \int (z_\alpha - \langle z_\alpha \rangle)(z_\beta - \langle z_\beta \rangle)\rho(z)d^{2n}z \tag{5.18}$$

and

$$(\Delta Z_\alpha)^2 = \Delta(Z_\alpha, Z_\alpha) = \int (z_\alpha - \langle z_\alpha \rangle)^2 \rho(z)d^{2n}z, \tag{5.19}$$

where the moments $\langle z_\alpha^k \rangle$ are the averages given by

$$\langle z_\alpha^k \rangle = \int z_\alpha^k \rho(z)d^{2n}z. \tag{5.20}$$

The quantities $Z_1, Z_2, \ldots, Z_{2n}$ should be viewed as random variables whose values are the phase-space coordinates $z_1, z_2, \ldots, z_n$. Since the integral of ρ is equal to one, formulae (7.31) and (7.32) can be rewritten in the familiar form

$$\Delta(Z_\alpha, Z_\beta) = \langle z_\alpha z_\beta \rangle - \langle z_\alpha \rangle \langle z_\beta \rangle, \tag{5.21}$$

$$(\Delta Z_\alpha)^2 = \Delta(Z_\alpha, Z_\alpha) = \langle z_\alpha^2 \rangle - \langle z_\alpha \rangle^2. \tag{5.22}$$

The quantities (7.31), (7.32), and (5.20) are well defined in view of condition (A.3): the integrals above are all absolutely convergent in view of the trivial estimates

$$\left| \int z_\alpha \rho(z) d^{2n}z \right| \leq \int (1 + |z|^2) |\rho(z)| d^{2n}z < \infty,$$

$$\left| \int z_\alpha z_\beta \rho(z) d^{2n}z \right| \leq \int (1 + |z|^2) |\rho(z)| d^{2n}z < \infty.$$

Definition 75. We will call the symmetric $2n \times 2n$ matrix

$$\sum = (\Delta(Z_\alpha, Z_\beta))_{1 \leq j, k \leq 2n}$$

the covariance matrix associated with ρ. When Σ is invertible, the phase space ellipsoid

$$W_\Sigma : \tfrac{1}{2} \Sigma^{-1} z \cdot z \leq \hbar \qquad (5.23)$$

is called the covariance (or Wigner) ellipsoid.

For instance, when $n = 1$, the covariance matrix is

$$\sum = \begin{pmatrix} \Delta X^2 & \Delta(X, P) \\ \Delta(P, X) & \Delta P^2 \end{pmatrix},$$

where by definition the variances ΔX^2 and ΔP^2 and the covariances $\Delta(X, P) = \Delta(P, X)$ are the quantities

$$\Delta X^2 = \iint x^2 \rho(x, p) d^n p d^n x - \left(\iint x \rho(x, p) d^n p d^n x \right)^2,$$

$$\Delta P^2 = \iint p^2 \rho(x, p) d^n p d^n x - \left(\iint p \rho(x, p) d^n p d^n x \right)^2,$$

$$\Delta(X, P) = \iint x p \rho(z) d^n p d^n x - \iint x \rho(z) d^n p d^n x \iint p \rho(z) d^n p d^n x.$$

When the averages (first moments) vanish, these equations become

$$\Delta X^2 = \iint x^2 \rho(x, p) dp dx, \quad \Delta P^2 = \iint p^2 \rho(x, p) dp dx \qquad (5.24)$$

and

$$\Delta(X, P) = \iint x p \rho(z) dp dx - \iint x \rho(z) dp dx \iint p \rho(z) dp dx. \qquad (5.25)$$

We are now going to show that the Robertson–Schrödinger inequalities can be written in a pleasant compact form, making the role played by the covariance matrix explicit. Let us begin with the case $n = 1$. The covariance matrix is here the 2×2 matrix

$$\Sigma = \begin{pmatrix} \Delta X^2 & \Delta(X,P) \\ \Delta(P,X) & \Delta P^2 \end{pmatrix};$$

we claim that the inequality

$$\Delta X^2 \Delta P^2 \geq \Delta(X,P)^2 + \tfrac{1}{4}\hbar^2$$

is equivalent to the condition

$$\Sigma + \frac{i\hbar}{2}J \geq 0,$$

where ≥ 0 means "is positive semidefinite". In fact, the trace of $\Sigma + \frac{i\hbar}{2}J$ being the positive number $\Delta X^2 + \Delta P^2 \geq 0$, the condition $\Sigma + \frac{i\hbar}{2}J \geq 0$ is equivalent to $\det(\Sigma + \frac{i\hbar}{2}J) \geq 0$ that is to

$$\Delta X^2 \Delta P^2 - (\Delta(X,P)^2 + \tfrac{1}{4}\hbar^2) \geq 0$$

which is precisely the Robertson–Schrödinger inequality. Notice that, in particular, the covariance matrix Σ must be positive definite. That this result extends to arbitrary dimension n was proven by Narcowich [218–220]. For related results see [217, 266, 267, 303].

Proposition 76 (Narcowich). *Assume that Σ is a quantum covariance matrix, i.e., that $\rho(z)$ is the Wigner transform of some function $\psi \in L^2(\mathbb{R}^n)$.*

(i) *Then Σ is positive definite (which we write $\Sigma > 0$); in particular Σ is invertible.*

(ii) *The Robertson–Schrödinger inequalities*

$$(\Delta X_j)^2(\Delta P_j)^2 \geq \Delta(X_j, P_j)^2 + \tfrac{1}{4}\hbar^2 \tag{5.26}$$

($1 \leq j \leq n$) are equivalent to the condition

$$\Sigma + \frac{i\hbar}{2}J \geq 0 \tag{5.27}$$

(i.e., $\Sigma + \frac{i\hbar}{2}J$ is positive semidefinite).

We remark that the complex matrix $\Sigma + \frac{i\hbar}{2}J$ is Hermitian because Σ is for $n = 1$ symmetric and $J^* = J^T = -J$. One proves (see the references above) that the condition $\Sigma + \frac{i\hbar}{2}J \geq 0$ implies that a quantum covariance matrix Σ must always be positive definite. Here is an example showing what can happen if one renounces to positive definiteness. Assume $n = 2$ and choose

$$\Sigma = \begin{pmatrix} \hbar/2 & -\hbar/2 & 0 & 0 \\ -\hbar/2 & \hbar/2 & 0 & 0 \\ 0 & 0 & \hbar/2 & 0 \\ 0 & 0 & 0 & \hbar/2 \end{pmatrix}.$$

We have $(\Delta X_1)^2 = (\Delta X_2)^2 = \hbar^2/4$ and $(\Delta P_1)^2 = (\Delta P_2)^2 = \hbar^2/4$, and also $\Delta(X_1, P_1) = \Delta(X_2, P_2) = 0$ so that the Robertson–Schrödinger inequalities (5.26) are trivially satisfied (they reduce to equalities). The matrix $\Sigma + \frac{i\hbar}{2}J$ is nevertheless indefinite, and can thus not be a quantum covariance matrix. The satisfaction of the Robertson–Schrödinger inequalities is thus not sufficient to decide whether a covariance matrix is quantum, or classical (see the discussion in [128]).

5.2. Mixed Quantum States

We have so far been assuming that the quantum system under consideration was in a well-known state characterized by a wavefunction ψ. Suppose instead that we have the choice between a finite or infinite number of states, described by functions $\psi_1, \psi_2, \ldots$, each ψ_j having a probability α_j to be the "true" description. We will then talk about a *mixed state*. We will see that the most adequate way of describing mathematically a mixed state consists in using the notion of density matrix, which can be expressed in terms of a weighted sum of Wigner functions, each of them representing the quasi-probability density for the mixed state to be one of the ψ_j. We will refine our study of pure and mixed quantum states in Chapters 6 and 7.

5.2.1. Trace class operators

We will give a detailed and rigorous study of trace class operators and their relation with Hilbert–Schmidt operators in next chapter (also see [112, Chapter 12]).

From now on we assume that $\mathcal{H} = L^2(\mathbb{R}^n)$ and that A is a Weyl operator $\widehat{A} \overset{\text{Weyl}}{\longleftrightarrow} a$. It is usual in the physics literature to define the trace of $\widehat{A}$

by the formula

$$\text{Tr}(\widehat{A}) = \int K_A(x, x) d^n x, \tag{5.28}$$

where K_A is the kernel of the operator $\widehat{A}$; this definition is obviously the extension to the infinite-dimensional case of the definition of the trace of a matrix as the sum of its diagonal elements. Needless to say, this formula does not follow directly from the definition of a trace class operator! In fact, even when the integral in (5.28) is absolutely convergent, this formula has no reason to be true in general (this will be discussed in detail in next chapter). On the positive side, if a trace class operator $\widehat{A}$ has a kernel K satisfying $\int |K(x, x)| d^n x < \infty$ then we are "almost sure" that formula (5.28) holds. But this is not, as Dubin *et al.* [67] note, a charter allowing carefree calculations!

Here is another "obvious" property: when $\widehat{A}$ is a Weyl operator $\widehat{A} \overset{\text{Weyl}}{\longleftrightarrow} a$ one then infers from (5.28) that the trace is expressed in terms of the Weyl symbol by the formula

$$\text{Tr}(\widehat{A}) = \left(\frac{1}{2\pi\hbar}\right)^n \int a(z) d^{2n} z \tag{5.29}$$

(which has no reason to be correct in general!). In fact, in view of formula (4.44) in Chapter 4 the kernel $K_{\widehat{A}}$ of $\widehat{A}$ is related to its Weyl symbol a by the Fourier transform

$$K_{\widehat{A}}(x, y) = \left(\frac{1}{2\pi\hbar}\right)^n \int e^{\frac{i}{\hbar} p \cdot (x-y)} a\left(\frac{1}{2}(x+y), p\right) d^n p$$

so that

$$K_{\widehat{A}}(x, x) = \left(\frac{1}{2\pi\hbar}\right)^n \int a(x, p) d^n p$$

hence formula (5.29). One should however keep in mind that formula (5.29) is correct only so far (5.28) is!

Another useful relation is the one giving the trace of a product of two operators: assuming that $\widehat{A}\widehat{B}$ also is of trace class, then

$$\text{Tr}(\widehat{A}\widehat{B}) = \left(\frac{1}{2\pi\hbar}\right)^n \int a(z) b(z) d^{2n} z. \tag{5.30}$$

Let us give a semi-heuristic proof of this formula. Setting $\widehat{C} = \widehat{A}\widehat{B}$ we have

$$\mathrm{Tr}(\widehat{A}\widehat{B}) = \left(\frac{1}{2\pi\hbar}\right)^n \int c(z) d^{2n}z;$$

in view of formula (4.61) in Proposition 60 the symbol c is explicitly given by the formula

$$c(z) = \left(\frac{1}{4\pi\hbar}\right)^{2n} \iint e^{\frac{i}{2\hbar}\sigma(z',z'')} a\left(z + \frac{1}{2}z'\right) b\left(z - \frac{1}{2}z''\right) d^{2n}z' d^{2n}z''.$$

Performing the change of variables $u = z + \frac{1}{2}z'$, $v = z - \frac{1}{2}z''$ we have $d^{2n}z' d^{2n}z'' = 4^{2n} d^{2n}u d^{2n}v$ and hence

$$c(z) = \left(\frac{1}{\pi\hbar}\right)^{2n} \iint e^{\frac{i}{2\hbar}\sigma(u-z,v-z)} a(u)b(v) d^{2n}u d^{2n}v$$

$$= \left(\frac{1}{\pi\hbar}\right)^{2n} \iint e^{\frac{2i}{\hbar}\sigma(z,u-v)} (e^{\frac{2i}{\hbar}\sigma(u,v)} a(u)b(v)) d^{2n}u d^{2n}v.$$

Integrating $c(z)$ yields

$$\int c(z) d^{2n}z = \left(\frac{1}{\pi\hbar}\right)^{2n} \iint \left(\int e^{\frac{2i}{\hbar}\sigma(z,u-v)} d^{2n}z\right)$$

$$\times e^{\frac{2i}{\hbar}\sigma(u,v)} a(u)b(v) d^{2n}u d^{2n}v$$

and, using the Fourier inversion formula

$$\int e^{\frac{2i}{\hbar}\sigma(z,u-v)} d^{2n}z = \int e^{\frac{2i}{\hbar} Jz\cdot(u-v)} d^{2n}z = (2\pi\hbar)^{2n} \delta(2u - 2v)$$

we get

$$\int c(z) d^{2n}z = 2^{2n} \iint \delta(2u - 2v) e^{\frac{2i}{\hbar}\sigma(u,v)} a(u)b(v) d^{2n}u d^{2n}v$$

$$= 2^{2n} \iint \delta(2u - 2v) a(u)b(v) d^{2n}u d^{2n}v$$

$$= \int a(u)b(u) d^{2n}u;$$

formula (5.30) follows in view of formula (5.29). Of course these manipulations are not justified in the "proof" above; in particular we have not discussed the convergence of the involved integrals, so it is correct only if we assume that the symbols a and b are of a very particular type (e.g., rapidly decreasing). We will give in next chapter a rigorous and fully justified proof of formula (5.30).

Needless to say, the "derivations" above are formal and one should be extremely cautious when using the "formulas" thus obtained (strictly speaking they are only valid if $\widehat{A} \overset{\text{Weyl}}{\longleftrightarrow} a$ and $\widehat{B} \overset{\text{Weyl}}{\longleftrightarrow} b$ are Hilbert–Schmidt operators; see [112, Chapters 12 and 13]).

We finally note the following invariance property of the trace of a Weyl operator under conjugation with metaplectic operators: assume that $\widehat{A} \overset{\text{Weyl}}{\longleftrightarrow} a$ of trace class on $L^2(\mathbb{R}^n)$ and $\widehat{S} \in \text{Mp}(2n, \mathbb{R})$; then $\widehat{S}\widehat{A}\widehat{S}^{-1} \overset{\text{Weyl}}{\longleftrightarrow} a \circ S^{-1}$ ($S = \pi^{\text{Mp}}(\widehat{S})$) is also of trace class and has same trace as $\widehat{A}$:

$$\text{Tr}(\widehat{S}\widehat{A}\widehat{S}^{-1}) = \text{Tr}(\widehat{A}). \tag{5.31}$$

This immediately follows from the equality

$$\int a(S^{-1}z)d^{2n}z = \int a(z)d^{2n}z$$

(because $\det S = 1$) provided we believe in formula (5.29). In fact, we will see in next chapter that formula (5.31) holds when $\widehat{S} \in \text{Mp}(2n, \mathbb{R})$ is replaced with any operator U that is unitary on $L^2(\mathbb{R}^n)$.

5.2.2. The density matrix

Density operators are central objects in quantum mechanics, because they are identified with the "mixed states" of a quantum system. Density operators are also called "density matrices" in physics; we will use both terms interchangeably. They contain, as a particular case, the "pure states" which are usually described by the wave function.

Let us begin by defining the notion of density operator in terms of trace class operators (which will be rigorously defined in Chapter 6).

Definition 77. A density matrix (or operator) on a Hilbert space $\mathcal{H}$ is a bounded operator $\widehat{\rho}$ on $\mathcal{H}$ having the following properties:

(i) $\widehat{\rho}$ is self-adjoint and semi-definite positive: $\widehat{\rho} = \widehat{\rho}^{\dagger}$, $\widehat{\rho} \geq 0$;
(ii) $\widehat{\rho}$ is of trace class and $\text{Tr}(\widehat{\rho}) = 1$.

We will see in Chapter 6 that every trace class operators is the product of two Hilbert–Schmidt operators; such operators being compact, it follows that trace class operators are compact.

Here is a first example of a density operator. Let us assume that we are in the presence of a well-defined quantum state, represented by an element $\psi \neq 0$ of $\mathcal{H}$. Such a state is called a *pure state* in quantum mechanics. It is no restriction to assume that ψ is normalized, that is $||\psi|| = 1$, so that the mean value of $\widehat{A}$ in the state ψ is

$$\langle \widehat{A} \rangle_\psi = \langle \psi | \widehat{A}\psi \rangle. \tag{5.32}$$

Consider now the projection operator

$$\widehat{\rho}_\psi : L^2(\mathbb{R}^n) \longrightarrow \{\alpha\psi : \alpha \in \mathbb{C}\} \tag{5.33}$$

of $L^2(\mathbb{R}^n)$ on the "ray" generated by ψ. For each $\phi \in L^2(\mathbb{R}^n)$ we have

$$\widehat{\rho}_\psi \phi = \alpha\psi, \ \alpha = \langle \phi | \psi \rangle. \tag{5.34}$$

We will call $\widehat{\rho}_\psi$ the *pure density matrix* associated with ψ; it is a trace-class operator with trace equal to one.

We are going to see that the pure density operator $\widehat{\rho}_\psi$ is in this case a Weyl operator whose symbol is (up to a factor) just the Wigner transform of ψ. Let us now assume that $\mathcal{H}$ is the space $L^2(\mathbb{R}^n)$ of square integrable functions on configuration space $\mathbb{R}^n$.

Proposition 78. *Let $\widehat{\rho}_\psi$ be the density operator on $L^2(\mathbb{R}^n)$ associated with a pure state ψ by (5.34).*

(i) *The Weyl symbol of $\widehat{\rho}_\psi$ is $(2\pi\hbar)^n \rho_\psi$ where*

$$\rho_\psi(z) = W\psi(z) \tag{5.35}$$

is the Wigner transform of the state $\widehat{\rho}_\psi$.

(ii) *Let $\widehat{A} = \mathrm{Op}^{\mathrm{W}}(a)$ be a self-adjoint operator. If $||\psi|| = 1$ then the expectation value $\langle \widehat{A} \rangle_\psi$ of $\widehat{A}$ in the state ψ is given by*

$$\langle \widehat{A} \rangle_\psi = \mathrm{Tr}(\widehat{\rho}_\psi \widehat{A}). \tag{5.36}$$

Proof. (i) In view of formula (4.43) the Weyl symbol a_ψ of $\widehat{\rho}_\psi$ is given by

$$(2\pi\hbar)^n \rho_\psi(x,p) = \int_{\mathbb{R}^n} e^{-\frac{i}{\hbar}p \cdot y} K_{\widehat{\rho}_\psi} \left(x + \frac{1}{2}y, x - \frac{1}{2}y \right) dy$$

$$= \int_{\mathbb{R}^n} e^{-\frac{i}{\hbar}p \cdot y} \psi \left(x + \frac{1}{2}y \right) \psi^* \left(x + \frac{1}{2}y \right) d^n y,$$

that is $\rho_\psi(z) = W\psi(z)$ as claimed. (ii) In view of formula (5.11) we have

$$\langle \widehat{A} \rangle_\psi = \int a(z) W\psi(z) d^{2n}z$$

so formula (5.38) follows from (i) using the expression (5.30) giving the trace of the compose of two Weyl operators.   □

Proposition 79. *Let $\widehat{\rho}_\psi$ be the density operator associated with a pure state ψ by (5.34).*

(i) *The Weyl symbol of $\widehat{\rho}_\psi$ is $(2\pi\hbar)^n \rho_\psi$ where*

$$\rho_\psi(z) = W\psi(z) \tag{5.37}$$

is the Wigner transform of the state $\widehat{\rho}_\psi$.

(ii) *Let $\widehat{A} \overset{Weyl}{\longleftrightarrow} a$ be self-adjoint. If $||\psi||_{L^2} = 1$ then the expectation value $\langle \widehat{A} \rangle_\psi$ of $\widehat{A}$ in the state ψ is given by*

$$\langle \widehat{A} \rangle_\psi = \mathrm{Tr}(\widehat{\rho}_\psi \widehat{A}). \tag{5.38}$$

Proof. (i) In view of formula (4.43) the Weyl symbol a_ψ of $\widehat{\rho}_\psi$ is given by

$$(2\pi\hbar)^n \rho_\psi(x,p) = \int_{\mathbb{R}^n} e^{-\frac{i}{\hbar}p \cdot y} K_{\widehat{\rho}_\psi}\left(x + \frac{1}{2}y, x - \frac{1}{2}y\right) dy$$

$$= \int_{\mathbb{R}^n} e^{-\frac{i}{\hbar}p \cdot y} \psi\left(x + \frac{1}{2}y\right) \psi^*\left(x + \frac{1}{2}y\right) d^n y,$$

that is $\rho_\psi(z) = W\psi(z)$ as claimed. (ii) In view of formula (5.11) we have

$$\langle \widehat{A} \rangle_\psi = \int a(z) W\psi(z) d^{2n}z$$

so formula (5.38) follows from (i) using the expression (5.30) giving the trace of the compose of two Weyl operators.   □

We will prove in next chapter the following essential result, which relates the theory of mixed quantum states to the Wigner formalism.

Proposition 80. *An operator $\widehat{\rho}$ on $L^2(\mathbb{R}^n)$ is a density matrix if and only if there exists a sequence $(\lambda_j)_j$ of numbers $\lambda_j \geq 0$ with $\Sigma_j \lambda_j = 1$ and a sequence (ψ_j) of functions in $L^2(\mathbb{R}^n)$ such that the Weyl symbol ρ*

of $\widehat{\rho}$ is given by

$$\rho = \sum_j \lambda_j W \psi_j. \tag{5.39}$$

It is clear that $\widehat{\rho}$ is a density matrix since trace-class operators form a vector space, $\widehat{\rho}$ is indeed of trace class and its trace is one since $\mathrm{Tr}(\widehat{\rho}_j) = 1$ and the α_j sum to one. That $\widehat{\rho} = \widehat{\rho}^\dagger$ is obvious, and the positivity of $\widehat{\rho}$ follows from the fact that $\alpha_j \geq 0$ for each j. The function ρ is often called the Wigner transform of the density matrix $\widehat{\rho}$.

Proposition 80 immediately yields a proof of the Robertson–Schrödinger inequalities in Section 5.1.2 since the function ρ is a quasi-distribution in the sense introduced there.

5.2.3. Uncertainty principle for operators

Let $\widehat{\rho}$ be a density matrix. Recall that the Robertson–Schrödinger inequalities (5.26) are

$$(\Delta X_j)^2_{\widehat{\rho}} (\Delta P_j)^2_{\widehat{\rho}} \geq \Delta(X_j, P_j)^2_{\widehat{\rho}} + \tfrac{1}{4}\hbar^2. \tag{5.40}$$

We are going to generalize these inequalities to larger classes of operators.

Let $\widehat{A}$ and $\widehat{B}$ be two Weyl operators; we assume that the expectation values

$$\langle \widehat{A} \rangle_{\widehat{\rho}} = \mathrm{Tr}(\widehat{\rho}\widehat{A}), \ \langle \widehat{A}^2 \rangle_{\widehat{\rho}} = \mathrm{Tr}(\widehat{\rho}\widehat{A}^2) \tag{5.41}$$

(and similar expressions for $\widehat{B}$) exist and are finite. Setting

$$(\Delta\widehat{A})^2_{\widehat{\rho}} = \langle \widehat{A}^2 \rangle_{\widehat{\rho}} - \langle \widehat{A} \rangle^2_{\widehat{\rho}}, (\Delta\widehat{B})^2_{\widehat{\rho}} = \langle \widehat{B}^2 \rangle_{\widehat{\rho}} - \langle \widehat{B} \rangle^2_{\widehat{\rho}},$$

$$\Delta(\widehat{A}, \widehat{B})_{\widehat{\rho}} = \tfrac{1}{2}\langle \widehat{A}\widehat{B} + \widehat{B}\widehat{A} \rangle_{\widehat{\rho}} - \langle \widehat{A} \rangle_{\widehat{\rho}}\langle \widehat{B} \rangle_{\widehat{\rho}},$$

we have the following result.

Proposition 81. *Let $\widehat{A} \overset{Weyl}{\longleftrightarrow} a$ and $\widehat{B} \overset{Weyl}{\longleftrightarrow} b$ be two (formally) self-adjoint Weyl operators on $L^2(\mathbb{R}^n)$ for which the expectation values (6.72)* *are defined. We have*

$$|\langle \widehat{A}\widehat{B}\rangle_{\widehat{\rho}}|^2 = \Delta(\widehat{A}, \widehat{B})^2_{\widehat{\rho}} - \tfrac{1}{4}\langle[\widehat{A}, \widehat{B}]\rangle^2_{\widehat{\rho}}, \qquad (5.42)$$

where $[\widehat{A}, \widehat{B}] = \widehat{A}\widehat{B} - \widehat{B}\widehat{A}$ *and hence*

$$(\Delta\widehat{A})^2_{\widehat{\rho}}(\Delta\widehat{B})^2_{\widehat{\rho}} \geq \Delta(\widehat{A}, \widehat{B})^2_{\widehat{\rho}} - \tfrac{1}{4}\langle[\widehat{A}, \widehat{B}]\rangle^2_{\widehat{\rho}} . \qquad (5.43)$$

Proof. Replacing $\widehat{A}$ and $\widehat{B}$ by $\widehat{A} - \langle \widehat{A}\rangle_{\widehat{\rho}}$ and $\widehat{B} - \langle \widehat{B}\rangle_{\widehat{\rho}}$ we may assume that $\langle \widehat{A}\rangle_{\widehat{\rho}} = \langle \widehat{B}\rangle_{\widehat{\rho}} = 0$ so that (5.42) and (5.43) reduce to, respectively,

$$|\langle \widehat{A}\widehat{B}\rangle_{\widehat{\rho}}|^2 = \tfrac{1}{2}\langle \widehat{A}\widehat{B} + \widehat{B}\widehat{A}\rangle^2_{\widehat{\rho}} - \tfrac{1}{4}\langle[\widehat{A}, \widehat{B}]\rangle^2_{\widehat{\rho}} \qquad (5.44)$$

and

$$\langle \widehat{A}^2\rangle_{\widehat{\rho}}\langle \widehat{B}^2\rangle_{\widehat{\rho}} \geq \tfrac{1}{2}\langle \widehat{A}\widehat{B} + \widehat{B}\widehat{A}\rangle^2_{\widehat{\rho}} - \tfrac{1}{4}\langle[\widehat{A}, \widehat{B}]\rangle^2_{\widehat{\rho}}. \qquad (5.45)$$

Writing $\widehat{A}\widehat{B} = \tfrac{1}{2}(\widehat{A}\widehat{B} + \widehat{B}\widehat{A}) + \tfrac{1}{2}(\widehat{A}\widehat{B} - \widehat{B}\widehat{A})$ we have,

$$\langle \widehat{A}\widehat{B}\rangle_{\widehat{\rho}} = \tfrac{1}{2}\langle \widehat{A}\widehat{B} + \widehat{B}\widehat{A}\rangle_{\widehat{\rho}} + \tfrac{1}{2}\langle \widehat{A}\widehat{B} - \widehat{B}\widehat{A}\rangle_{\widehat{\rho}}.$$

Now, $\Delta(\widehat{A}, \widehat{B})_{\widehat{\rho}}$ is a real number, and $\langle[\widehat{A}, \widehat{B}]\rangle_{\widehat{\rho}}$ is pure imaginary (because $[\widehat{A}, \widehat{B}]^* = -[\widehat{A}, \widehat{B}]$ since $\widehat{A}$ and $\widehat{B}$ are self-adjoint), hence formula (5.44). We next observe that

$$\langle \widehat{A}\widehat{B}\rangle_{\widehat{\rho}} = \sum_{j \in \mathcal{J}} \alpha_j(\widehat{A}\widehat{B}\psi_j|\psi_j)_{L^2} = \sum_{j \in \mathcal{J}} \alpha_j(\widehat{B}\psi_j|\widehat{A}\psi_j)_{L^2}; \qquad (5.46)$$

applying the Cauchy–Schwarz inequality to each scalar product $(\widehat{B}\psi_j|\widehat{A}\psi_j)_{L^2}$ we get

$$|\langle \widehat{A}\widehat{B}\rangle_{\widehat{\rho}}|^2 \leq \sum_{j \in \mathcal{J}} \alpha_j||\widehat{B}\psi_j||_{L^2}\,||\widehat{A}\psi_j||_{L^2}. \qquad (5.47)$$

Since $\langle \widehat{A}\rangle^2_{\widehat{\rho}} = \langle \widehat{B}\rangle^2_{\widehat{\rho}} = 0$ we have

$$||\widehat{A}\psi_j|| = \langle \widehat{A}^2\rangle^{1/2}_{\psi_j} = (\Delta\widehat{A})^2_{\psi_j}, \;\; ||\widehat{B}\psi_j|| = \langle \widehat{B}^2\rangle^{1/2}_{\psi_j} = (\Delta\widehat{B})^2_{\psi_j}$$

and the inequality (6.73) is thus equivalent to

$$|\langle \widehat{A}\widehat{B}\rangle_{\widehat{\rho}}| \leq \sum_{j \in \mathcal{J}} \alpha_j\langle \widehat{A}^2\rangle^{1/2}_{\psi_j}\langle \widehat{B}^2\rangle^{1/2}_{\psi_j}.$$

Writing $\alpha_j = (\sqrt{\alpha_j})^2$ the Cauchy–Schwarz inequality for sums yields

$$|\langle \widehat{A}\widehat{B}\rangle_{\widehat{\rho}}| \leq \left(\sum_{j \in \mathcal{J}} \alpha_j \langle \widehat{A}^2 \rangle_{\psi_j}^{1/2} \right) \left(\sum_{j \in \mathcal{J}} \alpha \langle \widehat{B}^2 \rangle_{\psi_j}^{1/2} \right) = \langle \widehat{A}^2 \rangle_{\widehat{\rho}} \langle \widehat{B}^2 \rangle_{\widehat{\rho}}$$

hence the inequality (5.45) using formula (5.44). □

Choosing for $\widehat{A}$ the operator of multiplication by x_j and $\widehat{B} = -i\hbar \partial/\partial x_j$ we get the usual Robertson–Schrödinger inequalities

$$(\Delta X_j)_{\widehat{\rho}}^2 (\Delta P_j)_{\widehat{\rho}}^2 \geq \Delta(X_j, P_j)_{\widehat{\rho}}^2 + \tfrac{1}{4}\hbar^2 \tag{5.48}$$

for $1 \leq j \leq n$.

Corollary 82. *Assume that $\widehat{A}$ and $\widehat{B}$ are Hermitian operators on $L^2(\mathbb{R}^n)$ such that $[\widehat{A}, \widehat{B}] = i\hbar$; then*

$$(\Delta \widehat{A})_{\widehat{\rho}}^2 (\Delta \widehat{B})_{\widehat{\rho}}^2 \geq \Delta(\widehat{A}, \widehat{B})_{\widehat{\rho}}^2 + \tfrac{1}{4}\hbar^2. \tag{5.49}$$

If $\widehat{\rho}$ represents a pure state ψ contained in the common definition domain of $\widehat{A}\widehat{B}$ and $\widehat{B}\widehat{A}$, then we have equality if and only if the vectors $(\widehat{A} - \langle \widehat{A}\rangle)\psi$ and $(\widehat{B} - \langle \widehat{B}\rangle)\psi$ are collinear.

Proof. The inequality (5.49) immediately follows from the inequality (5.43). Assume that we have equality in (5.49). It is sufficient to consider the case $\langle \widehat{A}\rangle_\psi = \langle \widehat{B}\rangle_\psi = 0$. In view of formula $\langle \widehat{A}\widehat{B}\rangle_\psi = \langle \widehat{A}\widehat{B}\psi|\psi\rangle = \langle \widehat{B}\psi|\widehat{A}\psi\rangle$ this means that the Cauchy–Schwarz inequality reduces to an equality, which implies that $\widehat{A}\psi$ and $\widehat{B}\psi$ are collinear. □

5.3. Symplectic Formulation of the Uncertainty Principle

Both the Heisenberg and the Robertson–Schrödinger uncertainty relations are expressed in terms of coordinates (more precisely, in terms of conjugate pairs). This is of course somewhat annoying, since we would prefer an intrinsic expression: for obvious physical reasons, the way we perceive indeterminacy should not depend on the coordinate systems we use. In this section we are going to see that symplectic geometry allows us to give a very concise formulation, which is independent of any choice of (symplectic) coordinates. In fact, it is a topological formulation, which is preserved under arbitrary symplectic transformations, linear or not. It can thus be seen as a universal formulation of quantum indeterminacy. For this purpose we have to introduce a few mathematical concepts closely related to Gromov's non-squeezing theorem.

5.3.1. Symplectic capacities

We will call *symplectic ellipsoid* with radius R the image of a ball $B^{2n}(z_0, R))$ centered at $z_0 \in \mathbb{R}^{2n}$ by a symplectic transformation $S \in \mathrm{Sp}(2n, \mathbb{R})$.

Definition 83. Let Ω be any subset of the phase space $\mathbb{R}^{2n}$. The linear Gromov width of Ω is the number $c_{\mathrm{LG}}(\Omega) = \pi R^2$ where R is the supremum of the radii of all symplectic ellipsoids contained inside the set Ω.

It immediately follows from this definition that we have

$$c_{\mathrm{LG}}[S(B^{2n}(z_0, R))] = c_{\mathrm{LG}}(B^{2n}(z_0, R)) = \pi R^2. \qquad (5.50)$$

It also follows from the squeezing theorem that we have

$$c_{\mathrm{G}}(Z_j^{2n}(R)) = \pi R^2, \qquad (5.51)$$

where $Z_j^{2n}(R)$ is the cylinder $x_j^2 + p_j^2 \leq R^2$. Clearly $c_{\mathrm{LG}}(\mathbb{R}^{2n}) = +\infty$. The linear Gromov width moreover has the following general properties:

- *Monotonicity*: $\Omega \subset \Omega'$ then $c_{\mathrm{LG}}(\Omega) \leq c_{\mathrm{LG}}(\Omega')$;
- *Symplectic invariance*: $c_{\mathrm{LG}}(\Omega) = c_{\mathrm{LG}}(S(\Omega))$ for every $S \in \mathrm{Sp}(2n, \mathbb{R})$;
- *Conformality*: $c_{\mathrm{LG}}(\lambda \Omega) = \lambda^2 c_{\mathrm{LG}}(\Omega)$ for every $\lambda > 0$.

The monotonicity property of the Gromov width is obvious; to prove its symplectic invariance it suffices to observe that we have $S'(B^{2n}(z_0, R)) \subset \Omega$ for some $S' \in \mathrm{Sp}(2n, \mathbb{R})$ if and only if $SS'(B^{2n}(z_0, R)) \subset S(\Omega)$. The conformality property follows from the linearity of S.

These properties, together with the explicit relations (5.50) and (5.51) show that the Gromov width is a *linear symplectic capacity*; linear symplectic capacities are particular cases of the general notion of symplectic capacity, which is defined as follows: it is a mapping $c : \Omega \to [0, \infty)$ having the following properties:

(**SC1**) *Invariance under symplectomorphisms*:

$$c(f(\Omega)) = c(\Omega) \quad \textit{if } f \in \mathrm{Symp}(2n, \mathbb{R}); \qquad (5.52)$$

(SC2) *Monotonicity:*

$$c(\Omega) \leq c(\Omega') \quad if \ \Omega \subset \Omega'; \tag{5.53}$$

(SC3) *Conformality:*

$$c(\lambda\Omega) = \lambda^2 c(\Omega) \quad for \ \lambda > 0; \tag{5.54}$$

(SC4) *Normalization:*

$$c(B^{2n}(R)) = \pi R^2 = c(Z_j^{2n}(R)), \tag{5.55}$$

where $Z_j^{2n}(R)$ is the cylinder with radius R based on the x_j, p_j plane and centered at the origin.

There are infinitely many symplectic capacities; in particular the Gromov width, defined by

$$c_G(\Omega) = \sup_{f \in \mathrm{Symp}(2n,\mathbb{R})} \{\pi R^2 : f(B^{2n}(R)) \subset \Omega\}$$

is a symplectic capacity; it is actually the smallest of all symplectic capacities, in the sense that if c is any symplectic capacity, then $c_G(\Omega) \leq c(\Omega)$ for all subsets Ω of $\mathbb{R}^{2n}$. While the verification of the properties **(SC1)**–**(SC3)** for c_G are relatively straightforward (and so is the proof of the equality $c_G(B^{2n}(R)) = \pi R^2$), the proof of the normalization condition $c_G(Z_j^{2n}(R)) = \pi R^2$ requires the full strength of Gromov's theorem. This explains why the proof of the existence of a symplectic capacity is never a trivial endeavor: the existence of a single such capacity automatically implies Gromov's non-squeezing theorem.

When condition **(SC1)** is weakened by replacing it with the following condition:

(SC1$'$) *Invariance under* $\mathrm{Sp}(2n, \mathbb{R})$ *and translations:*

$$c(S(\Omega)) = c(\Omega) \quad if \ S \in \mathrm{Sp}(2n, \mathbb{R}); \tag{5.56}$$

$$c(\Omega + z_0) = c(\Omega) \quad if \ z_0 \in \mathbb{R}^2, \tag{5.57}$$

one says that c is a linear symplectic capacity; for instance this is the case of the linear Gromov width c_{LG}.

For a detailed study of symplectic capacity we refer to [131]; also see [108, 112, 159].

5.3.2. The case of ellipsoids

It turns out that while there are infinitely many symplectic capacities, a remarkable fact is that they all agree on ellipsoids.

Let M be a real $m \times m$ symmetric matrix. Elementary linear algebra tells us that all the eigenvalues $\lambda_1, \lambda_2, \ldots, \lambda_n$ of M are real, and that M can be diagonalized using an orthogonal transformation. Williamson's theorem provides us with a symplectic variant of this result. It says that every symmetric and positive definite $2n \times 2n$ matrix M can be diagonalized using symplectic matrices, and this in a very particular way. We begin by noting that eigenvalues of JM are those of the antisymmetric matrix $M^{1/2}JM^{1/2}$ and are thus of the type $\pm i\lambda_j^\sigma(M)$ with $\lambda_j^\sigma(M) > 0$. We will always arrange the λ_j^σ in *decreasing* order:

$$\lambda_1^\sigma(M) \geq \lambda_2^\sigma(M) \geq \cdots \geq \lambda_n^\sigma(M). \tag{5.58}$$

This ordered set of numbers $(\lambda_1^\sigma(M), \lambda_2^\sigma(M), \ldots, \lambda_n^\sigma(M))$ is called the symplectic spectrum of M and is denoted by $\mathrm{Spec}_\sigma(M)$. The symplectic spectrum has the following obvious properties:

- $\mathrm{Spec}_\sigma(\alpha M) = (\alpha\lambda_1^\sigma(M), \ldots, \alpha\lambda_n^\sigma(M))$ for every $\alpha > 0$;
- $\mathrm{Spec}_\sigma(M^{-1}) = ((\lambda_n^\sigma(M))^{-1}, \ldots, (\lambda_1^\sigma(M))^{-1})$.

A very useful result, which has many applications in symplectic geometry, is Williamson's [300] symplectic diagonalization theorem. It says that ever positive definite matrix can be diagonalized using a symplectic matrix.

Proposition 84 (Williamson). *Let M be a real symmetric $2n \times 2n$ positive definite matrix M with symplectic spectrum $(\lambda_1^\sigma, \ldots, \lambda_n^\sigma)$. There exists (a non-unique) matrix $S \in \mathrm{Sp}(2n, \mathbb{R})$ such that*

$$S^T M S = \begin{pmatrix} \Lambda^\sigma & 0 \\ 0 & \Lambda^\sigma \end{pmatrix} \tag{5.59}$$

where $\Lambda^\sigma = \mathrm{diag}(\lambda_1^\sigma, \ldots, \lambda_n^\sigma)$.

For modern proofs, see [83, 159, 220], or [108, 112]. Williamson's diagonalization result generalizes to the multidimensional case the elementary observation that every 2×2 real symmetric matrix $\Sigma = \begin{pmatrix} a & b \\ b & c \end{pmatrix}$ with $a > 0$ and $ac - b^2 > 0$ can be written $\Sigma = S^T D S$ where

$$S = \begin{pmatrix} \sqrt{a/d} & b/\sqrt{ad} \\ 0 & \sqrt{d/a} \end{pmatrix}, \quad D = \begin{pmatrix} d & 0 \\ 0 & d \end{pmatrix}$$

and $d = \sqrt{ac - b^2}$.

Using Williamson's diagonalization result one proves the following.

Proposition 85. *Let M be as above and $\mathrm{Spec}_\sigma(M) = (\lambda_1^\sigma, \dots, \lambda_n^\sigma)$ its symplectic spectrum. Let $z_0 \in \mathbb{R}^{2n}$. The symplectic capacity of the ellipsoid:*

$$\Omega_{M,z_0} = \{z : M(z - z_0)^2 \le 1\} \tag{5.60}$$

is given by

$$c(\Omega_{M,z_0}) = c_{\mathrm{LG}}(\Omega_{M,z_0}) = \pi/\lambda_{\max}^\sigma, \tag{5.61}$$

where $\lambda_{\max}^\sigma = \lambda_1^\sigma$ is the largest symplectic eigenvalue of M.

The proof is an application of Gromov's non-squeezing theorem; we will not give it here and refer to [108, 112, 131]; the argument consists in using Williamson's theorem to reduce the ellipsoid (5.60) to the case

$$\sum_{j=1}^n \lambda_1^\sigma(M)(x_j^2 + p_j^2) \le 1$$

thus making the proof a trivial consequence of Gromov's non-squeezing theorem. Note that the equality $c(\Omega_{M,z_0}) = c_{\mathrm{LG}}(\Omega_{M,z_0})$ in (5.61) says that it doesn't matter whether one uses linear or nonlinear symplectic capacities when studying ellipsoids.

5.3.3. The strong uncertainty principle

We will use the following consequence of Proposition 85.

Corollary 86. *Let Σ be a symmetric positive definite $2n \times 2n$ real matrix. The symplectic capacity of the ellipsoid*

$$W_\Sigma = \{z \in \mathbb{R}^{2n} : \tfrac{1}{2}\Sigma^{-1}z^2 \le \hbar\} \tag{5.62}$$

is given by the formula

$$c(W_\Sigma) = 2\pi\lambda_{\min}^\sigma(\Sigma), \tag{5.63}$$

where $\lambda_{\min}(\Sigma)$ is the smallest symplectic eigenvalue of the matrix Σ.

Proof. Set $M = \tfrac{1}{2}\Sigma^{-1}$. We have $c(W_\Sigma) = \pi/\lambda_{\max}^\sigma(M)$ where $\lambda_{\max}^\sigma(M)$ is the largest symplectic eigenvalue of M. Since $\Sigma = \tfrac{1}{2}M^{-1}$ the symplectic spectrum of Σ is $(\tfrac{1}{2}(\lambda_n^\sigma(M))^{-1}, \dots, \tfrac{1}{2}(\lambda_1^\sigma(M))^{-1})$ hence the result using formula (5.61). $\square$

We will see later in this book that the covariance matrix of a quantum density matrix always is positive definite and that it satisfies the important condition

$$\sum + \frac{i\hbar}{2} J \geq 0 \tag{5.64}$$

(which means that all the eigenvalues of the Hermitian matrix $\sum + \frac{i\hbar}{2} J$ are ≥ 0). The ellipsoid (5.62) is called in this context the covariance ellipsoid (or Wigner ellipsoid). The following geometric lemma is the key to our formulation of the Robertson–Schrödinger uncertainty principle in terms of symplectic capacities.

Proposition 87. *Let $\sum$ be as above. The quantum condition (5.64) is equivalent to the following condition on the covariance ellipsoid:*

$$c(W_\Sigma) \geq \Pi\hbar. \tag{5.65}$$

Proof. The characteristic polynomial of $\sum + \frac{i\hbar}{2} J \geq 0$ is the product $P(t) = P_1(t) \cdots P_n(t)$ where

$$P_j(t) = t^2 - (\lambda_j^\sigma)^{-1} t + \tfrac{1}{4}(\lambda_j^\sigma)^{-2} - \eta^2.$$

The eigenvalues of the matrix $\frac{1}{2} M^{-1} + i\eta J$ are thus the real numbers $\frac{1}{2}[(\lambda_j^\sigma)^{-1} \pm 2\eta]$ hence that matrix is non-negative if and only if $\lambda_j^\sigma \leq \frac{1}{2}|\eta|^{-1}$ for every j, that is if and only if $\lambda_{\max}^\sigma \leq \frac{1}{2}|\eta|^{-1}$; in view of formula (5.61) in Proposition 85 this is equivalent to $c(W_\Sigma) = \pi/\lambda_{\max}^\sigma \geq 2\pi|\eta|$. □

An interesting point is that the equivalent conditions (5.64) and (5.65) imply satisfy the Robertson–Schrödinger inequalities

$$(\Delta X_j)^2 (\Delta P_j)^2 \geq \Delta(X_j, P_j)^2 + \tfrac{1}{4}\hbar^2 \tag{5.66}$$

for $1 \leq j \leq n$; however the latter do not imply the strong uncertainty principle (5.64) as follows the discussion after Proposition 76. We will come back to these delicate points in Chapters 6 and 7.

5.3.4. Hardy's uncertainty principle

A "folk theorem" is that a function ψ and its Fourier transform $F\psi$ cannot be simultaneously too sharply localized because this would violate the uncertainty principle. The mathematician Hardy [146] made this statement precise already in 1933 by showing that if there exist constants C_a and C_b such that $\psi \in L^2(\mathbb{R})$ and its Fourier transform satisfy

$$|\psi(x)| \leq C_a e^{-\frac{a}{2\hbar}x^2}, \quad |F\psi(p)| \leq C_b e^{-\frac{b}{2\hbar}p^2} \tag{5.67}$$

for $a, b > 0$, then the following holds true:

- *If $ab > 1$ then $\psi = 0$.*
- *If $ab = 1$ we have $\psi(x) = Ce^{-\frac{a}{2\hbar}x^2}$ for some complex constant C.*
- *If $ab < 1$ then (5.67) is satisfied by every function of the type $\psi(x) = Q(x)e^{-\frac{a}{2\hbar}x^2}$ where Q is a polynomial.*

It is not very difficult to extend Hardy's uncertainty principle to the multidimensional case and one finds the following proposition [129, 131].

Proposition 88. *Let A and B be two real positive definite matrices and $\psi \in L^2(\mathbb{R}^n)$, $\psi \neq 0$. Assume that*

$$|\psi(x)| \leq C_A e^{-\frac{1}{2\hbar}Ax^2} \quad \text{and} \quad |F\psi(p)| \leq C_B e^{-\frac{1}{2\hbar}Bp^2} \tag{5.68}$$

for some constants $C_A, C_B > 0$. Then:

(i) *The eigenvalues λ_j, $j = 1, \ldots, n$, of AB are ≤ 1.*
(ii) *If $\lambda_j = 1$ for all j, then $\psi(x) = Ce^{-\frac{1}{2\hbar}Ax^2}$ for some complex constant C.*
(iii) *If $\lambda_j < 1$ for some index j then each function $\psi(x) = Q(x)e^{-\frac{1}{2\hbar}Ax^2}$ (Q a polynomial) satisfies the Hardy inequalities (5.68).*

This extended Hardy uncertainty principle can be restated using symplectic capacities; in fact the conditions (5.68) imply that the ellipsoid

$$\Omega_{A,B} = \{z = (x, p) : Ax^2 + Bp^2 \leq \hbar\}$$

satisfies $c(\otimes_{A,B}) \geq \pi\hbar = \frac{1}{2}h$. To see this we rewrite the equation of $\Omega_{A,B}$ as $Mz^2 \leq \hbar$ with $M = \begin{pmatrix} A & 0 \\ 0 & B \end{pmatrix}$. Let $(\lambda_1^\sigma, \lambda_2^\sigma, \ldots, \lambda_n^\sigma)$ be the symplectic spectrum of M; using the results above (formula (5.61)) we have $c(\Omega_{A,B}) = \pi\hbar/\lambda_1^\sigma$. One then verifies that $\lambda_j^\sigma = \sqrt{\lambda_j}$ where the λ_j are the eigenvalues of AB, hence we must have $\lambda_j \leq 1$ that is $c(\Omega_{A,B}) \geq \pi\hbar$.

Hardy's uncertainty principle has an important application to the Wigner function; it quantifies the intuitive statement that a Wigner function cannot be too sharply concentrated around a point.

Proposition 89. *Let $\psi \in L^2(\mathbb{R}^n)$ and assume that there exist a constant $C > 0$ and a positive-definite symmetric matrix M such that*

$$W\psi(z) \leq Ce^{-\frac{1}{\hbar}Mz \cdot z} \tag{5.69}$$

for all z. Then we must have

$$c(W_\Sigma) \geq \tfrac{1}{2}h \tag{5.70}$$

where W_Σ is the Wigner ellipsoid corresponding to the choice $\Sigma = \frac{\hbar}{2}M^{-1}$.

Proof. (Cf. [129].) In view of Williamson's theorem, we can find $S \in \mathrm{Sp}(2n, \mathbb{R})$ such that

$$W\psi(S^{-1}z) \leq Ce^{-\frac{1}{\hbar}(\Lambda x^2 + \Lambda p^2)}, \tag{5.71}$$

where $\Lambda^\sigma = \mathrm{diag}(\lambda_1^\sigma, \ldots, \lambda_n^\sigma)$ and $\mathrm{Spec}_\sigma(M) = (\lambda_1^\sigma, \ldots, \lambda_n^\sigma)$. In view of the symplectic covariance property (formula (4.55) in Proposition 58) we have $W\psi(S^{-1}z) = W(\widehat{S}\psi)(z)$ for any $\widehat{S} \in \mathrm{Mp}(2n, \mathbb{R})$ with projection S. Since $\widehat{S}\psi \in L^2(\mathbb{R}^n)$ and $c(W_\Sigma)$ is a symplectic invariant, we may thus assume that $\widehat{S}$ is the identity operator (and hence $S = I$); this choice reduces the estimate to $W\psi(z) \leq Ce^{-\frac{1}{\hbar}Mz^2}$ to the diagonal case

$$W\psi(z) \leq Ce^{-\frac{1}{\hbar}(\Lambda^\sigma x^2 + \Lambda^\sigma p^2)}. \tag{5.72}$$

Integrating this inequality in x and p, respectively, we get

$$|\psi(x)| \leq C_1 e^{-\frac{1}{\hbar}\Lambda^\sigma x^2} \quad \text{and} \quad |F\psi(p)| \leq C_1 e^{-\frac{1}{\hbar}\Lambda^\sigma x^2} \tag{5.73}$$

for some constant $C_1 > 0$. In view of the multidimensional Hardy uncertainty principle the eigenvalues λ_j of $(\Lambda^\sigma)^2$ must be ≤ 1; since $\lambda_j = (\lambda_j^\sigma)^2$ we must thus have $\lambda_1^\sigma \leq 1$; in view of (5.61) this is equivalent to the inequality $c(\mathcal{B}_M) \geq \frac{1}{2}h$. $\qquad\square$

Chapter 6

Quantum States and the Density Matrix

In the present chapter, we review the theory of the density matrix from a rigorous point of view; this will give us the opportunity to examine the validity of some of the trace formulas we have given in last chapter. This rigor is motivated and needed by the applications to the density matrix with variable h we have in mind, and that will be addressed in next chapter. For more on the topics of trace class and Hilbert–Schmidt operators, see for instance [112, §12.1]; [262, Appendix 3].

6.1. The Density Matrix

The formalism of density operators and matrices was introduced by John von Neumann [226] in 1927 and independently, by Lev Landau and Felix Bloch in 1927 and 1946, respectively. Ugo Fano was one of the first to put the theory of the density matrix in a rigorous form in his foundational paper [76].

6.1.1. Quantum states

A quantum system is said to be in a pure state if we have complete knowledge about that system, meaning we know exactly which state it is in. Pure states can be prepared using *maximal tests* (see [238, §§2–3]): suppose we are dealing with a quantum system and let N be the maximum number of *different* outcomes that can be obtained in a test of that system. If such, a test has exactly N different outcomes, it is called a maximal

test. The quantum system under consideration is in a pure state if it is
prepared in such a way that it certainly yields a predictable outcome in that
maximal test; the outcomes in any other test have well-defined probabilities
which do not depend on the procedure used for the preparation. A pure
state can thus be identified by specifying the complete experiment that
characterizes it uniquely (see [76]). One usually writes a pure state using
Dirac's ket notation $|\psi\rangle$; for all practical purposes it is convenient to use
the wavefunction ψ defined by $\psi(x) = \langle x|\psi\rangle$; it is a normalized element of a
certain Hilbert space $\mathcal{H}$, which is usually identified in the case of continuous
variables with $L^2(\mathbb{R}^n)$ (the square integrable functions). When doing this
the state is identified with the linear span of the function ψ, that is the ray
$\mathbb{C}\psi = \{\lambda\psi : \lambda \in \mathbb{C}\}$. The pure state $|\psi\rangle$ can then be identified with the
orthogonal projection $\widehat{\rho}_\psi$ of $\mathcal{H}$ on the subspace $\mathbb{C}\psi$. This projection, which
is of rank one, is denoted by $|\psi\rangle\langle\psi|$ in quantum mechanics; it is analytically
given by the formula

$$\widehat{\rho}_\psi\phi = |\psi\rangle\langle\psi|\phi\rangle = \langle\psi|\phi\rangle\psi,$$

where $\langle\psi|\phi\rangle$ is identified with the inner product in $\mathcal{H}$. Most tests are
however not maximal, and most preparations do not produce pure states,
in which case we only have a partial knowledge of the quantum system
under consideration. The information on such a system is less than a maximum, with reference to the lack of a complete experiment with a uniquely
predetermined outcome. The state of the system is nevertheless fully
identified by any data adequate to predict the statistical results of all
conceivable observations on the system [76]. When this is the case one says
that the system is in a *mixed state*. Mixed states are *classical* probability
mixtures of pure states; however, different distributions of pure states can
generate physically indistinguishable mixed states. A quantum mixed state
can be viewed as the datum of a set of pairs $\{(\psi_j, \alpha_j)\}$ where each ψ_j is a
(normalized and square integrable) pure state and α_j a classical probability;
these probabilities sum up to one: $\sum_j \alpha_j = 1$. A *caveat*: one should not
confuse the mixed state $\{(\psi_j, \alpha_j)\}$ with the superposition $\psi = \sum_j \alpha_j\psi_j$
which is always a *pure* state!

6.1.2. Definition of trace class operators

We will use several times the Bessel equality

$$\sum_j \langle\psi|\psi_j\rangle\langle\phi|\psi_j\rangle^* = \langle\psi|\phi\rangle, \tag{6.1}$$

which is valid for every orthonormal basis (ψ_j) of the Hilbert space $\mathcal{H}$; it reduces to the familiar relation

$$\sum_j |\langle \psi | \psi_j \rangle|^2 = ||\psi||^2 \tag{6.2}$$

when $\psi = \phi$.

The way to do things correctly consists in using the mathematical definition of trace class operators; the latter is very general (and hence very powerful).

Definition 90. A bounded operator $\widehat{A}$ on a Hilbert space $\mathcal{H}$ is of *trace class* if there exist two orthonormal bases (ψ_j) and (ϕ_j) of $\mathcal{H}$ (indexed by the same set) such that

$$\sum_j |\langle \psi_j | \widehat{A} \phi_j \rangle| < \infty. \tag{6.3}$$

The set of trace class operators on $\mathcal{H}$ is denoted by $\mathcal{L}_1(\mathcal{H})$. The trace of a trace class operator is defined by the absolutely convergent series

$$\mathrm{Tr}(\widehat{A}) = \sum_j \langle \psi_j | \widehat{A} \psi_j \rangle, \tag{6.4}$$

where (ψ_j) is an arbitrary orthonormal basis of $\mathcal{H}$.

For the definitions above to make sense we have to show that if condition (6.3) holds for one pair of orthonormal bases, then it also holds for *all* pairs of orthonormal bases, and that if (ψ_j) and (ϕ_j) are two such pairs then

$$\sum_j \langle \psi_j | \widehat{A} \psi_j \rangle = \sum_j \langle \phi_j | \widehat{A} \phi_j \rangle \tag{6.5}$$

both series being absolutely convergent. The proof of these properties is not difficult; it consists in expanding each base using the vectors of the other. In fact (cf. [112, Chapter 12]), writing

$$\psi_i' = \sum_j \langle \psi_j | \psi_i' \rangle \psi_j, \ \ \phi_\ell' = \sum_k \langle \phi_k | \phi_\ell' \rangle \phi_k,$$

we have $\widehat{A}\phi_\ell' = \sum_k \langle \phi_k | \phi_\ell' \rangle \widehat{A}\phi_k$ and hence

$$\langle \psi_i' | \widehat{A} \phi_\ell' \rangle = \sum_{j,k} \langle \psi_i' | \psi_j \rangle \langle \phi_\ell' | \phi_k \rangle^* \langle \psi_j | \widehat{A} \phi_k \rangle. \tag{6.6}$$

Summing the left-hand side with respect to the indices i, ℓ and using the triangle inequality, we get

$$\sum_{i,\ell} |\langle \psi'_i | \widehat{A} \phi'_\ell \rangle| \leq \sum_{j,k} \left(\sum_{i,\ell} |\langle \psi'_i | \psi_j \rangle| \, |\langle \phi'_\ell | \phi_k \rangle| \right) |\langle \psi_j | \widehat{A} \phi_k \rangle|. \qquad (6.7)$$

In view of the trivial inequality $ab \leq \frac{1}{2}(a^2 + b^2)$ we have

$$\sum_{i,\ell} |\langle \psi'_i | \psi_j \rangle| \, |\langle \phi'_\ell | \phi_k \rangle| \leq \frac{1}{2} \sum_i |\langle \psi'_i | \psi_j \rangle|^2 + \frac{1}{2} \sum_\ell |\langle \phi'_\ell | \phi_k \rangle|^2$$

that is, since $\sum_i |\langle \psi'_i | \psi_j \rangle|^2 = ||\psi_j||^2 = 1$ and $\sum_\ell |\langle \phi'_\ell | \phi_k \rangle|^2 = ||\phi_k||^2 = 1$,

$$\sum_{i,\ell} |\langle \psi'_i | \widehat{A} \phi'_\ell \rangle| \leq \sum_{j,k} |\langle \psi_j | \widehat{A} \phi_k \rangle| < \infty, \qquad (6.8)$$

which proves that the condition (6.3) holds for all orthonormal bases if it holds for one. Let us now prove that formula (6.4) defining the trace is also independent of the choice of the orthonormal basis that is used. Assume that $\psi_i = \phi_i$ and $\psi'_i = \phi'_i$ for all indices i. Using the equality (6.6) we have

$$\langle \psi'_i | \widehat{A} \psi'_i \rangle = \sum_{j,k} \langle \psi'_i | \psi_j \rangle \langle \psi'_i | \psi_k \rangle^* \langle \psi_j | \widehat{A} \psi_k \rangle$$

and hence

$$\sum_i \langle \psi'_i | \widehat{A} \psi'_i \rangle = \sum_{j,k} \left(\sum_i \langle \psi'_i | \psi_j \rangle \langle \psi'_i | \psi_k \rangle^* \right) \langle \psi_j | \widehat{A} \psi_k \rangle.$$

In view of the Bessel equality (6.1)

$$\sum_i \langle \psi'_i | \psi_j \rangle \langle \psi'_i | \psi_k \rangle^* = \langle \psi_j | \psi_k \rangle = \delta_{jk}$$

so that

$$\sum_i \langle \psi'_i | \widehat{A} \psi'_i \rangle = \sum_j \langle \psi_j | \widehat{A} \psi_j \rangle,$$

which shows that the right-hand side of (6.4) indeed is independent of the choice of orthonormal basis; that the series is absolutely convergent follows from the inequality (6.8) taking $\psi_i = \phi_i$ for all indices i.

Notice that it is clear that the set $\mathcal{L}_1(\mathcal{H})$ of all trace-class operators on $\mathcal{H}$ is a vector space: that $\lambda\widehat{A} \in \mathcal{L}_1(\mathcal{H})$ if $\lambda \in \mathbb{C}$ and $\widehat{A} \in \mathcal{L}_1(\mathcal{H})$ is obvious. For $\widehat{A}, \widehat{B} \in \mathcal{L}_1(\mathcal{H})$ we have

$$\sum_i \langle \psi_i | (\widehat{A} + \widehat{B})\psi_i \rangle = \sum_i \langle \psi_i | \widehat{A}\psi_i \rangle + \sum_i \langle \psi_i | \widehat{B}\psi_i \rangle;$$

each sum on the right-hand side being absolutely convergent we have $\widehat{A} + \widehat{B} \in \mathcal{L}_1(\mathcal{H})$. We leave to the reader to verify that

$$\mathrm{Tr}(\widehat{A} + \widehat{B}) = \mathrm{Tr}\widehat{A} + \mathrm{Tr}\widehat{B} \quad \text{and} \quad \mathrm{Tr}(\lambda\widehat{A}) = \lambda\mathrm{Tr}(\widehat{A})$$

for every complex number λ. It follows that $\mathcal{L}_1(\mathcal{H})$ is a vector subspace of the algebra $\mathcal{B}(\mathcal{H})$ of bounded operators on $\mathcal{H}$. Also notice that it immediately follows from the definitions that $\widehat{A}$ is of trace class if and only if its adjoint $\widehat{A}^\dagger$ is: we have

$$\sum_j |\langle \phi_j | \widehat{A}^\dagger \chi_j \rangle| = \sum_j |\langle \widehat{A}\phi_j | \chi_j \rangle| = \sum_j |\langle \chi_j | \widehat{A}\phi_j \rangle|$$

and hence

$$\mathrm{Tr}(\widehat{A}^\dagger) = \mathrm{Tr}(\widehat{A})^*. \tag{6.9}$$

In particular the trace is real if the trace class operator $\widehat{A}$ is self-adjoint.

The following invariance property of the trace under unitary conjugation reflects the independence of trace class operators on changes of orthonormal bases.

Proposition 91. *Let $\widehat{A} \in \mathcal{L}_1(\mathcal{H})$ and $\widehat{U}$ be a unitary operator on $\mathcal{H}$. Then $\widehat{U}^\dagger\widehat{A}\widehat{U} \in \mathcal{L}_1(\mathcal{H})$ and we have*

$$\mathrm{Tr}(\widehat{U}^\dagger\widehat{A}\widehat{U}) = \mathrm{Tr}(\widehat{A}). \tag{6.10}$$

Proof. It is clear that $\widehat{U}^\dagger\widehat{A}\widehat{U}$ is a positive, bounded, and self-adjoint operator. The operator $\widehat{\rho}$ is of trace class if and only if $\sum_j \langle \psi_j | \widehat{\rho} | \psi_j \rangle < \infty$ for one (and hence every) orthonormal basis $(\psi_j)_j$ of $\mathcal{H}$. Since $\langle \psi_j | \widehat{U}^\dagger\widehat{A}\widehat{U} | \psi_j \rangle = \langle \widehat{U}\psi_j | \widehat{A}\widehat{U}\psi_j \rangle$ and the basis $(\widehat{U}\psi_j)_j$ also is orthonormal, the operator $\widehat{U}^\dagger\widehat{A}\widehat{U}$ is of trace class. The trace formula (6.10) follows from the orthonormal basis independence of the trace. $\qquad\square$

We have the following metaplectic covariance property.

Corollary 92. *For every* $\widehat{S} \in \mathrm{Mp}(n)$ *and* $\widehat{A} \in \mathcal{L}_1(L^2(\mathbb{R}^n))$, *we have*

$$\mathrm{Tr}(\widehat{S}^{-1}\widehat{A}\widehat{S}) = \mathrm{Tr}(\widehat{A}).$$

Proof. The metaplectic operators are unitary hence $\widehat{S}^\dagger = \widehat{S}^{-1}$.   □

An important functional property of self-adjoint trace class operators (which is shared by the Hilbert–Schmidt operators we will study in a moment) is that they are *compact operators*. This means that if $\widehat{A} \in \mathcal{L}_1(\mathcal{H})$ then for every bounded sequence (θ_j) (not necessarily a basis!) the sequence $(\widehat{A}\theta_j)_j$ contains a convergent subsequence. We will not prove this here; see [27, §22.4; 180, or 262, Appendix 3].

6.1.3. The density matrix

These mathematical preliminaries allow us to *define rigorously* the notion of density matrix. We recall that an operator $\widehat{A}$ on a Hilbert space $\mathcal{H}$ is positive (and we write $\widehat{A} \geq 0$) if we have $\langle \psi | \widehat{A}\psi \rangle \geq 0$ for all $\psi \in \mathcal{H}$. When the Hilbert space $\mathcal{H}$ is complex, the condition $\widehat{A} \geq 0$ automatically implies self-adjointness.

Let us check that the definition of a density matrix just given is compatible with the one given in previous chapters. Let $\{(\psi_j, \alpha_j)\}$ be a mixed state and consider the operator

$$\widehat{\rho} = \sum_j \alpha_j \widehat{\rho}_j, \tag{6.11}$$

where $\widehat{\rho}_j = |\psi_j\rangle\langle\psi_j|$ is the orthogonal projection on the ray $\mathbb{C}\psi_j = \{\lambda\psi_j : \lambda \in \mathbb{C}\}$.

Proposition 93. *The operator* $\widehat{\rho}$ *defined by formula (6.11) is a density matrix in the sense of the definition above.*

The set of density matrices has interesting properties that can be stated using the language of convex geometry.

Proposition 94. *Let* $\mathcal{B}(\mathcal{H})$ *be the algebra of bounded operators on the Hilbert space* $\mathcal{H}$.

(i) *The density matrices on a Hilbert space* $\mathcal{H}$ *form a convex subset of the space* $\mathcal{B}(\mathcal{H})$ *of bounded operators on* $\mathcal{H}$.

(ii) *The extreme points of this set are the rank-one projections, which correspond to the pure states on* $\mathcal{H}$.

Proof. The boundedness of $\hat{\rho}$ follows from the obvious inequalities

$$||\hat{\rho}\psi|| \leq \sum_j \alpha_j ||\hat{\rho}_j \psi|| \leq \sum_j \alpha_j ||\psi|| = ||\psi||.$$

The orthogonal projections $\hat{\rho}_j$ being self-adjoint operators so is $\hat{\rho}$. Let us show that $\hat{\rho} \geq 0$. We have $\hat{\rho}_j \psi = \langle \psi_j | \psi \rangle \psi_j$ hence

$$\langle \psi | \hat{\rho}_j \psi \rangle = \langle \psi_j | \psi \rangle \langle \psi_j | \psi \rangle = |\langle \psi_j | \psi \rangle|^2 \qquad (6.12)$$

so that $\hat{\rho}_j \geq 0$; it follows that $\hat{\rho} \geq 0$ as well. Let us show that $\hat{\rho}$ is of trace class and has trace unity. For this it suffices to show that $\hat{\rho}_j$ is a trace class operator with trace equal to one since we then have

$$\operatorname{Tr}\hat{\rho} = \sum_j \alpha_j = 1.$$

Let (ϕ_k) be an orthonormal basis (ϕ_k) of $\mathcal{H}$. We have $\hat{\rho}_j \phi_k = \langle \psi_j | \phi_k \rangle \psi_j$ hence $\langle \phi_k | \hat{\rho}_j \phi_k \rangle = |\langle \psi_j | \phi_k \rangle|^2$ and thus, using Bessel's equality (6.2),

$$\sum_k \langle \phi_k | \hat{\rho}_j \phi_k \rangle = \sum_k |\langle \psi_j | \phi_k \rangle|^2 = ||\psi_j||^2 = 1,$$

that is $\operatorname{Tr}(\hat{\rho}_j) = 1$ as claimed. □

6.1.4. Functional analysis and the spectral theorem

An especially useful expansion of a density operator is obtained using elementary functional analysis (the spectral theorem for compact operators). Recall that each eigenvalue of a self-adjoint compact operator (except possibly zero) has finite multiplicity (see any book on elementary functional analysis, e.g., [27] or [180]).

Theorem 95. *A bounded linear self-adjoint operator $\hat{\rho}$ on a complex Hilbert space $\mathcal{H}$ is of trace class if and only if there exist a sequence of real numbers $\lambda_j \geq 0$ and an orthonormal basis (ψ_j) of $\mathcal{H}$ such that for all $\psi \in \mathcal{H}$*

$$\hat{\rho}\psi = \sum_j \lambda_j \langle \psi_j | \psi \rangle \psi_j, \qquad (6.13)$$

that is $\hat{\rho} = \sum_j \lambda_j \hat{\rho}_j$ where $\hat{\rho}_j$ is the orthogonal projection on the ray $\mathbb{C}\psi_j$. In particular, $\hat{\rho}$ is a density matrix if and only if $\lambda_j \geq 0$ for every j and $\sum_j \lambda_j = 1$; the vector ψ_j is the eigenvector corresponding to the eigenvalue λ_j.

Proof. This is a classical result from the theory of compact self-adjoint operators on a Hilbert space; see any introductory book on functional analysis, for instance [27]. That ψ_j is an eigenvector corresponding to λ_j is clear: since $\langle \psi_k | \psi_j \rangle = \delta_{kj}$ we have

$$\widehat{\rho}\psi_j = \sum_k \lambda_k \langle \psi_k | \psi_j \rangle \psi_k = \lambda_j \psi_j. \qquad \square$$

This result shows that, as mentioned earlier, different mixtures can lead to the same density matrix. Such mixtures are said to be *equivalent*. In particular, every mixture is thus equivalent to a mixture of orthogonal (and hence independent) pure states.

Corollary 96. *A bounded self-adjoint operator $\widehat{\rho}$ on $\mathcal{H}$ is a density matrix if and only if its eigenvalues λ_j are ≥ 0 for every j and sum up to one: $\sum_j \lambda_j = 1$; we then have*

$$\widehat{\rho}\psi = \sum_j \lambda_j \widehat{\rho}_j \psi = \sum_j \lambda_j \langle \psi_j | \psi \rangle \psi_j, \qquad (6.14)$$

where ψ_j is the eigenvector corresponding to the eigenvalue λ_j.

It also follows from the theorem above that the trace of the square $\widehat{\rho}^2$ of the density matrix always lies in the interval $[0, 1]$.

Corollary 97. *If $\widehat{\rho}$ is a density matrix, then $\mathrm{Tr}(\widehat{\rho}^2) \leq 1$ with equality if and only if $\widehat{\rho}$ represents a pure state.*

Proof. We have

$$\widehat{\rho}^2 = \left(\sum_j \lambda_j \widehat{\rho}_j \right)^2 = \sum_{j,k} \lambda_j \lambda_k \widehat{\rho}_j \widehat{\rho}_k = \sum_j \lambda_j^2 \widehat{\rho}_j$$

the second equality because $\widehat{\rho}_j \widehat{\rho}_k = 0$ if $j \neq k$ since ψ_j and ψ_k are then orthogonal, and $\widehat{\rho}_j^2 = \widehat{\rho}_j$. Since $\lambda_j^2 \leq \lambda_j \leq 1$ we have

$$\mathrm{Tr}(\widehat{\rho}^2) = \sum_j \lambda_j^2 \leq 1.$$

The equality $\sum_j \lambda_j^2 = 1$ can only occur if all the coefficients λ_j are equal to zero, except one which is equal to one. Thus $\mathrm{Tr}(\widehat{\rho}^2) = 1$ if and only if $\widehat{\rho}$ represents a pure state. $\qquad \square$

The number $\mathrm{Tr}(\widehat{\rho}^2)$ is called the *purity* of the quantum state represented by the density matrix $\widehat{\rho}$. Another way of measuring the purity of a state is to use the von Neumann entropy $S(\widehat{\rho})$. By definition

$$S(\widehat{\rho}) = -\sum_j \lambda_j \ln \lambda_j \tag{6.15}$$

(with the convention $0 \ln 0 = 0$). One often uses the suggestive notation

$$S(\widehat{\rho}) = -\mathrm{Tr}(\widehat{\rho} \ln \widehat{\rho}). \tag{6.16}$$

Notice that the von Neumann entropy $S(\widehat{\rho})$ is zero if and only if $\widehat{\rho}$ is a pure state. Perhaps the most directly useful property of trace class operators (and hence of density matrices) is Proposition 98 below; it says that if we compose a trace class operator with any bounded operator we obtain again a trace class operator.

Proposition 98. *The set $\mathcal{L}_1(\mathcal{H})$ of all trace class operators on $\mathcal{H}$ is both a vector subspace of $\mathcal{B}(\mathcal{H})$ and a two-sided ideal in $\mathcal{B}(\mathcal{H})$: if $\widehat{A} \in \mathcal{L}_1(\mathcal{H})$ and $\widehat{B} \in \mathcal{B}(\mathcal{H})$ then $\widehat{A}\widehat{B} \in \mathcal{L}_1(\mathcal{H})$ and $\widehat{B}\widehat{A} \in \mathcal{L}_1(\mathcal{H})$ and we have*

$$\mathrm{Tr}(\widehat{B}\widehat{A}) = \mathrm{Tr}(\widehat{A}\widehat{B}). \tag{6.17}$$

Proof. That $\mathcal{L}_1(\mathcal{H})$ is a vector space is clear using formula (6.3): if for every $\chi_j \in \mathcal{H}$

$$\sum_j |\langle \widehat{A}_1 \phi_j | \chi_j \rangle| < \infty \quad \text{and} \quad \sum_j |\langle \widehat{A}_2 \phi_j | \chi_j \rangle| < \infty,$$

then we also have, using the triangle inequality,

$$\sum_j |\langle (\widehat{A}_1 + \widehat{A}_2) \phi_j | \chi_j \rangle| \leq \sum_j |\langle \widehat{A}_1 \phi_j | \chi_j \rangle| + \sum_j |\langle \widehat{A}_2 \phi_j | \chi_j \rangle| < \infty$$

so that $\widehat{A}_1 + \widehat{A}_2 \in \mathcal{L}_1(\mathcal{H})$; that $\lambda \widehat{A} \in \mathcal{L}_1(\mathcal{H})$ if $\widehat{A} \in \mathcal{L}_1(\mathcal{H})$ and $\lambda \in \mathbb{C}$ is clear. Let us show that $\widehat{B}\widehat{A} \in \mathcal{L}_1(\mathcal{H})$ if $\widehat{A} \in \mathcal{L}_1(\mathcal{H})$ and $\widehat{AB}$ is a bounded operator on $\mathcal{H}$. Recall that the boundedness of $\widehat{B}$ is equivalent to the existence of a number $C \geq 0$ such that $||\widehat{AB}\psi|| \leq C||\psi||$ for all $\psi \in \mathcal{H}$. Let now (ψ_j) and (ϕ_j) be two orthonormal bases of $\mathcal{H}$; writing $\langle \widehat{B}\widehat{A}\psi_j | \phi_j \rangle = \langle \widehat{A}\psi_j | \widehat{B}^\dagger \phi_j \rangle$ and applying Bessel's equality (6.1) to $\langle \widehat{A}\psi_j | \widehat{B}^\dagger \phi_j \rangle$ we get

$$\langle \widehat{B}\widehat{A}\psi_j | \phi_j \rangle = \sum_k \langle \widehat{A}\psi_j | \phi_k \rangle \langle \widehat{B}^\dagger \phi_j | \psi_k \rangle^*. \tag{6.18}$$

In view of the Cauchy–Schwarz inequality, we have, since $||\widehat{B}^\dagger \phi_j|| \leq C$,

$$|\langle \widehat{B}^\dagger \phi_j | \psi_k \rangle^*| \leq ||\widehat{B}^\dagger \phi_j|| \, ||\psi_k|| \leq C,$$

and hence

$$|\langle \widehat{B}\widehat{A}\psi_j | \phi_j \rangle| \leq \sum_k |\langle \widehat{A}\psi_j | \phi_k \rangle \langle \widehat{B}^\dagger \phi_j | \psi_k \rangle^*| \leq C \sum_k |\langle \widehat{A}\psi_j | \phi_k \rangle|.$$

Summing this inequality with respect to the index j yields, since $\widehat{A}$ is of trace class,

$$\left| \sum_j \langle \widehat{B}\widehat{A}\psi_j | \phi_j \rangle \right| \leq C \sum_{j,k} |\langle \widehat{A}\psi_j | \phi_k \rangle| < \infty,$$

and hence $\widehat{B}\widehat{A}$ is indeed of trace class as claimed. That $\widehat{A}\widehat{B}$ also is of trace class is immediate noting that we can write $\widehat{A}\widehat{B} = (\widehat{B}^\dagger \widehat{A}^\dagger)^\dagger$. There remains to prove the trace equality (6.17). Choosing $(\psi_j) = (\phi_j)$ the equality (6.18) becomes

$$\langle \widehat{B}\widehat{A}\psi_j | \psi_j \rangle = \sum_k \langle \widehat{A}\psi_j | \psi_k \rangle \langle \widehat{B}^\dagger \psi_j | \psi_k \rangle^*$$

and, similarly,

$$\langle \widehat{A}\widehat{B}\psi_j | \psi_j \rangle = \sum_k \langle \widehat{B}\psi_j | \psi_k \rangle \langle \widehat{A}^\dagger \psi_j | \psi_k \rangle^* = \sum_k \langle \widehat{A}\psi_k | \psi_j \rangle \langle \widehat{B}^\dagger \psi_k | \psi_j \rangle^*.$$

Summing this equality over j we get

$$\mathrm{Tr}(\widehat{A}\widehat{B}) = \sum_j \langle \widehat{B}\widehat{A}\psi_j | \psi_j \rangle = \sum_{j,k} \langle \widehat{A}\psi_j | \psi_k \rangle \langle \widehat{B}^\dagger \psi_j | \psi_k \rangle^*,$$

$$\mathrm{Tr}(\widehat{B}\widehat{A}) = \sum_j \langle \widehat{A}\widehat{B}\psi_j | \psi_j \rangle = \sum_{j,k} \langle \widehat{A}\psi_k | \psi_j \rangle \langle \widehat{B}^\dagger \psi_k | \psi_j \rangle^*,$$

and hence $\mathrm{Tr}(\widehat{A}\widehat{B}) = \mathrm{Tr}(\widehat{B}\widehat{A})$ since the sums of both right-hand sides are the same. $\qquad\Box$

6.2. Hilbert–Schmidt Operators

Closely related to trace class operators are the Hilbert–Schmidt operators; they are essential tools for understanding the structure and properties of density matrices. Hilbert–Schmidt operators are particular cases of the more general notion of *Schatten classes*; see [44] for a discussion of these spaces of operators.

6.2.1. The trace norm

An operator $\widehat{A}$ on a Hilbert space $\mathcal{H}$ is called a *Hilbert–Schmidt operator* if there exists an orthonormal basis (ψ_j) of $\mathcal{H}$ such that

$$\sum_j \langle \widehat{A}\psi_j | \widehat{A}\psi_j \rangle = \sum_j ||\widehat{A}\psi_j||^2 < \infty. \tag{6.19}$$

In particular such an operator is bounded on $\mathcal{H}$. As in the case of trace class operators one shows that if this property holds for one orthonormal basis then it holds for all, and that the number $\Sigma_j ||\widehat{A}\psi_j||^2$ is independent of the choice of such a basis: let in fact (ϕ_j) be a second orthonormal basis, and write $\widehat{A}\psi_j = \Sigma_k \langle \phi_j | \widehat{A}\psi_j \rangle \phi_k$. Then, by repeated use of the Bessel equality (6.1),

$$\sum_j ||\widehat{A}\psi_j||^2 = \sum_{j,k} |\langle \phi_j | \widehat{A}\psi_j \rangle|^2$$

$$= \sum_{j,k} |\langle \widehat{A}^\dagger \phi_j | \psi_j \rangle|^2$$

$$= \sum_k ||\widehat{A}^\dagger \phi_k||^2.$$

Choosing $(\phi_j) = (\psi_j)$ we thus get

$$\sum_j ||\widehat{A}\psi_j||^2 = \sum_k ||\widehat{A}^\dagger \psi_k||^2 < \infty$$

showing that $\widehat{A}^\dagger$ is Hilbert–Schmidt if and only if $\widehat{A}$ is. We may thus replace $\widehat{A}$ with $\widehat{A}^\dagger$ in the sequence of equalities above which yields, since $(\widehat{A}^\dagger)^\dagger = \widehat{A}$,

$$\sum_j ||\widehat{A}^\dagger \psi_j||^2 = \sum_k ||\widehat{A}\phi_k||^2$$

hence our claim.

Hilbert–Schmidt operators form a vector space $\mathcal{L}_2(\mathcal{H})$ and

$$||\widehat{A}||_{\mathrm{HS}} = \left(\sum_j ||\widehat{A}\psi_j||^2 \right)^{1/2} \tag{6.20}$$

defines a norm on this space; this norm is associated with the scalar product

$$\langle \widehat{A} | \widehat{B} \rangle_{\mathrm{HS}} = \mathrm{Tr}(\widehat{A}^\dagger \widehat{B}) = \sum_j \langle \widehat{A}\psi_j | \widehat{B}\psi_j \rangle. \tag{6.21}$$

If $\widehat{A}$ and $\widehat{B}$ are Hilbert–Schmidt operators then $\lambda\widehat{A}$ is trivially a Hilbert–Schmidt operator and $||\lambda\widehat{A}||_{\mathrm{HS}} = |\lambda|\,||\widehat{A}||_{\mathrm{HS}}$ for every $\lambda \in \mathbb{C}$; on the other hand, by the triangle inequality,

$$\sum_j ||(\widehat{A} + \widehat{B})\psi_j||^2 \le \sum_j ||\widehat{A}\psi_j||^2 + \sum_j ||\widehat{B}\psi_j||^2 < \infty$$

for every orthonormal basis (ψ_j) hence $\widehat{A} + \widehat{B}$ is also a Hilbert–Schmidt operator and we have

$$||\widehat{A} + \widehat{B}||_{\mathrm{HS}}^2 \le ||\widehat{A}||_{\mathrm{HS}}^2 + ||\widehat{B}||_{\mathrm{HS}}^2 \le (||\widehat{A}||_{\mathrm{HS}} + ||\widehat{B}||_{\mathrm{HS}})^2$$

and hence

$$||\widehat{A} + \widehat{B}||_{\mathrm{HS}} \le ||\widehat{A}||_{\mathrm{HS}} + ||\widehat{B}||_{\mathrm{HS}}.$$

Finally, $||\widehat{A}||_{\mathrm{HS}} = 0$ is equivalent to $\widehat{A}\psi_j = 0$ for every index j that is to $\widehat{A} = 0$.

The space $\mathcal{L}_2(\mathcal{H})$ is complete for that norm, and hence a Banach space (it is actually even a Hilbert space when equipped with the scalar product (6.21). In addition $\mathcal{L}_2(\mathcal{H})$ is closed under multiplication (and hence an algebra).

It turns out that the space $\mathcal{L}_2(\mathcal{H})$ is a two-sided deal in the algebra of $\mathcal{B}(\mathcal{H})$ of bounded operators: if $\widehat{A} \in \mathcal{L}_2(\mathcal{H})$ and $\widehat{B} \in \mathcal{B}(\mathcal{H})$ then $\widehat{A}\widehat{B} \in \mathcal{L}_2(\mathcal{H})$ and $\widehat{B}\widehat{A} \in \mathcal{L}_2(\mathcal{H})$. Let us show that $\widehat{B}\widehat{A} \in \mathcal{L}_2(\mathcal{H})$. We have, denoting by $||\widehat{B}||$ the operator norm of $\widehat{B}$,

$$||\widehat{B}\widehat{A}||_{\mathrm{HS}}^2 = \sum_j ||\widehat{B}\widehat{A}\psi_j||^2 \le ||\widehat{B}|| \left(\sum_j ||\widehat{A}\psi_j||^2 \right) < \infty.$$

Applying the same argument to $\widehat{A}\widehat{B} = (\widehat{B}^\dagger\widehat{A}^\dagger)^\dagger$ shows that $\widehat{A}\widehat{B} \in \mathcal{L}_2(\mathcal{H})$ as well.

6.2.2. Hilbert–Schmidt and trace class

An essential property is that every trace class operator is the product of two Hilbert–Schmidt operators (and hence itself a Hilbert–Schmidt operator).

Proposition 99. (i) *A bounded operator $\widehat{A}$ on $\mathcal{H}$ is of trace class if and only if it is the product of two Hilbert–Schmidt operators:* $\mathcal{L}_1(\mathcal{H}) = (\mathcal{L}_2(\mathcal{H}))^2$.

(ii) *A trace class operator $\widehat{A}$ on $\mathcal{H}$ is itself a Hilbert–Schmidt operator:* $\mathcal{L}_1(\mathcal{H}) \subset \mathcal{L}_2(\mathcal{H})$.

(iii) *A trace class operator $\widehat{A}$ is positive if and only if there exists a Hilbert–Schmidt operator $\widehat{B}$ on $\mathcal{H}$ such that $\widehat{A} = \widehat{B}^\dagger \widehat{B}$.*

Proof. In what follows (ψ_j) is an orthonormal basis in $\mathcal{H}$. (i) Assume that $\widehat{A} = \widehat{B}\widehat{C}$ where $\widehat{B}$ and $\widehat{C}$ are both Hilbert–Schmidt operators. We have, using respectively the triangle and the Cauchy–Schwarz inequalities,

$$\left| \sum_j \langle \psi_j | \widehat{A}\psi_j \rangle \right| \leq \sum_j |\langle \widehat{B}^\dagger \psi_j | \widehat{C}\psi_j \rangle| \leq \sum_j \|\widehat{B}^\dagger \psi_j\| \, \|\widehat{C}\psi_j\|;$$

in view of the trivial inequality

$$\sum_j \|\widehat{B}^\dagger \psi_j\| \, \|\widehat{C}\psi_j\| \leq \frac{1}{2} \left(\sum_j \|\widehat{B}^\dagger \psi_j\|^2 + \|\widehat{C}\psi_j\|^2 \right),$$

we get, since $\widehat{B}$ and $\widehat{C}$ are both Hilbert–Schmidt operators,

$$\left| \sum_j \langle \psi_j | \widehat{A}\psi_j \rangle \right| \leq \frac{1}{2} \left(\sum_j \|\widehat{B}^\dagger \psi_j\|^2 + \|\widehat{C}\psi_j\|^2 \right) < \infty$$

proving that $\widehat{A}$ is indeed of trace class. Assume, conversely, that $\widehat{A} \in \mathcal{L}_1(\mathcal{H})$. In view of the polar decomposition theorem there exists a unitary operator $\widehat{U}$ on $\mathcal{H}$ such that $\widehat{A} = \widehat{U}(\widehat{A}^\dagger\widehat{A})^{1/2}$. Setting $\widehat{B} = \widehat{U}(\widehat{A}^\dagger\widehat{A})^{1/4}$ and $\widehat{C} = (\widehat{A}^\dagger\widehat{A})^{1/4}$ we have $\widehat{A} = \widehat{B}\widehat{C}$; let us show that $\widehat{C}$ and $\widehat{B}$ are Hilbert–Schmidt operators. We have,

$$\sum_j |\langle \widehat{C}\psi_j | \widehat{C}\psi_j \rangle| = \sum_j |\langle \widehat{C}^\dagger\widehat{C}\psi_j | \psi_j \rangle|$$

$$= \sum_j |\langle (\widehat{A}^\dagger\widehat{A})^{1/2}\psi_j | \psi_j \rangle| < \infty$$

because $(\widehat{A}^\dagger\widehat{A})^{1/2}$ is of trace class (Proposition 98), hence $\widehat{C} \in \mathcal{L}_2(\mathcal{H})$. It follows that $\widehat{B} = \widehat{U}\widehat{C} \in \mathcal{L}_2(\mathcal{H})$ as well. (ii) We have seen that every trace class operator is a product $\widehat{A} = \widehat{B}\widehat{C}$ of two Hilbert–Schmidt operators. In view of the algebra property (i) of $\mathcal{L}_2(\mathcal{H})$ the operator $\widehat{A}$ is itself Hilbert–Schmidt operator. (iii) An operator $\widehat{A} = \widehat{B}^\dagger \widehat{B}$ is always positive since

$$\langle \widehat{B}^\dagger \widehat{B}\psi | \psi \rangle = \langle \widehat{B}\psi | \widehat{B}\psi \rangle \geq 0$$

for all $\psi \in \mathcal{H}$. Suppose conversely that $\widehat{A} \in \mathcal{L}_1(\mathcal{H})$ and $\widehat{A} \geq 0$. By the same argument as in the proof of (ii) we now have $\widehat{A} = (\widehat{A}^\dagger \widehat{A})^{1/2}$; it suffices thus to take $\widehat{B} = \widehat{C} = (\widehat{A}^\dagger \widehat{A})^{1/4}$ and to note that $\widehat{B}^\dagger = \widehat{B}$. $\qquad \square$

6.2.3. The case of $L^2(\mathbb{R}^n)$

Let us now specialize to the case where $\mathcal{H} = L^2(\mathbb{R}^n)$. In this case Hilbert–Schmidt operators are exactly those operators that have a square integrable kernel; as a consequence a density matrix also has square integrable kernel.

Proposition 100. *Let $\widehat{A}$ be a bounded operator on $L^2(\mathbb{R}^n)$.*

(i) *It is a Hilbert–Schmidt operator if and only if there exists a function $K \in L^2(\mathbb{R}^n \times \mathbb{R}^n)$ such that*

$$\widehat{A}\psi(x) = \int K(x, y)\psi(y)d^n y. \qquad (6.22)$$

(ii) *Every trace class operator (and hence every density matrix) on $L^2(\mathbb{R}^n)$ can be represented in this way with a kernel $K \in L^2$ $(\mathbb{R}^n \times \mathbb{R}^n)$.*

Proof. (i) *The condition is necessary.* Let (ψ_j) be an orthonormal basis in $L^2(\mathbb{R}^n)$. Let $\widehat{A} \in \mathcal{L}_2(L^2(\mathbb{R}^n))$; we have $\widehat{A}\psi_i = \sum_j \langle \psi_j | \widehat{A}\psi_i \rangle \psi_j$ and hence

$$\widehat{A}\psi = \sum_i \langle \psi_i | \psi \rangle \widehat{A}\psi_i = \sum_{i,j} \langle \psi_i | \psi \rangle \langle \psi_j | \widehat{A}\psi_i \rangle \psi_j$$

which we can rewrite, using the definition

$$\langle \psi_i | \psi \rangle = \int \psi_i^*(y)\psi(y)d^n y$$

of the inner product as

$$\widehat{A}\psi(x) = \sum_{i,j} \langle \psi_i | \psi \rangle \langle \psi_j | \widehat{A}\psi_i \rangle \psi_j(x)$$

$$= \sum_{i,j} \langle \psi_i | \psi \rangle \langle \psi_j | \widehat{A}\psi_i \rangle \psi_j(x)$$

$$= \sum_{i,j} \langle \psi_j | \widehat{A}\psi_i \rangle \int \psi_j(x)\psi_i^*(y)\psi(y)d^n y.$$

This is now (6.22) with

$$K(x,y) = \sum_{i,j} \langle \psi_j | \widehat{A}\psi_i \rangle \psi_j(x)\psi_i^*(y).$$

Let us show that $K \in L^2(\mathbb{R}^n \times \mathbb{R}^n)$. Remarking that the tensor products $(\psi_j \otimes \psi_i^*)$ form an orthonormal basis in $L^2(\mathbb{R}^n \times \mathbb{R}^n)$ we have

$$\int |K(x,y)|^2 d^n x d^n y \le \sum_{i,j} |\langle \psi_j | \widehat{A}\psi_i \rangle|^2;$$

applying the Bessel equality (6.1) to $|\langle \psi_j | \widehat{A}\psi_i \rangle|^2$ we get

$$\sum_{i,j} |\langle \psi_j | \widehat{A}\psi_i \rangle|^2 = \sum_i |\langle \widehat{A}\psi_i | \widehat{A}\psi_i \rangle|^2 < \infty$$

since $\widehat{A}$ is a Hilbert–Schmidt operator. It follows that $K \in L^2(\mathbb{R}^n \times \mathbb{R}^n)$ as claimed. *The condition is sufficient.* Assume that the kernel K of $\widehat{A} \in \mathcal{B}(L^2(\mathbb{R}^n))$ belongs to $L^2(\mathbb{R}^n \times \mathbb{R}^n)$. Since $(\psi_j \otimes \psi_i^*)$ is an orthonormal basis in $L^2(\mathbb{R}^n \times \mathbb{R}^n)$, we can find numbers c_{ij} such that $\sum_{i,j} |c_{ij}|^2 < \infty$ and

$$K(x,y) = \sum_{i,j} c_{ij}\psi_j(x) \otimes \psi_i^*(y).$$

Define now the operator $\widehat{A}$ by the equality (6.22); we have

$$\widehat{A}\psi(x) = \sum_{i,j} c_{ij}\psi_j(x) \int \psi_i^*(y)\psi(y)d^n y = \sum_{i,j} c_{ij}\langle \psi_i | \psi \rangle \psi_j(x)$$

and hence, since the basis $(\psi_j)_j$ is orthonormal,

$$\widehat{A}\psi_k(x) = \sum_{i,j} c_{ij}\langle \psi_i | \psi_k \rangle \psi_j(x) = \sum_j c_{kj}\psi_j(x).$$

so that

$$\sum_k ||\widehat{A}\psi_k||^2 = \sum_{j,k} |c_{kj}|^2 < \infty$$

and $\widehat{A}$ is thus a Hilbert–Schmidt operator. (ii) In view of property the algebra property (see (i) in Proposition 99) a trace class operator is *a fortiori* a Hilbert–Schmidt operator. The claim follows in view of the statement (i). □

It is essential, both for theoretical and practical (tomography) purposes to identify the phase space object corresponding to a density matrix.

6.3. The Phase Space Picture

From now on we assume that the Hilbert space $\mathcal{H}$ is $L^2(\mathbb{R}^n)$, the space of complex-valued square integrable functions on $\mathbb{R}^n$ (we are thus dealing with quantum systems with n degrees of freedom). We begin by shortly reviewing the Weyl correspondence; for details and complements see [112, 120] or [83, 189].

6.3.1. The Weyl correspondence

Recall that a function $K(x, y)$ defined on $\mathbb{R}^n \times \mathbb{R}^n$ is the kernel of an operator $\widehat{A}$ if we have

$$\widehat{A}\psi(x) = \int K(x, y)\psi(y)d^n y \qquad (6.23)$$

for all $\psi \in L^2(\mathbb{R}^n)$. A deep theorem from functional analysis ("Schwartz's kernel theorem") tells us that every continuous linear operator from spaces of test functions to the tempered distributions[a] can be represented in this way, the integral in (6.23) being possibly replaced by a distributional bracket. By definition, the Weyl symbol of the operator $\widehat{A}$ is the function

$$a(x, p) = \int e^{-\frac{i}{\hbar}p\cdot y} K\left(x + \frac{1}{2}y, x - \frac{1}{2}y\right)d^n y; \qquad (6.24)$$

---

[a]In this context we use the word "distribution" in the sense of L. Schwartz's "generalized functions".

this formula is easily inverted using an inverse Fourier transform in p, yielding the expression of the kernel in terms of the symbol:

$$K(x,p) = \left(\frac{1}{2\pi\hbar}\right)^n \int e^{\frac{i}{\hbar}p(x-y)} a\left(\frac{1}{2}(x+y),p\right) d^n p. \qquad (6.25)$$

These two formulas uniquely define the kernel and the symbol in terms of each other, and imply the "Weyl correspondence (or transform)", which expresses unambiguously the operator $\widehat{A}$ in terms of the symbol a:

$$\widehat{A}\psi(x) = \left(\frac{1}{2\pi\hbar}\right)^n \iint e^{\frac{i}{\hbar}p(x-y)} a\left(\frac{1}{2}(x+y),p\right) \psi(y) d^n y d^n p; \qquad (6.26)$$

one often writes $\widehat{A} = \text{Op}_{\text{W}}(a)$, and this notation is unambiguous because the symbol of $\widehat{A}$ is uniquely determined by (6.24). Formula (6.26) can be rewritten in several different ways; one common expression is

$$\widehat{A}\psi(x) = \left(\frac{1}{\pi\hbar}\right)^n \int a(z_0)\widehat{\Pi}(z_0)\psi(x) d^{2n} z_0, \qquad (6.27)$$

where $\widehat{\Pi}(z_0) = \widehat{\Pi}(x_0,p_0)$ is the reflection operator already introduced in Chapter 4:

$$\widehat{\Pi}(x_0,p_0)\psi(x) = e^{\frac{2i}{\hbar}p_0(x-x_0)}\psi(2x_0 - x). \qquad (6.28)$$

The usefulness of the Weyl correspondence in quantum mechanics comes from the fact that it associates to real symbols self-adjoint operators. In fact, more generally:

$$\widehat{A} = \text{Op}^{\text{W}}(a) \implies \widehat{A}^\dagger = \text{Op}^{\text{W}}(a^*).$$

We refer to [108, 112, 120] for detailed discussions of the Weyl correspondence from the point of view outlined above; Littlejohn's well-cited paper [189] contains a very nice review of the topic with applications to semiclassical approximations.

The essential point to understand now is that the Wigner function $\rho = W_{\widehat{\rho}}(x,p)$ we are going to define below is (up to an unimportant constant factor) the *Weyl symbol* of the operator $\widehat{\rho}$: the Wigner function is thus a *dequantization* of $\widehat{\rho}$, that is a phase space function obtained from this operator.[b] Also notice that it is the first time Planck's constant appears

---

[b]It is perhaps a little bit daring to speak about "dequantization" in this context since the Wigner function is not a classical object.

in a quite explicit way; we could have *a priori* replaced $\hbar$ with any other real parameter η: this change would not have consequence for the involved mathematics (but it would of course change the physics!). We will come back to this essential point in Chapter 7.

Definition of the Wigner function of a density matrix

To a density matrix $\widehat{\rho}$ on $L^2(\mathbb{R}^n)$ one associates in standard texts its Wigner function (also called Wigner distribution). It is the function $W_{\widehat{\rho}}$ of the variables $x = (x_1, \ldots, x_n)$ and of the conjugate momenta $p = (p_1, \ldots, p_n)$ usually defined in quantum physics texts by the expression

$$W_{\widehat{\rho}}(x, p) = \left(\frac{1}{\pi\hbar}\right)^n \int e^{-\frac{2i}{\hbar}px'} \langle x + x'|\widehat{\rho}|x - x'\rangle d^n x', \qquad (6.29)$$

where $|x\rangle$ is an eigenstate of the operator $\widehat{x} = (\widehat{x}_1, \ldots, \widehat{x}_n)$ (where $\widehat{x}_j = $ multiplication by x_j). Performing the change of variables $x \mapsto y = 2x'$ we can rewrite this definition in the equivalent form

$$W_{\widehat{\rho}}(x, p) = \left(\frac{1}{2\pi\hbar}\right)^n \int e^{-\frac{i}{\hbar}py} \left\langle x + \frac{1}{2}y|\widehat{\rho}|x - \frac{1}{2}y \right\rangle d^n y; \qquad (6.30)$$

this has some practical advantages when one uses the Wigner–Weyl–Moyal formalism. In spite of their formal elegance, formulas (6.29), (6.30) are at first sight somewhat obscure and need to be clarified, especially if one wants to work analytically with them. Assume first that $\widehat{\rho}$ represents a pure state: $\widehat{\rho} = \widehat{\rho}_\psi = |\psi\rangle\langle\psi|$ where $\psi \in L^2(\mathbb{R}^n)$ is normalized. We get, using the relations $\psi(x) = \langle x|\psi\rangle$ and $\psi^*(x) = \langle\psi|x\rangle$,

$$\langle x + x'|\widehat{\rho}|x - x'\rangle = \langle x + x'|\psi\rangle\langle\psi|x - x'\rangle$$
$$= \psi(x + x')\psi^*(x - x')$$

and hence $W_{\widehat{\rho}_\psi}(x, p) = W\psi(x, p)$ where

$$W\psi(x, p) = \left(\frac{1}{\pi\hbar}\right)^n \int e^{-\frac{2i}{\hbar}px'} \psi(x + x')\psi^*(x - x') d^n x' \qquad (6.31)$$

is the usual Wigner function (or Wigner distribution, or Wigner transform) of $\psi \in L^2(\mathbb{R}^n)$ (see [158, 299]; for the mathematical theory

[108, 112, 120]); equivalently

$$W\psi(x,p) = \left(\frac{1}{2\pi\hbar}\right)^n \int e^{-\frac{i}{\hbar}py}\psi\left(x+\frac{1}{2}y\right)\psi^*\left(x-\frac{1}{2}y\right)d^n y. \quad (6.32)$$

In the general case, where $\widehat{\rho} = \sum_j \alpha_j |\psi_j\rangle\langle\psi_j|$ is a convex sum of operators of the type above one immediately gets, by linearity, the expression

$$W_{\widehat{\rho}}(x,p) = \sum_j \alpha_j W\psi_j(x,p). \quad (6.33)$$

A very important result we will prove later on (Proposition 104) (but use immediately!) is the following:

The Wigner function of a mixed state is square integrable: we have $\rho = W_{\widehat{\rho}} \in L^2(\mathbb{R}^{2n})$.

One also often uses the cross-Wigner transform of a pair of square integrable functions. It is given by

$$W(\psi,\phi)(x,p) = \left(\frac{1}{2\pi\hbar}\right)^n \int e^{-\frac{i}{\hbar}p\cdot y}\psi\left(x+\frac{1}{2}y\right)\phi^*\left(x-\frac{1}{2}y\right)d^n y.$$
$$(6.34)$$

It naturally appears as an interference term when calculating the Wigner function of a sum; in fact, using definition (6.32) one immediately checks that

$$W(\psi+\phi) = W\psi + W\phi + 2\mathrm{Re}W(\psi,\phi). \quad (6.35)$$

Notice that $W(\psi,\phi)$ is in general a complex number and that

$$W(\psi,\phi)^* = W(\phi,\psi). \quad (6.36)$$

The cross-Wigner function has many applications; in particular it allows to reformulate the notion of weak-value as an interference between the past and the future in the time-symmetric approach to quantum mechanics (see [93]).

The Weyl symbol of a density matrix

In the case of density matrices we have the following proposition.

Proposition 101. *Let $\widehat{\rho}$ be a density matrix on $L^2(\mathbb{R}^n)$:*

$$\widehat{\rho} = \sum_j \alpha_j |\psi_j\rangle\langle\psi_j| \quad \text{with } \alpha_j \geq 0 \quad \text{and} \quad \sum_j \alpha_j = 1.$$

The Weyl symbol of $\widehat{\rho}$ is $a = (2\pi\hbar)^n \rho$ where

$$\rho(x,p) = W_{\widehat{\rho}}(x,p) = \sum_j \alpha_j W\psi_j(x,p) \tag{6.37}$$

is the Wigner function of $\widehat{\rho}$.

Proof. The action of the projection $\widehat{\rho}_j = |\psi_j\rangle\langle\psi_j|$ on a vector $\psi \in L^2(\mathbb{R}^n)$ is given by

$$\widehat{\rho}_j\psi(x) = \langle\psi_j|\psi\rangle\psi_j(x) = \int \psi_j^*(y)\psi(y)\psi_j(x)d^n y,$$

and hence the kernel of $\widehat{\rho}_j$ is the function

$$K_j(x,y) = \psi_j(x)\psi_j^*(y).$$

It follows, using formula (6.24), that the Weyl symbol of $\widehat{\rho}_j$ is

$$a_j(x,p) = \int e^{-\frac{i}{\hbar}py}\psi_j\left(x + \frac{1}{2}y\right)\psi_j^*\left(x - \frac{1}{2}y\right)d^n y$$

$$= (2\pi\hbar)^n W\psi_j(x,p).$$

Formula (6.37) follows by linearity. $\qquad\square$

Statistical interpretation of the Wigner function

The importance of the Wigner function of a density matrix comes from the fact that we can use it as a substitute for an ordinary probability density for calculating averages (it is precisely for this purpose Wigner introduced his eponymous transform in [299]). We have already seen that for all $\psi \in L^1(\mathbb{R}^n) \cap L^2(\mathbb{R}^n)$ the marginal properties

$$\int W\psi(x,p)d^n p = |\psi(x)|^2, \quad \int W\psi(x,p)d^n x = |F\psi(p)|^2 \tag{6.38}$$

hold, and hence, in particular,

$$\iint W\psi(x,p)d^npd^nx = 1 \quad \text{if } ||\psi|| = 1. \tag{6.39}$$

In the second equality (6.38)

$$F\psi(p) = \left(\frac{1}{2\pi\hbar}\right)^{n/2} \int e^{-\frac{i}{\hbar}p \cdot x}\psi(x)d^nx \tag{6.40}$$

is the $\hbar$-Fourier transform of ψ. One should be aware of the fact that while $W\psi$ is always real (and hence so is $\rho = W_{\widehat{\rho}}$) as can be easily checked by taking the complex conjugates of both sides of the equality (6.32,) it takes negative values for all ψ which are not Gaussian functions. (This is the celebrated "Hudson theorem" [164].) A *caveat*: this result is only true for the Wigner function $W\psi$ of a single function ψ; the case of a general distribution $\rho = \Sigma_j\alpha_jW\psi_j$ is much subtler, and will be discussed later.

Let us introduce the following terminology: we call an observable $\widehat{A}$ a "good observable" for the density matrix $\widehat{\rho}$ if its Weyl symbol a (i.e., the corresponding classical observable) satisfies $a\rho \in L^1(\mathbb{R}^{2n})$, that is

$$\int |a(z)\rho(z)|d^{2n}z < \infty \tag{6.41}$$

(ρ the Wigner function of $\widehat{\rho}$; we are using the shorthand notation $z = (x,p)$, $d^{2n}z = d^nxd^np$). We assume in addition that a is real so that $\widehat{A}$ is Hermitian. Notice that "goodness" is guaranteed if the symbol a is square integrable, because the Cauchy–Schwarz inequality then implies that

$$\left(\int |a(z)\rho(z)|d^{2n}z\right)^2 \leq \int \rho(z)^2d^{2n}z \int |a(z)|^2d^{2n}z < \infty$$

since ρ is square integrable (as mentioned above, see Proposition 104).

Proposition 102. *Let $\widehat{\rho}$ be a density matrix on $L^2(\mathbb{R}^n)$ and ρ its Wigner function. The average value of every good observable $\widehat{A}$ with respect to $\widehat{\rho}$ is then finite and given by the formula*

$$\langle\widehat{A}\rangle_{\widehat{\rho}} = \int a(z)\rho(z)d^{2n}z. \tag{6.42}$$

Proof. By linearity it suffices to consider the case where $\widehat{\rho} = |\psi\rangle\langle\psi|$ so that $\rho = W\psi$; this reduces the proof of formula (6.42) to that of the simpler equality

$$\langle \widehat{A} \rangle_\psi = \int a(x,p)W\psi(x,p)d^n x d^n p. \tag{6.43}$$

Replacing in the equality above $W\psi(x,p)$ by its integral expression (6.32) yields

$$\langle \widehat{A} \rangle_\psi = \left(\frac{1}{2\pi\hbar} \right)^n \iint a(x,p)$$
$$\times \left(\int e^{-\frac{i}{\hbar}py} \psi \left(x + \frac{1}{2}y \right) \psi^* \left(x - \frac{1}{2}y \right) d^n y \right) d^n x d^n p.$$

Since we assume that the "goodness" assumption (6.41) is satisfied, we can use Fubini's theorem and rewrite this equality as a double integral:

$$\langle \widehat{A} \rangle_\psi = \left(\frac{1}{2\pi\hbar} \right)^n \iint a(x,p)e^{-\frac{i}{\hbar}py} \psi \left(x + \frac{1}{2}y \right) \psi^* \left(x - \frac{1}{2}y \right) d^n y d^n x d^n p.$$

Let us perform the change of variables $x' = x + \frac{1}{2}y$ and $y' = x - \frac{1}{2}y$; we have $x = \frac{1}{2}(x' + y')$ and $y = x' - y'$ and hence, using definition (6.26) of the Weyl operator $\widehat{A}$,

$$\langle \widehat{A} \rangle_\psi = \left(\frac{1}{2\pi\hbar} \right)^n \iint e^{-\frac{i}{\hbar}p(x'-y')} a \left(\frac{1}{2}(x' + y'),p \right) \psi(x')\psi^*(y')d^n y' d^n x' d^n p$$

$$= \left(\frac{1}{2\pi\hbar} \right)^n \int \left(\iint e^{-\frac{i}{\hbar}p(x'-y')} a \left(\frac{1}{2}(x' + y'),p \right) \right.$$
$$\left. \times \psi^*(y')d^n y' d^n p \right) \psi(x')d^n x'$$

and hence

$$\langle \widehat{A} \rangle_\psi = \int \widehat{A}\psi^*(x')\psi(x')d^n x' = \langle \widehat{A}\psi|\psi \rangle,$$

which we set out to prove. $\square$

We remark that the identity (6.43) can be extended to the cross-Wigner function (6.34); in fact, adapting the proof of (6.43) one sees that if ψ and ϕ are square integrable, then

$$\langle \psi|\widehat{A}\phi \rangle = \int a(z)W(\psi,\phi)(z)d^{2n}z. \tag{6.44}$$

6.3.2. The displacement operator and the ambiguity function

In this subsection we review a few properties of the Wigner function which are perhaps not all so well known in quantum mechanics; these properties are important because they give an insight into some of the subtleties of the Weyl formalism. We also define a related transform, the ambiguity function.

Redefinition of the Wigner function

Recall that the reflection operator is explicitly given by the formula

$$\widehat{\Pi}(x_0, p_0)\psi(x) = e^{\frac{2i}{\hbar} p_0 \cdot (x - x_0)} \psi(2x_0 - x).$$

It can be used to define the Wigner function in a very concise way. In fact, for every $\psi \in L^2(\mathbb{R}^n)$ we have

$$W\psi(z_0) = \left(\frac{1}{\pi\hbar} \right)^n \langle \psi | \widehat{\Pi}(z_0)\psi \rangle. \tag{6.45}$$

This is easy to verify: we have, by definition of $\widehat{\Pi}(x_0, p_0)$,

$$\langle \psi | \widehat{\Pi}(z_0)\psi \rangle = \int e^{\frac{2i}{\hbar} p_0 \cdot (x - x_0)} \psi(2x_0 - x)\psi^*(x) d^n x;$$

setting $y = 2(x_0 - x)$ we have $x = x_0 - \frac{1}{2}y$, $2x_0 - x = x_0 + \frac{1}{2}y$, and $d^n x = 2^{-n} d^n y$ hence

$$\langle \psi | \widehat{\Pi}(z_0)\psi \rangle = 2^{-n} \int e^{-\frac{i}{\hbar} p_0 \cdot y} \psi\left(x_0 + \frac{1}{2}y \right) \psi^*\left(x_0 - \frac{1}{2}y \right) d^n y$$

which proves (6.45), taking definition (6.32) of the Wigner function into account. Formula (6.45) shows quite explicitly that, up to the factor $(\pi\hbar)^{-n}$, the Wigner function is the probability amplitude for the state $|\psi\rangle$ to be in the state $|\widehat{\Pi}(x_0, p_0)\psi\rangle$; this was actually already observed by Grossmann [138] and Royer [245] in the mid-1970s.

6.3.3. The Moyal identity

An important equality satisfied by the Wigner function is *Moyal's identity* (sometimes also called the "orthogonality relation").

Proposition 103. *The cross-Wigner transform satisfies the "Moyal identity"*

$$\langle\langle W(\psi,\psi')|W(\phi,\phi')\rangle\rangle = \left(\frac{1}{2\pi\hbar}\right)^n \langle\psi|\phi\rangle\langle\psi'|\phi'\rangle^* \qquad (6.46)$$

for all $(\psi,\phi) \in L^2(\mathbb{R}^n) \times L^2(\mathbb{R}^n)$; *here* $\langle\langle\cdot|\cdot\rangle\rangle$ *is the inner product on* $L^2(\mathbb{R}^{2n})$.

Proof. Let us set

$$A = (2\pi\hbar)^{2n}\langle\langle W(\psi,\phi)|W(\psi',\phi')\rangle\rangle.$$

By definition of the cross-Wigner transform we have

$$A = \int e^{-\frac{i}{\hbar}p\cdot(y-y')}\psi\left(x+\frac{1}{2}y\right)\psi'\left(x+\frac{1}{2}y'\right)^*$$

$$\times \phi\left(x-\frac{1}{2}y\right)^*\phi'\left(x-\frac{1}{2}y'\right)d^nyd^ny'd^nxd^np.$$

Observing that

$$\int e^{-\frac{i}{\hbar}p\cdot(y-y')}d^np = (2\pi\hbar)^n\delta(y-y')$$

we have

$$A = (2\pi\hbar)^n\int \psi\left(x+\frac{1}{2}y\right)\psi'\left(x-\frac{1}{2}y\right)^*\phi\left(x+\frac{1}{2}y\right)^*$$

$$\times \phi'\left(x-\frac{1}{2}y\right)d^nyd^ny'd^nx.$$

Setting $x' = x + \frac{1}{2}y$ and $y' = x - \frac{1}{2}y$ we have $d^nx'd^ny' = d^nxd^ny$ hence

$$A = (2\pi\hbar)^n\left(\int \psi(x')\psi'(x')^*d^nx'\right)\left(\int \phi(y')^*\phi'(y')d^ny'\right)$$

which proves Moyal's identity (6.46). $\qquad\qquad\square$

It immediately follows from Moyal's identity that we have

$$\int W\psi(z)W\phi(z)d^{2n}z = \left(\frac{1}{2\pi\hbar}\right)^n |\langle\psi|\phi\rangle|^2 \qquad (6.47)$$

for all $\psi, \phi \in L^2(\mathbb{R}^n)$. In particular,

$$\int W\psi(z)^2 d^{2n}z = \left(\frac{1}{2\pi\hbar}\right)^n ||\psi||^4. \qquad (6.48)$$

This formula implies the following interesting fact which is not immediately obvious: consider the spectral decomposition (6.13) of a density operator in Theorem 95:

$$\widehat{\rho}\psi = \sum_j \lambda_j \langle\psi_j|\psi\rangle\psi_j$$

here the λ_j are the eigenvalues of $\widehat{\rho}$ and the corresponding eigenvectors ψ_j form an orthonormal basis of $\mathcal{H}$. When $\mathcal{H} = L^2(\mathbb{R}^n)$ the corresponding Wigner function is therefore

$$\widehat{\rho}\psi = \sum_j \lambda_j W\psi_j. \qquad (6.49)$$

It follows from Moyal's identity (6.47) that the $W\psi_j$ form an orthonormal system of vectors in the Hilbert space $L^2(\mathbb{R}^{2n})$ (but not a basis as is easily seen by "dimension count").

As a consequence of the Moyal identity we prove the fact that the Wigner function of a density matrix is square integrable.

Proposition 104. *Let $\{(\psi_j, \alpha_j)\}$ be a mixed state ($\psi_j \in L^2(\mathbb{R}^n)$, $\alpha_j \geq 0$, $\sum_j \alpha_j = 1$). The Wigner function $\rho = W_{\widehat{\rho}}$ is square integrable: $\rho \in L^2(\mathbb{R}^{2n})$.*

Proof. Since $L^2(\mathbb{R}^{2n})$ is a vector space it is sufficient to consider the pure case, that is to prove that $W\psi \in L^2(\mathbb{R}^{2n})$ if $\psi \in L^2(\mathbb{R}^{2n})$. But this immediately follows from Moyal's identity (6.48). $\qquad\square$

The Moyal identity can be extended to the cross-Wigner function (6.34); recall that for $\psi, \phi \in L^2(\mathbb{R}^n)$ it is defined by

$$W(\psi, \phi)(x, p) = \left(\frac{1}{2\pi\hbar}\right)^n \int e^{-\frac{i}{\hbar}p\cdot y}\psi\left(x + \frac{1}{2}y\right)\phi^*\left(x - \frac{1}{2}y\right)d^n y.$$

In fact, for all $\psi, \psi', \phi, \phi' \in L^2(\mathbb{R}^n)$ we have

$$\int W(\psi, \psi')^*(z)W(\phi, \phi')(z)d^{2n}z = \left(\frac{1}{2\pi\hbar}\right)^n \langle\psi|\phi\rangle\langle\psi'|\phi'\rangle^* \qquad (6.50)$$

(see for instance [112, 120]). Denoting by $\langle\langle\cdot|\cdot\rangle\rangle$ the inner product on $L^2(\mathbb{R}^{2n})$ this identity can be written as

$$\int |W(\psi, \psi')(z)|^2 d^{2n}z = \left(\frac{1}{2\pi\hbar}\right)^n ||\psi||^2 ||\psi'||^2. \qquad (6.51)$$

An important remark: one can prove [112, 120], using this generalized Moyal identity, that if vectors ψ_j form an orthonormal basis of $L^2(\mathbb{R}^n)$ then the vectors $(2\pi\hbar)^{n/2}W(\psi_j, \psi_k)$ form an orthonormal basis of the space $L^2(\mathbb{R}^{2n})$ (that these vectors are orthonormal is clear from (6.46)).

The ambiguity function

A transform closely related to the Wigner function and well known from signal analysis (especially radar theory) is the *ambiguity function* $A\psi$ (it is also called the "auto-correlation function"). It can be introduced in several equivalent ways; we begin by defining it explicitly by a formula: for $\psi \in L^2(\mathbb{R}^n)$

$$A\psi(x, p) = \left(\frac{1}{2\pi\hbar}\right)^n \int e^{-\frac{i}{\hbar}p\cdot y}\psi\left(y + \frac{1}{2}x\right)\psi^*\left(y - \frac{1}{2}x\right)d^n y. \qquad (6.52)$$

Comparing with the definition

$$W\psi(x, p) = \left(\frac{1}{2\pi\hbar}\right)^n \int e^{-\frac{i}{\hbar}p\cdot y}\psi\left(x + \frac{1}{2}y\right)\psi^*\left(x - \frac{1}{2}y\right)d^n y$$

of the Wigner function one cannot help being surprised by the similarity of both definitions. In fact, it is easy to show by performing an elementary change of variables that if ψ is an even function (that is $\psi(-x) = \psi(x)$ for all $x \in \mathbb{R}^n$) then $W\psi$ and $A\psi$ are related by

$$A\psi(x, p) = 2^{-n}W\psi(\tfrac{1}{2}x, \tfrac{1}{2}p). \qquad (6.53)$$

There are two complementary "natural" ways to define the ambiguity function. The first is to observe that the Wigner function and the ambiguity function are symplectic Fourier transforms of each other:

$$A\psi = F_\sigma W\psi \text{ and } W\psi = F_\sigma A\psi; \tag{6.54}$$

they are of course equivalent since $F_\sigma^{-1} = F_\sigma$. For a proof, see [112, 120]. There is still another way to define the ambiguity function. Let $\widehat{D}(z_0) = \widehat{D}(x_0, p_0)$ be the *displacement operator* (it is also called the Glauber–Sudarshan displacement operator, or the Heisenberg operator, or the Heisenberg–Weyl operator). It is defined by

$$\widehat{D}(z_0)\psi(x) = e^{\frac{i}{\hbar}(p_0 x - \frac{1}{2}p_0 x_0)}\psi(x - x_0). \tag{6.55}$$

This operator is the time-one propagator for the Schrödinger equation associated with the classical translation Hamiltonian $\sigma(z, z_0) = x_0 p - p_0 x$ (see the discussions in [108, 112, 119, 189]); this observation motivates the notation

$$\widehat{D}(z_0) = e^{-\frac{i}{\hbar}\sigma(\widehat{z}, z_0)} = e^{-\frac{i}{\hbar}(x_0\widehat{p} - p_0\widehat{x})} \tag{6.56}$$

often found in the literature. We are using here the coordinate expression of the displacement operator; we leave it to the reader as an exercise to check that $\widehat{D}(z_0)$ coincides with the operator

$$D(\alpha) = \exp\left[\frac{i}{\hbar}(\alpha a^\dagger - \alpha^* a)\right]$$

commonly used in quantum optics (a and $a^\dagger$ are the annihilation and creation operators; see [241] for a discussion of these notational issues). The displacement operator is related to the reflection operator $\widehat{\Pi}(z_0)$ by the simple formula

$$\widehat{\Pi}(z_0) = \widehat{D}(z_0)\Pi\widehat{D}(z_0)^\dagger, \tag{6.57}$$

where Π is the parity operator $\Pi\psi(x) = \psi(-x)$. That the operators $\widehat{D}(z_0)$ correspond to translations in phase space quantum mechanics is illustrated by the following important relation satisfied by the Wigner transform:

$$W(\widehat{D}(z_0)\psi)(z) = W\psi(z - z_0) \tag{6.58}$$

(it is easily proven by a direct computation, see [108, 112, 120, 189]). Using the displacement operator, the ambiguity function is given by

$$A\psi(z_0) = \left(\frac{1}{2\pi\hbar}\right)^n \langle\widehat{D}(z_0)\psi|\psi\rangle; \tag{6.59}$$

one verifies by a direct calculation using (6.55) that one recovers the first analytical definition (6.52).

The displacement operators play a very important role, not only in quantum mechanics, but also in related disciplines such as harmonic analysis, signal theory, and time-frequency analysis. They can be viewed as a representation of the canonical commutation relations (the Schrödinger representation of the Heisenberg group); this is related to the fact these operators satisfy

$$\widehat{D}(z_0)\widehat{D}(z_1) = e^{\frac{i}{\hbar}\sigma(z_0,z_1)}\widehat{D}(z_1)\widehat{D}(z_0) \tag{6.60}$$

and also

$$\widehat{D}(z_0 + z_1) = e^{-\frac{i}{2\hbar}\sigma(z_0,z_1)}\widehat{D}(z_0)\widehat{D}(z_1). \tag{6.61}$$

The second formula shows that the displacement operators form a projective representation of the phase space translation group. In addition to being used to define the ambiguity function, the displacement operators allow one to define Weyl operators in terms of their "twisted symbol" (sometimes also called "covariant symbol"), which is by definition the symplectic Fourier transform

$$a_\sigma(z) = F_\sigma a(z) \tag{6.62}$$

of the ordinary symbol a. Let in fact $\widehat{A} = \mathrm{Op_W}(a)$, that is

$$\widehat{A} = \left(\frac{1}{\pi\hbar}\right)^n \int a(z_0)\widehat{\Pi}(z_0)d^{2n}z_0$$

(formula 6.27). Using the displacement operator $\widehat{D}(z_0)$ in place of the reflection operator $\widehat{\Pi}(z_0)$ we have

$$\widehat{A} = \left(\frac{1}{2\pi\hbar}\right)^n \int a_\sigma(z_0)\widehat{D}(z_0)d^{2n}z_0 \tag{6.63}$$

(see [108, 112, 120, 189]). This formula has many applications; it is essential in the study of the positivity properties of trace class operators as we will see in a moment. Notice that formula (6.63) is Weyl's original definition [295] in disguise: making the change of variables $z_0 \mapsto -Jz_0$ in this formula one gets, noting that $a_\sigma(-Jz_0) = Fa(z_0)$ and

$$\widehat{D}(-Jz_0) = e^{-\frac{i}{\hbar}(x_0\widehat{x}+p_0\widehat{p})}e^{-\frac{i}{\hbar}(x_0\widehat{p}-p_0\widehat{x})},$$

$$\widehat{A} = \left(\frac{1}{2\pi\hbar}\right)^n \iint Fa(x,p)e^{\frac{i}{\hbar}(x\widehat{x}+p\widehat{p})}d^npd^nx,$$

which is the formula originally proposed by Weyl [295], in analogy with the Fourier inversion formula (see the discussion in [119, 120]).

6.3.4. Calculating traces rigorously: Kernels and symbols

It is tempting to redefine the trace of a Weyl operator $\widehat{A} = \mathrm{Op_W}(a)$ by the formula

$$\mathrm{Tr}(\widehat{A}) = \left(\frac{1}{2\pi\hbar}\right)^n \int a(z)d^{2n}z \qquad (6.64)$$

as we did in the previous chapter. But doing this one should not forget that even if the operator $\widehat{A}$ is of trace class, formula (6.64) need not give the actual trace. First, the integral in the right-hand side might not be convergent; secondly even if it is we have to prove that it really yields the right result. We will discuss the validity of formula (6.64) and of other similar formulas below, but let us first prove some intermediary results.

We begin by discussing the Weyl symbols of Hilbert–Schmidt and trace class operators.

Proposition 105. *Let $\widehat{A} = \mathrm{Op^W}(a)$ be a Hilbert–Schmidt operator. Then $a \in L^2(\mathbb{R}^{2n})$ and we have*

$$\int |a(z)|^2 d^{2n}z = (2\pi\hbar)^{n/2} \iint K(x,y)d^nxd^ny. \qquad (6.65)$$

Conversely, every Weyl operator with symbol $a \in L^2(\mathbb{R}^{2n})$ is a Hilbert–Schmidt operator.

Proof. In view of Proposition 100 we have $K \in L^2(\mathbb{R}^n \times \mathbb{R}^n)$. Let us prove formula (6.65) when $K \in \mathcal{S}(\mathbb{R}^n \times \mathbb{R}^n)$; it will then hold by continuity for arbitrary $K \in L^2(\mathbb{R}^n \times \mathbb{R}^n)$ in view of the density of $\mathcal{S}(\mathbb{R}^n \times \mathbb{R}^n)$ in $L^2(\mathbb{R}^n \times \mathbb{R}^n)$. In view of formula (6.24) relating the kernel and the symbol of a Weyl operator we have

$$\int |a(x,p)|^2 dp = (2\pi\hbar)^n \int \left| K\left(x + \frac{1}{2}y, x - \frac{1}{2}y\right) \right|^2 d^n y. \qquad (6.66)$$

Integrating this equality with respect to the x variables we get, using Fubini's theorem,

$$\int |a(z)|^2 d^{2n} z = (2\pi\hbar)^n \int \left(\int \left| K\left(x + \frac{1}{2}y, x - \frac{1}{2}y\right) \right|^2 d^n y \right) d^n x$$

$$= (2\pi\hbar)^n \iint \left| K\left(x + \frac{1}{2}y, x - \frac{1}{2}y\right) \right|^2 d^n x d^n y.$$

Set now $x' = x + \frac{1}{2}y$ and $y' = x - \frac{1}{2}y$; we have $d^n x' d^n y' = d^n x d^n y$ and hence

$$\int |a(z)|^2 d^{2n} z = (2\pi\hbar)^n \iint |K(x', y')|^2 d^n x' d^n y' \qquad (6.67)$$

which we set out to prove. The converse is obvious since the condition $a \in L^2(\mathbb{R}^{2n})$ is equivalent to $K \in L^2(\mathbb{R}^n \times \mathbb{R}^n)$ in view of the inequality above. $\square$

6.3.5. Rigorous results

We now specialize our discussion to the case $\mathcal{H} = L^2(\mathbb{R}^n)$. Recall that we showed in Section 6.2 that the formula

$$\langle \widehat{A} | \widehat{B} \rangle_{\mathrm{HS}} = \mathrm{Tr}(\widehat{A}^\dagger \widehat{B}) \qquad (6.68)$$

defines an inner product on the ideal $\mathcal{L}_2(\mathcal{H})$ of Hilbert–Schmidt operators in $\mathcal{H}$ the associated "trace norm" being defined by

$$||\widehat{A}||_{\mathrm{HS}} = \mathrm{Tr}(\widehat{A}^\dagger \widehat{A})^{1/2}. \qquad (6.69)$$

Let us prove the following important result which justifies many "formulas" found in quantum mechanical texts when the symbols satisfy some integrability conditions.

Proposition 106. *Let $\widehat{A} = \mathrm{Op}^{\mathrm{W}}(a)$ and $\widehat{B} = \mathrm{Op}^{\mathrm{W}}(a)$ be Hilbert–Schmidt operators: $\widehat{A}, \widehat{B} \in \mathcal{L}_2(\mathcal{H})$.*

(i) *The trace class operator $\widehat{A}\widehat{B}$ has trace*

$$\mathrm{Tr}(\widehat{A}\widehat{B}) = \left(\frac{1}{2\pi\hbar}\right)^n \int a(z)b(z)d^{2n}z. \tag{6.70}$$

(ii) *The Hilbert–Schmidt inner product is given by the convergent integral*

$$\langle\widehat{A}|\widehat{B}\rangle_{\mathrm{HS}} = \left(\frac{1}{2\pi\hbar}\right)^n \int a^*(z)b(z)d^{2n}z \tag{6.71}$$

and hence $\|\widehat{A}\|^2_{\mathrm{HS}} = \mathrm{Tr}(\widehat{A}^\dagger\widehat{A})$ *is given by*

$$\|\widehat{A}\|^2_{\mathrm{HS}} = \left(\frac{1}{2\pi\hbar}\right)^n \int |a(z)|^2 d^{2n}z. \tag{6.72}$$

Proof. (i) We first observe that in view of Proposition 105 we have $a \in L^2(\mathbb{R}^n)$ and $b \in L^2(\mathbb{R}^n)$ hence the integrals in (6.70) and (6.71) are absolutely convergent. Let (ψ_j) be an orthonormal basis of $L^2(\mathbb{R}^n)$; by definition of the trace we have

$$\mathrm{Tr}(\widehat{A}\widehat{B}) = \sum_j \langle\psi_j|\widehat{A}\widehat{B}\psi_j\rangle = \sum_j \langle\widehat{A}^\dagger\psi_j|\widehat{B}\psi_j\rangle.$$

Expanding $\widehat{B}\psi_j$ and $\widehat{A}^\dagger\psi_j$ in the basis (ψ_j) we get

$$\widehat{B}\psi_j = \sum_k \langle\psi_k|\widehat{B}\psi_j\rangle\psi_k, \quad \widehat{A}^\dagger\psi_j = \sum_\ell \langle\widehat{A}\psi_\ell|\psi_j\rangle\psi_\ell,$$

and hence, using the Bessel equality (6.1),

$$\langle\widehat{A}^\dagger\psi_j|\widehat{B}\psi_j\rangle = \sum_k \langle\widehat{A}^\dagger\psi_j|\psi_k\rangle\langle\widehat{B}\psi_j|\psi_k\rangle^*. \tag{6.73}$$

In view of formula (6.44) we have

$$\langle\widehat{A}^\dagger\psi_j|\psi_k\rangle = \int a(z)W(\psi_j,\psi_k)(z)d^{2n}z,$$

$$\langle\widehat{B}\psi_j|\psi_k\rangle = \int b^*(z)W(\psi_j,\psi_k)(z)d^{2n}z,$$

where $W(\psi_j, \psi_k)$ is the cross-Wigner transform (6.34) of ψ_j, ψ_k; denoting by $\langle\langle\cdot|\cdot\rangle\rangle$ the inner product on $L^2(\mathbb{R}^{2n})$ these equalities can be rewritten

$$\langle\widehat{A}^\dagger \psi_j | \psi_k\rangle = \langle\langle a^*|W(\psi_j, \psi_k)\rangle\rangle,$$

$$\langle\widehat{B}\psi_j | \psi_k\rangle = \langle\langle b|W(\psi_j, \psi_k)\rangle\rangle,$$

and hence it follows from the extended Moyal identity (6.50) that

$$\operatorname{Tr}(\widehat{A}\widehat{B}) = \sum_{j,k}\langle\langle a^*|W(\psi_j, \psi_k)\rangle\rangle\langle\langle b|W(\psi_j, \psi_k)\rangle\rangle^*.$$

Since (ψ_j) is an orthonormal basis the vectors $(2\pi\hbar)^{n/2}W(\psi_j, \psi_k)$ also form an orthonormal basis (see the remark following formula (6.51)), hence the Bessel identity (6.1) allows us to write the equality above as

$$\operatorname{Tr}(\widehat{A}\widehat{B}) = \left(\frac{1}{2\pi\hbar}\right)^n \langle\langle a^*|b\rangle\rangle$$

which is formula (6.70).

(ii) It immediately follows from formula (6.70) using (6.68) and (6.69), recalling that if $\widehat{A} = \operatorname{Op}^W(a)$ then $\widehat{A}^\dagger = \operatorname{Op}^W(a^*)$. $\qquad\square$

Part (i) of the result above allows us — at last! — to express the trace of a Weyl operator in terms of its symbol provided that the latter is absolutely integrable.

Corollary 107. *Let $\widehat{A} = \operatorname{Op}^W(a)$ be a trace class operator (hence $a \in L^2(\mathbb{R}^n)$). If in addition we have $a \in L^1(\mathbb{R}^n)$, then*

$$\operatorname{Tr}(\widehat{A}) = \left(\frac{1}{2\pi\hbar}\right)^n \int a(z)d^{2n}z. \qquad (6.74)$$

Proof. It is equivalent to prove that the symplectic Fourier transform $a_\sigma = F_\sigma a$ satisfies

$$\operatorname{Tr}(\widehat{A}) = a_\sigma(0). \qquad (6.75)$$

Writing $\widehat{A} = \widehat{B}\widehat{C}$ where $\widehat{B}$ and $\widehat{C}$ are Hilbert–Schmidt operators we have

$$\operatorname{Tr}(\widehat{A}) = \left(\frac{1}{2\pi\hbar}\right)^n \int b(z)c(z)d^{2n}z,$$

hence it suffices to show that

$$a_\sigma(0) = \left(\frac{1}{2\pi\hbar}\right)^n \int b(z)c(z)d^{2n}z.$$

We have, in view of formula (4.62) giving the twisted symbol of the product of two Weyl operators,

$$a_\sigma(z) = \left(\frac{1}{2\pi\hbar}\right)^n \int e^{\frac{i}{2\hbar}\sigma(z,z')} b_\sigma(z - z') c_\sigma(z') d^{2n} z',$$

and hence, using the Plancherel identity

$$\int F_{\sigma,\eta} a(z) F_{\sigma,\eta} b(-z) d^{2n} z = \int a(z) b(z) d^{2n} z$$

for $F_{\sigma,\eta}$,

$$a_\sigma(0) = \left(\frac{1}{2\pi\hbar}\right)^n \int b_\sigma(-z') c_\sigma(z') d^{2n} z'$$

$$= \left(\frac{1}{2\pi\hbar}\right)^n \int b(z) c(z) d^{2n} z,$$

which proves the formula (6.75). $\square$

Chapter 7

Varying Planck's Constant

The quantum world is very sensitive to the value of Planck's constant h. For instance, as discussed by Yang [229], doubling the value of h would result in a radical change on the geometric sizes and apparent colors, the solar spectrum and luminosity, as well as the energy conversion between light and materials. In the present chapter, we study the mathematical consequences of (possibly infinitesimal) variations of Planck's constant on the quantum states, mixed or pure. Doing this we have in mind the (still controversial) hypothesis that Planck's constant could indeed have been varying after the Big Bang. If, as it is believed, h had been steadily decreasing since the emergence of the early universe it would imply that there is an ongoing transition from quantum to classical, mixed states becoming pure, and pure states evolving into classical states. We would thus have an emergence of the classical from the quantum. Reversing this scenario in time could also imply that the early Universe was much more "quantum" than it is today. These hypotheses should of course be confirmed by cosmological experimental data.

7.1. Variability of Physical Constants

Dirac had speculated in [63] that physical constants such as the gravitational constant G or the fine-structure constant α might be subject to change over time. In fact, the first investigations of time-varying constants were made by Lord Kelvin who was interested in possible time variations of the speed of light (see the discussion in [16]). In this section we provide a short summary of a few recent advances, with a special focus on the possible variability of Planck's constant h, which has been explicitly addressed by many authors; see for instance Pegg [236], Hutchin [165], Kentosh and Mohageg [175, 176].

7.1.1. Discussion of units

However, testing the constancy of a physical parameter means going to extraordinary lengths in terms of precision measurements, and is intimately related to choices of *unit systems*. The physicist M. Duff has remarked [68–70] that all the fundamental physical dimensions can be expressed using only one unit: mass. Duff first noticed the obvious, namely that length can be expressed as times using c, the velocity of light, as a conversion factor. One can therefore take $c = 1$, and measure lengths in seconds. The second step is to use the relation $E = h\nu$ which relates energy to frequency, the latter being the inverse of time. We can thus measure a time using the inverse of energy. But energy is equivalent to mass as shown by Einstein, so that a time can be measured by the inverse of a mass. Setting $c = h = 1$ we have thus reduced all the fundamental dimensions to one: mass. A further step consists in choosing a reference mass such that the gravitational constant is equal to one: $G = 1$. Summarizing, we have got a theoretical system of units in which $c = h = G = 1$. There are other ways to define irreducible unit systems; the topic of reduction of unit systems has actually been discussed many times also in the older literature. Already Stoney [271], noting that electric charge is quantized, derived units of length, time, and mass in 1881 by normalizing G, c, and e to unity; Planck suggested in 1898–1899 that it would suffice to use G, c, and h to define length, mass, and time units. His proposal led to what is called today Planck's length $\ell_P = \sqrt{Gh/c^3}$ and Planck mass $M_P = \sqrt{hc/G}$ and time $T_P = \sqrt{Gh/c^5}$.

7.1.2. The fine structure constant

Now, a very important — and somewhat mysterious — physical parameter is the fine-structure constant $\alpha = e^2/2\hbar c$; it is a dimensionless number whose inverse has the approximate value 137. This constant measures the strength of interactions between light and matter. It can be expressed as a combination of three physical constants: the charge e on an electron, the speed c of light, and Planck's constant h. And here is the crux: many scientists have suggested that the fine structure constant might not be constant, but could vary over time and space. It started with the discovery made at an uranium mine in Oklo (Gabon) of a natural nuclear reactor. The Oklo reactor is well known to give limits on variation of the fine-structure constant over the period since the reactor was running, which is ca. 1.8 billion years (see Barrow's book [15] for a detailed and well-written account of the "Oklo story"). These findings spurred attempts to measure variations of α using cosmological methods. In 1999, a team of astronomers using the Keck observatory in Hawaii reported that measurements of light absorbed by very distant galaxy-like objects in space called quasars — which are so far away that we see them today as they looked billions of years ago — suggest that the value of the fine-structure constant was once slightly different from what it is today. Experiments can in principle only put an upper bound on the relative change per year. For the fine-structure constant, this upper bound seems to be comparatively low, at roughly 10^{-17} per year.

The fact that α could be drifting is held as controversial by some physicists. The quest for testing this hypothesis is ongoing, and very interesting advances have been obtained by John Webb and his collaborators; among the most cited papers are [22, 215, 216, 286–288, 301]. But if true, it must mean that at least one of the three fundamental constants $e, \hbar, c$ that constitute it must vary (see for instance [17]).

Mohageg and Kentosh's experiment. To test whether Planck's constant is really constant, Mohageg and Kentosh [175] of California State University in Northridge turned to the same GPS systems that car drivers use to find their way home. GPS relies on the most accurate timing devices we currently possess: atomic clocks. These count the passage of time according to frequency of the radiation that atoms emit when their electrons jump between different energy levels. Kentosh and Mohageg looked through a year's worth of GPS data and found that the corrections depended in an unexpected way on a satellite's distance above the Earth. This small discrepancy could be due to atmospheric effects or random error, but it could also arise from a position-dependent Planck's constant. The reason why Kentosh and Mohageg asked whether h is really constant, however, is not just because it is a central number for modern physics, but because h also appears in the fine-structure constant discussed above.

So, what did they discover? Well, if there is any difference in h it would have to be really tiny. After careful analysis of the data from seven highly stable GPS satellites, Kentosh and Mohageg conclude that h is identical at different locations to an accuracy of seven parts in a thousand. In other words, if h were a one-meter measuring stick, two sticks in different places anywhere in the world do not differ by more than 7 mm.

Atomic clocks. Atomic clocks are among the most accurate scientific instruments ever built, their precision reaching up (for the moment) to a 10^{-18} fractional inaccuracy. This opens the intriguing prospect of using clocks to variations of constants of Nature. As pointed out to me by the physicist Freeman Dyson (private communication), any increase in the accuracy of atomic clocks will also lead to increased accuracy in the measurements of possible fluctuations of Planck's constant. Kentosh and Mohageg fixed on whether h depends on where (not when) you measure it. If h changes from place to place, so do the frequencies, and thus the "ticking rate", of atomic clocks. And any dependence of h on location would translate as a tiny timing discrepancy between different GPS clocks.

We will from now on work in units where the change in Planck's constant is made while keeping other systems of units fixed.

7.2. First Remarks

We begin by discussing from an elementary point of view some rather pedestrian consequences of a change of Planck's constant. We thereafter outline some (obvious) modifications of the Weyl–Wigner–Moyal formalism under such changes. We thereafter show that using Moyal's identity one can already obtain a few intriguing results.

For obvious reasons, we do not use a system of units in which $c = h = G = 1$.

7.2.1. Some elementary observations à *la* Gamow

In his entertaining recent paper [229] Yang, inspired by Gamow's Mr. Tompkins [85], elaborates on some consequences at the macroscopic level of a doubling of Planck's constant. We are following his text and argumentation here, assuming that Planck's constant h has undergone a relative change of δ, thus becoming

$$h^\delta = (1 + \delta)h;$$

we will suppose that δ is small: $|\delta| \ll 1$ and can take both positive and negative values.

Let us now consider a hydrogen atom; its Bohr radius (viewed as the average "distance" between the proton and the electron in Bohr's model) is given by

$$a_0 = \frac{4\pi\varepsilon_0\hbar^2}{m_e c^2},$$

where ε_0 is the permittivity of free space and m_e the mass of the electron. Assuming that masses and the velocity c of light are conserved this radius will become, when one replaces $\hbar$ with $\hbar^\delta$,

$$a_0^\delta = (1+\delta)^2 a_0 \approx (1+2\delta)a_0$$

hence the relative change of Bohr's radius is approximately 2δ. Extrapolating to other atoms, it is reasonable to assume that their size would also increase or decrease by a factor $1 + 2\delta$. Since all atoms would change approximately equally, we would not be able to detect this change of size. However, having supposed that the mass of the Earth remains the same, its radius R_E would change by a factor $(1+\delta)^4 \approx (1+4\delta)$ and become

$$R_E^\delta \approx R_E(1+4\delta).$$

Suppose for instance that $\delta = 2.5\%$; this leads to an increase of radius of the Earth into by 10%. Due to the conservation of angular momentum the spinning of the Earth would slow down, so that the length of a day would increase to approximately 27 h; since the masses of the Earth and the Sun are not affected and there would be less than 365 days in a year! The acceleration $g \approx 9.81$ ms^{-2} of gravity would become

$$g^\delta = \frac{GM_E}{(R_E^\delta)^2} \approx g(1-8\delta);$$

for $\delta = 2.5\%$ it would thus decrease by 20%; the density of the atmosphere near the surface of the Earth would thus become substantially smaller because of the weaker gravity.

7.2.2. Dependence on $\hbar$ of the Wigner function

Setting aside for the moment the debate on whether Planck's constant can vary or not, consider the following situation: we have an unknown quantum state, on which we perform a quorum of measurements in order to determine its density matrix $\hat{\rho}$. Now, one should be aware of the fact that this density matrix will not be determined directly by these experiments; what one does is to measure by, say, a homodyne quantum tomography, certain properties of that system. Quantum homodyne tomography originates from the observation by Vogel and Risken [282] that the probability distributions determined by homodyne detection are just the Radon transforms of the Wigner function of the density matrix, and that the latter allows us to infer the density matrix using the Weyl correspondence. The method works as follows: suppose now that we have been able to determine a statistical function $\rho(x, p)$ of the position and momentum variables (called "quadratures" in this context), yielding the properties of the quantum system under investigation. If we now identify this function $\rho(x, p)$ with the Wigner function

$$\rho(x, p) = \sum_j \alpha_j W\psi_j(x, p) \tag{7.1}$$

of the corresponding density matrix, this relation can be inverted and yields $\widehat{\rho}$; mathematically speaking $\widehat{\rho}$ is just, up to a factor, the Weyl operator

$$\widehat{\rho}\psi(x) = \int e^{\frac{i}{\hbar}p(x-y)} \rho(\tfrac{1}{2}(x+y), p)\psi(x)d^n y. \tag{7.2}$$

Now, two essential observations. The first is that when using this procedure we assume quite explicitly that we are using the Weyl–Wigner–Moyal formalism: we identify the function ρ with the Wigner function, and "quantize" it thereafter using the Weyl transform. This is very good, of course, but one should keep in mind that there are other possible representations in quantum mechanics; a physically very interesting one is for instance the Born–Jordan quantization scheme (see Appendix C). We will not investigate the implications of such a choice here. Let us instead emphasize another problem. Even if we place ourselves in the Weyl–Wigner–Moyal framework we are tacitly assuming that $\hbar = h/2\pi$ has a fixed value in time and space. If it happens that Planck's constant h has another value, η, at another location or at another time, formulas (7.1) and (7.2) would have to be replaced with the different expressions

$$\rho(x, p) = \sum_j \alpha_j W'\psi_j'(x, p), \tag{7.3}$$

where the usual Wigner function should be replaced with

$$W'\psi_j'(x, p) = \left(\frac{1}{2\pi\hbar'}\right)^n \int e^{-\frac{i}{\eta}py} \psi_j'\left(x + \frac{1}{2}y\right) \psi_j'\left(x + \frac{1}{2}y\right)^* d^n y$$

and the corresponding density operator with

$$\hat{\rho}'\psi(x) = \int e^{\frac{i}{\eta}p(x-y)}\rho\left(\frac{1}{2}(x+y),p\right)\psi(x)d^n y. \tag{7.4}$$

7.2.3. A consequence of Moyal's identity

Let η be a real parameter; we assume for the moment that $\eta > 0$. This parameter will play the role of a variable $\hbar = h/2\pi$. For any square integrable ψ, we define the η-Wigner function of ψ by replacing $\hbar$ by η in the usual definition:

$$W_\eta\psi(x,p) = \left(\frac{1}{2\pi\eta}\right)^n \int e^{-\frac{i}{\eta}py}\psi\left(x+\frac{1}{2}y\right)\psi^*\left(x-\frac{1}{2}y\right)d^n y. \tag{7.5}$$

Of course $W_\hbar\psi = W\psi$ (the usual Wigner function). The mathematical properties of $W_\eta\psi$ are of course the same that those of $W\psi$, replacing $\hbar$ everywhere with η. In particular, replacing the $\hbar$-Fourier transform (6.40) with the η-Fourier transform

$$F_\eta\psi(p) = \left(\frac{1}{2\pi\eta}\right)^{n/2} \int e^{-\frac{i}{\eta}px}\psi(x)d^n x, \tag{7.6}$$

the marginal properties (6.38) satisfied by the Wigner function become here

$$\int W_\eta\psi(x,p)d^n p = |\psi(x)|^2, \quad \int W_\eta\psi(x,p)d^n x = |F_\eta\psi(p)|^2 \tag{7.7}$$

(assuming that $\psi \in L^1(\mathbb{R}^n) \cap L^2(\mathbb{R}^n)$).

The Moyal identity we studied in Section 6.3.3 of Chapter 5 can be restated *mutatis mutandis* in the η-dependent case as

$$\int W_\eta\psi(z)W_\eta\phi(z)d^{2n} z = \left(\frac{1}{2\pi\eta}\right)^n |\langle\psi|\phi\rangle|^2 \tag{7.8}$$

for all square integrable functions ψ and ϕ. In particular

$$\int W_\eta\psi(z)^2 d^{2n} z = \left(\frac{1}{2\pi\eta}\right)^n ||\psi||^4. \tag{7.9}$$

Let us now address the following question: for ψ given, can we find ϕ such that $W_\eta\phi = W\psi$ for $\eta \neq \hbar$? The answer is "no"! More generally, we have the following proposition.

Proposition 108. *Let $\psi \in L^2(\mathbb{R}^n)$ be a normalized wavefunction.*

(i) *A pure state $|\psi\rangle$ does not remain a pure state if we vary $\hbar$. There does not exist any state $|\phi\rangle(\phi \in L^2(\mathbb{R}^n))$ such that $W_\eta \phi = W\psi$ if $\eta \neq \hbar$.*

(ii) *Assume that $|\psi\rangle$ becomes a mixed state when $\hbar$ is replaced with η. Then we must have $\eta \leq \hbar$.*

Proof. (i) Assume that $W_\eta \phi = W\psi$; then

$$\int W_\eta \phi(x,p) d^n p = \int W\psi(x,p) d^n p,$$

hence, using the marginal properties (6.38), $|\phi(x)|^2 = |\psi(x)|^2$ so that ϕ and ψ have same norm: $||\phi|| = ||\psi|| = 1$. On the other hand, using the Moyal identity (7.9), the equality $W\psi = W_\eta \phi$ implies that

$$\int W\psi(z)^2 d^{2n} z = \left(\frac{1}{2\pi\hbar}\right)^n ||\psi||^4,$$

$$\int W_\eta \phi(z) d^{2n} z = \left(\frac{1}{2\pi\eta}\right)^n ||\phi||^4,$$

hence we must have $\eta = \hbar$. (ii) Assume that there exist a sequence (ϕ_j) of (normalized) functions in $L^2(\mathbb{R}^n)$ and a sequence of positive constants α_j summing up to one such that $W\psi = \sum_j \alpha_j W_\eta \phi_j$. Proceeding as above we get, using again the marginal properties,

$$||\psi||^2 = \sum_j \alpha_j ||\phi_j||^2. \tag{7.10}$$

On the other hand, squaring $W\psi$ we have

$$(W\psi)^2 = \sum_{j,k} \alpha_j \alpha_k W_\eta \phi_j W_\eta \phi_k,$$

hence, integrating and using respectively the Moyal identity for $(W\psi)^2$ and $W_\eta \phi_j W_\eta \phi_k$, the Cauchy–Schwarz inequality and (7.10), we get

$$\left(\frac{1}{2\pi\hbar}\right)^n ||\psi||^4 = \left(\frac{1}{2\pi\eta}\right)^n \sum_{j,k} \alpha_j \alpha_k |\langle \phi_j | \phi_k \rangle|^2$$

$$\leq \left(\frac{1}{2\pi\eta}\right)^n \sum_{j,k} \alpha_j \alpha_k ||\phi_j||^2 ||\phi_k||^2$$

$$= \left(\frac{1}{2\pi\eta}\right)^n \left(\sum_j \alpha_j ||\phi_j||^2\right)^2$$

$$= \left(\frac{1}{2\pi\eta}\right)^n ||\psi||^4,$$

which implies that $\left(\frac{1}{2\pi\hbar}\right)^n \leq \left(\frac{1}{2\pi\eta}\right)^n$, that is $\eta \leq \hbar$ as claimed. $\square$

A *caveat*: property (ii) in the result above does not say that a pure state automatically becomes a mixed state if we decrease Planck's constant. It merely says that if a pure state becomes mixed, it can only happen if Planck's constant has decreased. We will discuss these delicate and difficult questions in next section.

7.3. The Quantum Bochner Theorem

Given a trace class operator it is notoriously difficult to decide whether it is positive or not (and hence a density matrix). In the present section, we give a general criterion which goes back to the work of Kastler [170], and Loupias and Miracle-Sole [191, 192]. It has been tackled by many authors, notable contributions being those of Narcowich [218–220], Narcowich and O'Connell [217], Bröckner and Werner [36], Werner [293], and Dias and Prata [54]. We begin by briefly recalling Bochner's theorem on the Fourier transform of probability density, and thereafter study its extension to the quantum case, which leads to the "KLM conditions" (the letters K, L, and M standing for Kastler, Loupias, and Miracle-Sole).

7.3.1. Bochner's classical theorem

Bochner's famous theorem about the Fourier transform of a probability density says that a (complex-valued) function f on $\mathbb{R}^m$, continuous at the origin and such that $f(0) = 1$ is the characteristic function of a probability density on $\mathbb{R}^m$ if and only if it is of *positive type*, that is, if for all choices of points $z_1, \ldots, z_N \in \mathbb{R}^m$ the symmetric $N \times N$ matrix

$$(f(z_j - z_k))_{1 \leq j,k \leq N} \tag{7.11}$$

is positive semidefinite (that is, its eigenvalues are all ≥ 0).

Let us introduce two modifications of the symplectic Fourier transform F_σ. First we allow the latter to depend on an arbitrary parameter $\eta \neq 0$

and set

$$F_{\sigma,\eta}f(z) = f_{\sigma,\eta}(z) = \left(\frac{1}{2\pi\eta}\right)^n \int e^{-\frac{i}{\eta}\sigma(z,z')}f(z')d^{2n}z'. \tag{7.12}$$

It coincides with F_σ when $\eta = \hbar$. We also define the reduced symplectic Fourier transform $F_\Diamond$ by

$$f_\Diamond(z) = F_\Diamond f(z) = \int e^{-i\sigma(z,z')}f(z')d^{2n}z'. \tag{7.13}$$

Obviously $F_\Diamond f$ and $F_{\sigma,\eta}f = f_{\sigma,\eta}$ are related by the simple formula

$$f_\Diamond(z) = (2\pi\eta)^n f_{\sigma,\eta}(\eta z). \tag{7.14}$$

With this notation Bochner's theorem on Fourier transforms of probability measures on phase space $\mathbb{R}^{2n}$ can be restated in the following way: a real function f on $\mathbb{R}^{2n}$ is a probability density if and only if its reduced symplectic Fourier transform $f_\Diamond$ is continuous, $\rho_\Diamond(0) = 1$, and for all choices of $z_1, \ldots, z_N \in \mathbb{R}^{2n}$ the $N \times N$ matrix Λ whose entries are the complex numbers $\rho_\Diamond(z_j - z_k)$ is positive semidefinite:

$$(f_\Diamond(z_j - z_k))_{1 \le j,k \le N} \ge 0. \tag{7.15}$$

When condition (7.15) is satisfied one says that the reduced symplectic Fourier transform $f_\Diamond$ is of *positive type*. We are going to generalize Bochner's condition to the case of the Wigner function of a density matrix of a mixed quantum state.

7.3.2. The notion of η-positivity

The notion of η-positivity, due to Kastler [170], generalizes Bochner's condition.

Definition 109. Let $f \in \mathcal{S}'(\mathbb{R}^m)$ and η be a real number; we say that f is of η-*positive type* if for every integer N the $N \times N$ matrix $(\Lambda_{jk})_{1 \le j,k \le N}$ with entries

$$\Lambda_{jk} = e^{-\frac{i\eta}{2}\sigma(z_j,z_k)}f(z_j - z_k)$$

is positive semidefinite for all choices of $(z_1, z_2, \ldots, z_N) \in (\mathbb{R}^m)^N$:

$$(\Lambda_{jk})_{1 \le j,k \le N} \ge 0. \tag{7.16}$$

The condition (7.16) is equivalent to the polynomial inequalities

$$\sum_{1 \le j,k \le N} \zeta_j^* \zeta_k e^{-\frac{i\eta}{2}\sigma(z_j,z_k)} f(z_j - z_k) \ge 0 \tag{7.17}$$

for all $N \in \mathbb{N}$, $\zeta_j, \zeta_k \in \mathbb{C}$, and $z_j, z_k \in \mathbb{R}^m$. If f is of η-positive type then it is also of $(-\eta)$-positive type as is immediately seen by taking the complex conjugate of the left-hand side of (7.17). Notice that when one chooses $\eta = 0$ one recovers the Bochner condition (7.15) hence 0-positivity is just positivity in the usual sense.

For future use let us see how these conditions can be written when the function f is the reduced symplectic Fourier transform of some symbol a defined on phase space $\mathbb{R}^{2n}$. First, it is clear that we must have

$$\sum_{1 \le j,k \le N} \zeta_j^* \zeta_k e^{-\frac{i\eta}{2}\sigma(z_j,z_k)} a_\Diamond(z_j - z_k) \ge 0. \tag{7.18}$$

When $\eta \ne 0$ we can rewrite these conditions using the symplectic η-Fourier transform: replacing (z_j, z_k) with $\eta^{-1}(z_k, z_j)$ and noting that $\sigma(z_k, z_j) = -\sigma(z_j, z_k)$ the conditions (7.16) are equivalent to

$$(\Lambda'_{jk}(z_j, z_k))_{1 \le j,k \le N} \ge 0, \tag{7.19}$$

where

$$\Lambda'_{jk}(z_j, z_k) = e^{\frac{i}{2\eta}\sigma(z_j,z_k)} a_{\sigma,\eta}(z_j - z_k). \tag{7.20}$$

The polynomial conditions (7.17) become in this case

$$\sum_{1 \le j,k \le N} \zeta_j^* \zeta_k e^{-\frac{i}{2\eta}\sigma(z_j,z_k)} a_{\sigma,\eta}(z_j - z_k) \ge 0. \tag{7.21}$$

The conditions (7.19) are in a sense very elegant, but they are — as we will see — not that easy to use in practice. Worse, they cannot be directly implemented numerically since they do not even consist in a discrete set of inequalities since the z_j are allowed to vary continuously in $\mathbb{R}^{2n}$. Fortunately they can be shown to be equivalent to a set of countable conditions (see below) which can be approximated to arbitrary accuracy by finite sets of conditions.

7.3.3. The KLM conditions

We are now going to prove an essential result (the "quantum Bochner theorem") originally due to Kastler [170], and Loupias and Miracle-Sole

[191, 192]; also see [231–233] for different points of view. The proof we will give (and which is partially based on the discussions in [217, 219, 293]) is much simpler than the original proof of Kastler who uses the theory of C^*-algebras. For this we will need a technical result from linear algebra ("Schur's product theorem"), which says that the Hadamard product (= entrywise product) of two positive semidefinite matrices is also positive semidefinite.

Lemma 110 (Schur). *Let $A = (A_{jk})_{1\leq j,k\leq N}$ and $B = (B_{jk})_{1\leq j,k\leq N}$ be two symmetric matrices with the same finite dimension N. If A and B are both positive semidefinite, then their Hadamard product*

$$A \circ B = (A_{jk}B_{jk})_{1\leq j,k\leq N}$$

is also positive semidefinite.

For a proof of this classical result from linear algebra, see for instance [12].

Let us now state and prove the quantum version of Bochner's theorem: Let $\widehat{\rho}$ be a self-adjoint trace class operator on $L^2(\mathbb{R}^n)$:

$$\widehat{\rho}\psi = \sum_j \alpha_j \langle \psi_j | \psi \rangle \psi_j.$$

Theorem 111. *Let $\rho = \sum_j \alpha_j W\psi_j$ be the Wigner function of $\widehat{\rho}$. We have $\widehat{\rho} \geq 0$ if and only if the reduced symplectic Fourier transform $\rho_\Diamond$ satisfies the KLM conditions: (i) $\rho_\Diamond$ is continuous and $\rho_\Diamond(0) = 1$ and (ii) $\rho_\Diamond$ is of η-positive type.*

Proof. Let us first show that the KLM conditions (i) and (ii) are necessary. Assume that $\widehat{\rho} \geq 0$; then

$$\rho = \sum_j \alpha_j W_\eta \psi_j \tag{7.22}$$

for a family of normalized functions $\psi_j \in L^2(\mathbb{R}^n)$, the coefficients α_j being ≥ 0 and summing up to one. It is thus sufficient to show that the Wigner transform $W_\eta \psi$ of an arbitrary $\psi \in L^2(\mathbb{R}^n)$ is of η-positive type.

This amounts to showing that for all $(z_1, \ldots, z_N) \in (\mathbb{R}^{2n})^N$ and all $(\zeta_1, \ldots, \zeta_N) \in \mathbb{C}^N$ we have

$$I_N(\psi) = \sum_{1 \leq j,k \leq N} \zeta_j \zeta_k^* e^{-\frac{i}{2\eta}\sigma(z_j,z_k)} F_{\sigma,\eta} W_\eta \psi(z_j - z_k) \geq 0 \qquad (7.23)$$

for every complex vector $(\zeta_1, \ldots, \zeta_N) \in \mathbb{C}^N$ and every sequence $(z_1, \ldots, z_N) \in (\mathbb{R}^{2n})^N$ (see condition (7.21)). Since the η-Wigner function $W_\eta \psi$ and the η-ambiguity A_η function are obtained from each other by the symplectic η-Fourier transform

$$F_{\sigma,\eta} a(z) = \left(\frac{1}{2\pi\eta}\right)^n \int e^{-\frac{i}{\eta}\sigma(z,z')} a(z') d^{2n} z'$$

we have

$$I_N(\psi) = \sum_{1 \leq j,k \leq N} \zeta_j \zeta_k^* e^{-\frac{i}{2\eta}\sigma(z_j,z_k)} A_\eta \psi(z_j - z_k).$$

Let us prove that

$$I_N(\psi) = \left(\frac{1}{2\pi\eta}\right)^n \| \sum_{1 \leq j \leq N} \zeta_j \widehat{D}_\eta(z_j) \psi \|^2, \qquad (7.24)$$

where $\widehat{D}_\eta(z_j)$ is the η-dependent version of the displacement operator $\widehat{D}(z_j)$:

$$\widehat{D}_\eta(z_j)\psi(x) = e^{\frac{i}{\eta}(p_j x - \frac{1}{2} p_j x_j)} \psi(x - x_j); \qquad (7.25)$$

the inequality (7.23) will follow. Taking into account the fact that $\widehat{D}_\eta(-z_k)^\dagger = \widehat{D}_\eta(z_k)$ and using the relation (6.61) which becomes here

$$\widehat{D}_\eta(z_0)\widehat{D}_\eta(z_1) = e^{\frac{i}{2\eta}\sigma(z_0,z_1)} \widehat{D}_\eta(z_0 + z_1) \qquad (7.26)$$

we have, expanding the square in the right-hand side of (7.24),

$$\left\| \sum_{1 \leq j \leq N} \zeta_j \widehat{D}_\eta(z_j)\psi \right\|^2 = \sum_{1 \leq j,k \leq N} \zeta_j \zeta_k^* \langle \widehat{D}_\eta(z_k)\psi | \widehat{D}_\eta(z_j)\psi \rangle$$

$$= \sum_{1 \leq j,k \leq N} \zeta_j \zeta_k^* \langle \widehat{D}_\eta(-z_j)\widehat{D}_\eta(z_k)\psi | \psi \rangle$$

$$= \sum_{1 \leq j,k \leq N} \zeta_j \zeta_k^* e^{-\frac{i}{2\eta}\sigma(z_j,z_k)} \langle \widehat{D}_\eta(z_k - z_j)\psi | \psi \rangle;$$

in view of formula (6.59), which becomes here

$$A_\eta \psi(z) = \left(\frac{1}{2\pi\eta}\right)^n \langle \widehat{D}_\eta(z)\psi|\psi\rangle$$

we thus have

$$\left\|\sum_{1\leq j\leq N} \zeta_j \widehat{D}_\eta(z_j)\psi\right\|^2 = (2\pi\eta)^n \sum_{1\leq j,k\leq N} \zeta_j \zeta_k^* e^{-\frac{i}{2\eta}\sigma(z_j,z_k)} A_\eta \psi(z_j - z_k)$$

proving the equality (7.24). Let us now show that, conversely, the conditions (i) and (ii) are sufficient, i.e., they imply that $\langle\psi|\widehat{\rho}\psi\rangle \geq 0$ for all $\psi \in L^2(\mathbb{R}^n)$; equivalently (see formula (6.42) in Theorem 10.3)

$$\int \rho(z)W_\eta\psi(z)d^{2n}z \geq 0 \qquad (7.27)$$

for $\psi \in L^2(\mathbb{R}^n)$. Let us set, as above,

$$\Lambda'_{jk} = e^{\frac{i}{2\eta}\sigma(z_j,z_k)} a_{\sigma,\eta}(z_j - z_k), \qquad (7.28)$$

where z_j and z_k are arbitrary vectors in $\mathbb{R}^{2n}$. To say that $a_{\sigma,\eta}$ is of η-positive type means that the matrix $\Lambda' = (\Lambda'_{jk})_{1\leq j,k\leq N}$ is positive semidefinite; choosing $z_k = 0$ and setting $z_j = z$ this means that every matrix $(a_{\sigma,\eta}(z))_{1\leq j,k\leq N}$ is positive semidefinite. Setting now

$$\Gamma_{jk} = e^{\frac{i}{2\eta}\sigma(z_j,z_k)} F_{\sigma,\eta} W_\eta \psi(z_j - z_k)$$

$$= e^{\frac{i}{2\eta}\sigma(z_j,z_k)} A_\eta \psi(z_j - z_k)$$

the matrix $\Gamma_{(N)} = (\Gamma_{jk})_{1\leq j,k\leq N}$ is positive semidefinite. Let us now write

$$M_{jk} = A_\eta\psi(z_j - z_k)\rho_{\sigma,\eta}(z_j - z_k);$$

we claim that the matrix $(M_{jk})_{1\leq j,k\leq N}$ is positive semidefinite as well. In fact, it is the Hadamard product of the positive semidefinite matrices $(M'_{jk})_{1\leq j,k\leq N}$ and $(M''_{jk})_{1\leq j,k\leq N}$ where

$$M'_{jk} = e^{\frac{i}{2\eta}\sigma(z_j,z_k)} A_\eta\psi(z_j - z_k),$$

$$M''_{jk} = e^{-\frac{i}{2\eta}\sigma(z_j,z_k)} \rho_{\sigma,\eta}(z_j - z_k),$$

and Lemma 110 implies that $(M_{jk})_{1 \le j,k \le N}$ is then also positive semidefinite. It follows from Bochner's theorem that the function b defined by

$$b_{\sigma,\eta}(z) = A_\eta \psi(z) \rho_{\sigma,\eta}(-z) = F_{\sigma,\eta} W \psi(z) \rho_{\sigma,\eta}(-z)$$

is a probability density; in particular we must have $b(0) \ge 0$. Integrating the equality above with respect to z we get, using the Plancherel formula,

$$\int F_{\sigma,\eta} a(z) F_{\sigma,\eta} b(-z) d^{2n} z = \int a(z) b(z) d^{2n} z, \qquad (7.29)$$

satisfied by the symplectic η-Fourier transform,

$$(2\pi\eta)^n b(0) = \int A_\eta \psi(z) \rho_{\sigma,\eta}(-z) d^{2n} z$$

$$= \int W_\eta \psi(z) \rho(z) d^{2n} z$$

hence the inequality (7.27) since $b(0) \ge 0$. $\qquad\qquad\square$

7.4. Application to Quantum States

7.4.1. Preliminaries; Narcowich's lemma

Let $\rho(z)$ be a real function on phase space $\mathbb{R}^{2n}$; it will be viewed as a quasi-probability density, i.e., we assume that

$$\int \rho(z) d^{2n} z = 1 \qquad (7.30)$$

and that the marginal identities (6.38) hold:

$$\int \rho(x,p) d^n p = |\psi(x)|^2, \quad \int \rho(x,p) d^n x = |F\psi(p)|^2.$$

To guarantee the existence of the covariances

$$\Delta(z_\alpha, z_\beta) = \int (z_\alpha - \langle z_\alpha \rangle)(z_\beta - \langle z_\beta \rangle) \rho(z) d^{2n} z \qquad (7.31)$$

and the variances

$$(\Delta z_\alpha)^2 = \Delta(z_\alpha, z_\alpha) = \int (z_\alpha - \langle z_\alpha \rangle)^2 \rho(z) d^{2n} z, \qquad (7.32)$$

we assume as in Chapter 5 that ρ satisfies the condition

$$\int (1 + |z|^2)|\rho(z)|d^{2n}z < \infty. \tag{7.33}$$

Let us define as before the covariance matrix

$$\sum = (\Delta(z_\alpha, z_\beta))_{1 \le \alpha, \beta \le 2n}.$$

Here is an example: let $\rho(z)$ be given by the formula

$$\rho(z) = (2\pi)^{-n}\sqrt{\det \Sigma^{-1}}e^{-\frac{1}{2}\Sigma^{-1}z^2}, \tag{7.34}$$

where Σ is a symmetric positive definite real $2n \times 2n$ matrix. A straight-forward calculation involving Gaussian integrals shows that Σ is precisely the associated covariance matrix. We will see later that $\rho(z)$ is the Wigner function of a quantum state if Σ satisfies a certain condition related to the uncertainty principle. However, $\rho(z)$ can always be viewed as a classical probability distribution.

Recall that we used in Chapter 5 Williamson's symplectic diagonalization theorem (Proposition 83): if M is a real symmetric positive definite $2n \times 2n$ matrix, we can find a (non-unique) $S \in \mathrm{Sp}(2n, \mathbb{R})$ such that

$$M = S^T \begin{pmatrix} \Lambda_\sigma & 0 \\ 0 & \Lambda_\sigma \end{pmatrix} S,$$

where Λ_σ is the diagonal matrix whose non-zero entries are the symplectic eigenvalues of the matrix M, i.e., the moduli of the eigenvalues $\pm i\lambda$ of the matrix JM.

In addition to Williamson's symplectic diagonalization result we will need the following technical result.

Lemma 112 (Narcowich). *Let $f(z)$ be a twice differentiable function defined on phase space. If f is of η-positive type with $\eta \ne 0$ the matrix $f'' = D^2 f$ of second derivatives of the function f in the variables z_α ($1 \le \alpha \le 2n$) satisfies*

$$-f''(0) + \frac{i\eta}{2}J \ge 0. \tag{7.35}$$

Proof. We sketch the proof and refer to Narcowich [218, Lemma 2.1] for computational details (also see [112, Lemma 310]). For arbitrary

$(\zeta_1, \ldots, \zeta_m) \in \mathbb{C}^m$ and $\varepsilon \in \mathbb{R}$ let us set

$$R(\varepsilon) = \sum_{j,k=1}^{m} \zeta_j^* \zeta_k e^{-\frac{i\varepsilon^2}{2\eta} \sigma(z_j, z_k)} f(\varepsilon(z_j - z_k)).$$

If f is of η-positive type we have $R(\varepsilon) \geq 0$ for every value of ε; choosing now the ζ_j such that $\sum_j \zeta_j = 0$ we have $R(0) = 0$ and $R''(0) \geq 0$. A straightforward but somewhat lengthy calculation now shows that

$$R''(0) = u^T(-2f''(0) + i\eta^{-1}J)u,$$

where $u = \sum_j \zeta_j z_j \in \mathbb{C}^{2n}$. The ζ_j and z_j being chosen arbitrarily we thus have $-2f''(0) + i\hbar\eta^{-1}J \geq 0$, which is condition (7.35). $\square$

7.4.2. A necessary (but not sufficient) condition for a state to be "quantum"

We are going to prove an essential result, which goes back to Narcowich [220]. It says that the covariance matrix of a quantum state must satisfy a certain condition which implies — but is stronger than — the Robertson–Schrödinger uncertainty principle

$$(\Delta X_j)^2 (\Delta P_j)^2 \geq \Delta(X_j, P_j)^2 + \tfrac{1}{4}\hbar^2$$

(see (5.2)).

We recall that it is assumed that the Wigner function ρ satisfies the condition

$$\int (1 + |z|^2) |\rho(z)| d^{2n} z < \infty. \tag{7.36}$$

This implies, among other things, that the Fourier transform (and hence also the symplectic Fourier transform) of ρ is twice continuously differentiable. In fact, writing

$$F\rho(z) = \left(\frac{1}{2\pi\hbar}\right)^n \int e^{-\frac{i}{\hbar}zz'} \rho(z') d^{2n} z'$$

we have

$$\partial_{z_\alpha} F\rho = -\frac{i}{\hbar} F(z_\alpha \rho), \quad \partial_{z_\alpha} \partial_{z_\beta} F\rho = \left(-\frac{i}{\hbar}\right)^2 F(z_\alpha z_\beta \rho)$$

and hence

$$|\partial_{z_\alpha} F\rho(z)| \leq \frac{1}{\hbar} \left| \int z_\alpha \rho(z) d^{2n} z \right| < \infty,$$

$$|\partial_{z_\alpha} \partial_{z_\beta} F\rho(z)| \leq \left(\frac{1}{\hbar}\right)^2 \left| \int\int z_\alpha z_\beta \rho(z) d^{2n} z \right| < \infty.$$

Proposition 113. *Suppose that the phase space function ρ with associated covariance matrix Σ is the η-Wigner transform of a density matrix $\widehat{\rho}$.*

(i) *We have s*

$$\Sigma + \frac{i\eta}{2} J \geq 0 \tag{7.37}$$

(i.e., the Hermitian matrix $\Sigma + \frac{i\eta}{2} J$ is positive semidefinite); this condition is equivalent to the inequality

$$|\eta| \leq 2\lambda_{\min}, \tag{7.38}$$

where $\lambda_{\min}$ is the smallest symplectic eigenvalue of Σ.

(ii) *If the conditions $(7.37) - (7.38)$ hold for one value of η, then they hold for every $\eta' < \eta$.*

Proof. (i) That $\Sigma + \frac{i\eta}{2} J$ is Hermitian is clear: since the adjoint of J is $-J$ we have

$$\left(\Sigma + \frac{i\eta}{2} J\right)^\dagger = \Sigma^\dagger + \left(\frac{i\eta}{2} J\right)^\dagger = \Sigma + \frac{i\eta}{2} J.$$

We next remark that $\Sigma = \Sigma_0$ where Σ_0 is the covariance matrix of $\rho_0(z) = \rho(z + \langle z \rangle_\rho)$: we have $\langle z \rangle_0 = 0$ and hence

$$\Delta(x_j, x_k)_0 = \int x_j x_k \rho_0(z) d^{2n} z = \Delta(x_j, x_k)_\rho;$$

similarly $\Delta(x_j, p_k)_0 = \Delta(x_j, p_k)_\rho$ and $\Delta(p_j, p_k)_0 = \Delta(p_j, p_k)_\rho$. It is thus sufficient to prove the result for the density operator $\widehat{\rho}_0$. We are going to use Lemma 112 with $f(z) = F_\sigma \rho_0(z) = \rho_{0,\sigma}(z)$. Observing that the symplectic Fourier transform $\rho_{0,\sigma}$ is twice continuously differentiable in view of the argument preceding the statement of the theorem, we have

$$\hbar^2 \rho_{0,\sigma}''(0) = (2\pi\hbar)^{-n} \begin{pmatrix} -\Sigma_{0,pp} & \Sigma_{0,xp} \\ \Sigma_{0,px} & -\Sigma_{0,xx} \end{pmatrix}$$

and hence

$$\eta^2 \rho_{0,\sigma}''(0) = \left(\frac{1}{2\pi\eta}\right)^n J\Sigma_0 J. \qquad (7.39)$$

Since $\hat{\rho}$ is a density matrix we have

$$M = -2\eta^{-1} J\Sigma J + iJ \geq 0;$$

the condition $M \geq 0$ being equivalent to $J^T M J \geq 0$ the inequality (7.37) follows. Let us finally show that the conditions (7.38) and (7.37) indeed are equivalent. Let $\Sigma = S^T D S$ be a symplectic diagonalization of Σ. Since $S^T J S = J$ condition (7.37) is equivalent to

$$D + \frac{i\eta}{2} J \geq 0, \quad D = \begin{pmatrix} \Lambda & 0 \\ 0 & \Lambda \end{pmatrix}.$$

The characteristic polynomial of $D + \frac{i\eta}{2} J$ is

$$P(\lambda) = \begin{vmatrix} \Lambda - \lambda I_n & \frac{i\eta}{2} I_n \\ -\frac{i\eta}{2} I_n & \Lambda - \lambda I_n \end{vmatrix}$$

$$= \det \left[(\Lambda - \lambda I_n)^2 - \frac{1}{4}\eta^2 I_n \right];$$

the matrix Λ being diagonal, the zeroes λ of $P(\lambda)$ are the solutions of the n equations $(\lambda_j - \lambda)^2 - \frac{1}{4}\eta^2 = 0$ that is $\lambda_j - \lambda = \pm \frac{1}{2}|\eta|$. Since $\lambda \geq 0$ we must have $\lambda_j \geq \frac{1}{2}|\eta|$ for all j hence (7.38). Property (ii) immediately follows from (7.38); it can also be proved directly: setting $\eta' = r\eta$ with $0 < r \leq 1$ we have

$$\sum + \frac{i\eta'}{2} J = (1-r) \sum + r\left(\sum + \frac{i\eta}{2} J \right) \geq 0. \qquad \Box$$

The relation of the result above with the uncertainty principle comes from the following observation (see [220; 112, Chapter 13; 131]): the relation

$$\sum + \tfrac{1}{2}i\eta J \geq 0; \tag{7.40}$$

(which we will call the "strong uncertainty principle") implies the Robertson–Schrödinger inequalities

$$(\Delta x_j)_{\hat{\rho}}^2 (\Delta p_j)_{\hat{\rho}}^2 \geq (\Delta(x_j, p_j)_{\hat{\rho}})^2 + \tfrac{1}{4}\eta^2 \tag{7.41}$$

$(j = 1, \ldots, n)$ and

$$(\Delta x_j)_{\psi}^2 (\Delta p_k)_{\psi}^2 \geq 0 \quad \text{if } j \neq k.$$

It is however essential to note that (7.40) and (7.41) are *not equivalent*. Recall the following counterexample from Chapter 5: the symmetric matrix

$$\sum = \begin{pmatrix} 1 & -1 & 0 & 0 \\ -1 & 1 & 0 & 0 \\ 0 & 0 & 1 & 0 \\ 0 & 0 & 0 & 1 \end{pmatrix}$$

is not a quantum covariance matrix in spite of the fact that the inequalities (7.41) are trivially satisfied for the choice $\eta = 1$ (they are in fact equalities). The matrix $\Sigma + iJ$ is nevertheless indefinite (its determinant is -1); the matrix Σ is not even invertible since $\det \Sigma = 0$.

It is also essential to remark that the condition (7.40) is *necessary* but *not sufficient* for a phase space function to be the Wigner function of a quantum state. Let us check this on the following example due to Narcowich and O'Connell [217] and which is further discussed in [127]. Consider the function $f(x, p)$ defined for $n = 1$ by

$$f(x, p) = (1 - \tfrac{1}{2}ax^2 - \tfrac{1}{2}bx^2)e^{-(a^2x^4 - b^2p^4)},$$

where a and b are positive constants such that $ab \geq \tfrac{1}{4}\hbar^2$. Define now

$$\rho(x, p) = \frac{1}{2\pi} \iint e^{-i(xx' + pp')} f(x', p')dp'dx';$$

since f is an even function ρ is real. In view of the Fourier inversion formula we have

$$f(x, p) = \iint e^{i(xx' + pp')} \rho(x', p')dp'dx'$$

and hence

$$\iint \rho(x,p)dpdx = f(0,0) = 1$$

so that $\rho(x,p)$ is a candidate for being the Wigner function of some density matrix. Calculating the covariance matrix Σ associated with ρ, one finds after some tedious but straightforward calculations (see [217, pp. 4–5]) that $\Sigma + \frac{i\hbar}{2}J \geq 0$. However, ρ cannot be the Wigner function of a density matrix $\widehat{\rho}$ for if this were the case we would have $\langle p^4 \rangle_{\widehat{\rho}} \geq 0$; but by definition of ρ

$$\langle p^4 \rangle_{\widehat{\rho}} = \iint p^4 \rho(x,p)dpdx = \frac{\partial^4 f}{\partial x^4}(0,0) = -24a^2 < 0.$$

7.5. Gaussian States

We focus now on the case where the states are Gaussian: we say that a pure or mixed quantum state is Gaussian if its Wigner function is of the type

$$\rho(z) = (2\pi)^{-n}\sqrt{\det \Sigma^{-1}}e^{-\frac{1}{2}\Sigma^{-1}(z-z_0)\cdot(z-z_0)}, \qquad (7.42)$$

where Σ (the "covariance matrix") is a positive definite $2n \times 2n$ matrix satisfying a certain condition that will be stated later (intuitively speaking Σ cannot be "too small" because then $\rho(z)$ would be too sharply peaked and thus violate the uncertainty principle of quantum mechanics). Gaussian states appear naturally in every quantum system which can be described or approximated by a quadratic Hamiltonian; because of their peculiarities these states play an exceptionally important role in quantum mechanics and optics (see [14]).

7.5.1. Definition and examples

We will define a *generalized Gaussian* as any complex function on $\mathbb{R}^n$ of the type

$$\psi_M^\hbar(x) = \left(\frac{1}{\pi\hbar}\right)^{n/4}(\det X)^{1/4}e^{-\frac{1}{2\hbar}M(x-x_0)^2}, \qquad (7.43)$$

where $M = X + iY$ is a complex symmetric $2n \times 2n$ invertible matrix; X and Y are real matrices such that $X = X^T > 0$ and $Y = Y^T$. The coefficient in front of the exponential is chosen so that $\psi_M^\hbar$ is normalized to unity: $\|\psi_M^\hbar\| = 1$.

Suppose that $X = I$ and $Y = 0$; then

$$\psi_M^\hbar(x) = \phi_0^\hbar(x) = \left(\frac{1}{\pi\hbar}\right)^{n/4} e^{-\frac{1}{2\hbar}|x|^2} \tag{7.44}$$

is the standard (or fiducial) coherent state.

Let $\phi_M(x) = e^{-\frac{1}{2\hbar}Mx^2}$ where $M = X + iY$ is a symmetric complex $n \times n$ matrix such that $X = \mathrm{Re}M > 0$. The Fourier transform

$$F\phi_M(p) = \left(\frac{1}{2\pi\hbar}\right)^{n/2} \int e^{-\frac{i}{\hbar}px}\phi_M(x)d^n x$$

is given by

$$F\phi_M(x) = (\det M)^{-1/2}\phi_{M^{-1}}(x), \tag{7.45}$$

where

$$(\det M)^{-1/2} = \lambda_1^{-1/2}\cdots\lambda_m^{-1/2};$$

the numbers $\lambda_1^{-1/2},\ldots,\lambda_n^{-1/2}$ are the square roots with positive real parts of the eigenvalues $\lambda_1^{-1},\ldots,\lambda_m^{-1}$ of M^{-1} (see, e.g., [83, Appendix A]). It follows that the Fourier transform of $\psi_M^\hbar$ is given by the formula

$$F\psi_M^\hbar(p) = \left(\frac{1}{\pi\hbar}\right)^{n/4}(\det X)^{1/4}(\det M)^{-1/2}\phi_{M^{-1}}(x). \tag{7.46}$$

7.5.2. The Wigner function of the Gaussian $\psi_M^\hbar$

We are following here almost verbatim our discussion in [112, §11.2.1].

Proposition 114. *The Wigner transform $W\psi_M^\hbar$ is the phase space Gaussian*

$$W\psi_M^\hbar(z) = \left(\frac{1}{\pi\hbar}\right)^n e^{-\frac{1}{\hbar}Gz^2}, \tag{7.47}$$

where G is the symplectic symmetric matrix

$$G = \begin{pmatrix} X + YX^{-1}Y & YX^{-1} \\ X^{-1}Y & X^{-1} \end{pmatrix}; \tag{7.48}$$

in fact $G = S^T S$ where

$$S = \begin{pmatrix} X^{1/2} & 0 \\ X^{-1/2}Y & X^{-1/2} \end{pmatrix} \tag{7.49}$$

is a symplectic matrix.

Proof. To simplify notation we set $C(X) = (\pi\hbar)^{n/4} (\det X)^{1/4}$. By definition of the Wigner transform, we have

$$W\psi_M^\hbar(z) = \left(\frac{1}{2\pi\hbar}\right)^n C(X)^2 \int e^{-\frac{i}{\hbar}p \cdot y} e^{-\frac{1}{2\hbar}F(x,y)} d^n y, \qquad (7.50)$$

where the phase F is defined by

$$F(x, y) = (X + iY)(x + \tfrac{1}{2}y)^2 + (X - iY)(x - \tfrac{1}{2}y)^2$$
$$= 2Xx^2 + 2iYx \cdot y + \tfrac{1}{2}Xy^2$$

so we can rewrite (7.50) as

$$W\psi_M^\hbar(z) = \left(\frac{1}{2\pi\hbar}\right)^n e^{-\frac{1}{\hbar}Xx^2} C(X)^2 \int e^{-\frac{i}{\hbar}(p+Yx) \cdot y} e^{-\frac{1}{4\hbar}Xy^2} d^n y.$$

Using the Fourier transformation formula (7.45) above with x replaced by $p + Yx$ and M by $\frac{1}{2}X$, we get

$$\int e^{-\frac{i}{\hbar}(p+Yx) \cdot y} e^{-\frac{1}{4\hbar}Xy^2} d^n y$$
$$= (2\pi\hbar)^{n/2} \left[\det(\tfrac{1}{2}X)\right]^{-1/2} C(X)^2 e^{-\frac{1}{\hbar}X^{-1}(p+Yx)^2}.$$

On the other hand we have

$$(2\pi\hbar)^{n/2} \left[\det\left(\frac{1}{2}X\right)\right]^{-1/2} C(X)^2 = \left(\frac{1}{\pi\hbar}\right)^n$$

and hence

$$W\psi_M^\hbar(z) = \left(\frac{1}{\pi\hbar}\right)^n e^{-\frac{1}{\hbar}Gz^2},$$

where

$$Gz^2 = (X + YX^{-1})x^2 + 2X^{-1}Yx \cdot p + X^{-1}p^2$$

so that G is given by (7.48). One immediately verifies that $G = S^T S$ where S is given by (7.49) and that $S^T J S = J$ hence $S \in \mathrm{Sp}(2n, \mathbb{R})$ as claimed. $\qquad \square$

In particular, when $\psi_M^\hbar$ is the standard coherent state (7.44) we recover the well-known formula

$$W\phi_0^\hbar(z) = \left(\frac{1}{\pi\hbar}\right)^n e^{-\frac{1}{\hbar}|z|^2}. \qquad (7.51)$$

As a consequence we have the following complete characterization of Gaussian pure states.

Corollary 115. *A Gaussian function* $\rho(z) = C_G e^{-G(z-z_0)^2/2\hbar}$ *(C_G a normalization constant and $G = G^T > 0$) is the Wigner distribution of a pure mixed state if and only if $G \in \mathrm{Sp}(2n, \mathbb{R})$.*

Proof. Assume first $z_0 = 0$. In view of Proposition 114 if $\rho(z) = W\psi_M^{\hbar}(z)$ then

$$\rho(z) = \left(\frac{1}{\pi\hbar}\right)^n e^{-\frac{1}{\hbar}Gz^2},$$

where G is symplectic. Assume conversely that G is symplectic. Since $G > 0$ we have, by the symplectic polar decomposition theorem, $G = S^T S$ for some $S \in \mathrm{Sp}(2n, \mathbb{R})$, hence $\rho(S^{-1}z) = (\pi\hbar)^{-n}e^{-|z|^2/\hbar}$, and $\rho \circ S^{-1}$ is thus the Wigner distribution $W\phi_0^{\hbar}$ of the fiducial coherent state $\phi_0^{\hbar}$ (formula (7.51). By the symplectic covariance property $\rho \circ S^{-1} \overset{\text{Weyl}}{\longleftrightarrow} \widehat{S}\widehat{\rho}\widehat{S}^{-1}$ (Proposition 57 in Chapter 4), hence $\rho \circ S^{-1}$ is also the Wigner distribution of $\widehat{S}\widehat{\rho}\widehat{S}^{-1}$; it follows that ρ is the Wigner distribution of the Gaussian state $\widehat{S}\phi_0^{\hbar}$. The case of general $z_0 \neq 0$ readily follows since we have $W\psi_M^{\hbar}(z - z_0) = W(\widehat{D}(z_0))\psi_M^{\hbar}(z)$ where $\widehat{D}(z_0)$ is the displacement operator. $\qquad\square$

A necessary and sufficient condition

We are going to address the following question:

For which values of η can the Gaussian function ρ be the η-Wigner function of a density operator?

Narcowich [218] was the first to address this question using techniques from harmonic analysis using the approach in Kastler's paper [170]; we give here a new and simpler proof using the multidimensional generalization of Hardy's uncertainty principle.

In what follows we denote by $F_\eta\psi$ the η-Fourier transform given, for $\psi \in L^2(\mathbb{R}^n)$, by

$$F_\eta\psi(p) = \left(\frac{1}{2\pi\eta}\right)^n \int e^{-\frac{i}{\eta}p\cdot x}\psi(x)d^nx. \tag{7.52}$$

We will need the following simple positivity result.

Lemma 116. *If R is a symmetric positive semidefinite $2n \times 2n$ matrix, then*

$$P_{(N)} = (Rz_j \cdot z_k)_{1 \le j,k \le N} \tag{7.53}$$

is a symmetric positive semidefinite $N \times N$ matrix for all $z_1, \ldots, z_N \in \mathbb{R}^{2n}$.

Proof. There exists a matrix L such that $R = L^*L$ (Cholesky decomposition). Denoting by $\langle z|z' \rangle = z \cdot \overline{z'}$ the inner product on $\mathbb{C}^{2n}$ we have, since the z_j are real vectors,

$$L^*z_j \cdot z_k = \langle L^*z_j|z_k \rangle = \langle z_j|Lz_k \rangle = z_j \cdot (Lz_k)^*,$$

hence $Rz_j \cdot z_k = Lz_j \cdot (Lz_k)^*$. It follows that

$$\sum_{1 \le j,k \le N} \lambda_j \lambda_k^* Rz_j \cdot z_k = \sum_{1 \le j \le N} \lambda_j Lz_j \left(\sum_{1 \le j \le N} \lambda_j Lz_j \right)^* \ge 0,$$

hence our claim. $\qquad\square$

We now have the tools needed to give a complete characterization of Gaussian η-Wigner functions. Recall from Proposition 113 that a necessary condition for a matrix Σ to be the covariance matrix of a quantum state is that it satisfies the condition $\Sigma + \frac{i\eta}{2}J \ge 0$. It turns out that in the Gaussian case this condition is also *sufficient*.

Proposition 117. *The Gaussian function*

$$\rho(z) = (2\pi)^{-n} \sqrt{\det \Sigma^{-1}} e^{-\frac{1}{2}\Sigma^{-1}z^2} \tag{7.54}$$

is the η-Wigner transform of a positive trace class operator if and only if it satisfies

$$|\eta| \le 2\lambda_{\min}, \tag{7.55}$$

where $\lambda_{\min}$ is the smallest symplectic eigenvalue of Σ; equivalently

$$\Sigma + \frac{i\eta}{2}J \ge 0. \tag{7.56}$$

Proof. Let us give a direct proof of the necessity of condition (7.55) for the Gaussian (7.54) to be the η-Wigner transform of a positive trace class operator. Let $\widehat{\rho} = (2\pi\eta)^n \mathrm{Op}_\eta^{\mathrm{W}}(\rho)$ and set $a(z) = (2\pi\eta)^n \rho(z)$. Let $\widehat{S} \in \mathrm{Mp}(2n, \mathbb{R})$; the operator $\widehat{\rho}$ is of trace class if and only if $\widehat{S}\widehat{\rho}\widehat{S}^{-1}$ is of trace class, in which case we have $\mathrm{Tr}(\widehat{\rho}) = \mathrm{Tr}(\widehat{S}\widehat{\rho}\widehat{S}^{-1})$. Choose $\widehat{S}$ with projection $S \in \mathrm{Sp}(2n, \mathbb{R})$ such that $\Sigma = S^T DS$ is a Williamson symplectic diagonalization of Σ as studied in Chapter 5. This choice reduces the proof to the case $\Sigma = D$, that is to

$$\rho(z) = (2\pi)^{-n}(\det \Lambda^{-1})e^{-\frac{1}{2}(\Lambda^{-1}x^2 + \Lambda^{-1}p^2)}. \tag{7.57}$$

Suppose now that $\widehat{\rho}$ is a density matrix; then there exist normalized functions $\psi_j \in L^2(\mathbb{R}^n)$ $(1 \leq j \leq n)$ such that

$$\rho(z) = \sum_j \alpha_j W_\eta \psi_j(z),$$

where the $\alpha_j > 0$ sum up to one. Integrating with respect to the p and x variables, respectively, the marginal conditions satisfied by the η-Wigner transform and formula (7.57) imply that we have

$$\sum_j \alpha_j |\psi_j(x)|^2 = (2\pi)^{-n/2}(\det \Lambda)^{1/2}e^{-\frac{1}{2}\Lambda^{-1}x^2},$$

$$\sum_j \alpha_j |F_\eta \psi_j(p)|^2 = (2\pi)^{-n/2}(\det \Lambda)^{1/2}e^{-\frac{1}{2}\Lambda^{-1}p^2}.$$

In particular, since $\alpha_j \geq 0$ for every $j = 1, 2, \ldots, n$,

$$|\psi_j(x)| \leq C_j e^{-\frac{1}{4}\Lambda^{-1}x^2}, \quad |F_\eta \psi_j(p)| \leq C_j e^{-\frac{1}{4}\Lambda^{-1}p^2},$$

here $C_j = (2\pi)^{-n/4}(\det \Lambda)^{1/4}/\alpha_j^{1/2}$. Applying the multidimensional Hardy uncertainty principle (Proposition 88 in Chapter 5) with $A = B = \frac{1}{2}\eta\Lambda^{-1}$ we must have $|\eta| \leq 2\lambda_j$ for all $j = 1, \ldots, n$ which is condition (7.55); this establishes the sufficiency statement. Let us finally show that, conversely, the condition (7.56) is sufficient. It is again no restriction to assume that Σ is the diagonal matrix $D = \left(\begin{smallmatrix}\Lambda & 0 \\ 0 & \Lambda\end{smallmatrix}\right)$; the symplectic Fourier transform of ρ is easily calculated and one finds that $\rho_\Diamond(z) = e^{-\frac{1}{4}Dz^2}$. Let $\Lambda_{(N)}$ be the $N \times N$ matrix with entries

$$\Lambda_{jk} = e^{-\frac{i\eta}{2}\sigma(z_j, z_k)}\rho_\Diamond(z_j - z_k);$$

a simple algebraic calculation shows that we have

$$\Lambda_{jk} = e^{-\frac{1}{4}Dz_j^2}e^{\frac{1}{2}(D+i\eta J)z_j \cdot z_k}e^{-\frac{1}{4}Dz_k^2}$$

and hence

$$\Lambda_{(N)} = \Delta_{(N)}\Gamma_{(N)}\Delta_{(N)}^*,$$

where $\Delta_{(N)} = \mathrm{diag}(e^{-\frac{1}{4}Dz_1^2}, \ldots, e^{-\frac{1}{4}Dz_N^2})$ and $\Gamma_{(N)} = (\Gamma_{jk})_{1 \leq j,k \leq N}$ with $\Gamma_{jk} = e^{\frac{1}{2}(D+i\eta J)z_j \cdot z_k}$. The matrix $\Lambda_{(N)}$ is thus positive semidefinite if and only if $\Gamma_{(N)}$ is, but this is the case in view of Lemma 116. $\qquad\square$

7.6. Discussion

The topic of "varying constants" has been somewhat controversial; however opposition to the reality of a variation of the fine-structure constant α seems to weaken due to the very precise repeated measurements made by Barrow and his collaborators at the Keck observatory. The indirect GPS measurements of possible variations of Planck's constant h due to Mohageg and his student Kentosh are perhaps somewhat more subject to controversy. However it is imperative to note that any measurement of a variation of a dimensionless constant (for instance the fine-structure constant) implies the variation of dimensional quantities. Quoting Barrow and Magueijo [17]:

"Suppose that evidence is found for varying dimensionless constants. Any theory explaining the phenomenon would necessarily have to make use of dimensional quantities. It would be a matter of choice as to which dimensional quantities were taken to vary. Any theory based on a choice to vary one-dimensional constant could always be reformulated as a theory based on another choice of varying constant."

While this seems to be an obvious statement, some physicists still do not accept it. It is however clear: assume that the fine structure constant $\alpha = e^2/\hbar c$ is found to have a slightly different value $\alpha(t_0)$ at some time t_0. At that time we have

$$\alpha(t_0) = \frac{e(t_0)^2}{\hbar(t_0)c(t_0)}$$

so *a priori* the electron charge e, the speed of light c and Planck's constant could be different. However, we are free to choose units so that two of three quantities remain constant — but not all three of them!. Hence, if α really varies, as suggested by experiments, we are free to freeze e and c and assume that Planck's constant is a function of time: $\hbar(t_0) \neq h$.

Using the rigorous theory of the density matrix, we have investigated some consequences of a possible change of Planck's constant. For the moment very little is known outside the Gaussian case. Let us mention that Dias and Prata have proved in [60] that if $|\psi\rangle$ is a *non-Gaussian* pure state then any variation of Planck's constant has dramatic consequences. They show that any change of h in such a state makes it to cease to be a pure quantum state (and it cannot become a classical state either since in view of Hudson's theorem its Wigner function $W\psi$ cannot be positive since ψ is not a Gaussian). Notice that it already is clear that $|\psi\rangle$ no longer can be a pure state in view of the argument in Proposition 108 where we shows that if $W\psi = W_\eta\phi$ then we must have $\eta = \hbar$. Dias and Prata's result is curious: while a Gaussian state (pure or mixed) remains a quantum state $\hbar$ decreases, why does the quantum character of a non-Gaussian pure state disappear so abruptly? This is one of the unsolved mysteries. Another mystery is what happens when we change the value of Planck's constant in an arbitrary mixed non-Gaussian states. The resolution of these difficult questions might help us to shed some light on the relation between the infinitely small and the infinitely big. And perhaps make us understand better our Universe.

Appendix A

The Symplectic Group

A.1. General Properties

We will work in the standard symplectic space $(\mathbb{R}^{2n}, \sigma)$ where

$$\sigma(z, z') = Jz \cdot z' = p \cdot x' - p' \cdot x$$

$(z = (x, p), z' = (x', p'))$; here

$$J = \begin{pmatrix} 0_{n \times n} & I_{n \times n} \\ -I_{n \times n} & 0_{n \times n} \end{pmatrix}$$

is the *standard symplectic matrix*.

Definition A.1. The group of all automorphisms s of $(\mathbb{R}^{2n}, \sigma)$ satisfying the condition

$$\sigma(Sz, Sz') = \sigma(z, z')$$

for all $z, z' \in \mathbb{R}^{2n}$ is denoted by $\mathrm{Sp}(2n, \mathbb{R})$ and called the "standard symplectic group".

Notice that $J^T = -J$ and $J^2 = -I$.

This condition is equivalent to

$$S^T J S = S J S^T = J,$$

where S^T is the transpose of S. Since

$$\det S^T J S = \det S^2 \det J = \det J,$$

it follows that $\det S = \pm 1$ (hence S is in particular invertible). It turns out, however, that

$$S \in \mathrm{Sp}(n) \Longrightarrow \det S = 1.$$

The group $\mathrm{Sp}(n)$ is stable under transposition: the condition $S \in \mathrm{Sp}(2n, \mathbb{R})$ is equivalent to $S^T J S = J$; since S^{-1} also is in $\mathrm{Sp}(2n, \mathbb{R})$ we have $(S^{-1})^T J S^{-1} = J$; taking the inverses of both sides of this equality we get $S J^{-1} S^T = J^{-1}$ that is $S J S^T = J$, so that $S^T \in \mathrm{Sp}(2n, \mathbb{R})$.

As symplectic basis of $(\mathbb{R}^{2n}, \sigma)$ being chosen, we can always write a symplectic matrix $S \in \mathrm{Sp}(2n, \mathbb{R})$ in block-matrix form

$$S = \begin{pmatrix} A & B \\ C & D \end{pmatrix}, \tag{A.1}$$

where the entries A, B, C, D are $n \times n$ matrices. It is easily seen, by a direct calculation, that the condition $S \in \mathrm{Sp}(2n, \mathbb{R})$ is equivalent to any of the following two conditions:

$$A^T C, \ B^T D \ \text{ symmetric, and } A^T D - C^T B \ = \ I, \tag{A.2}$$

$$A B^T, \ C D^T \ \text{ symmetric, and } A D^T - B C^T = I. \tag{A.3}$$

It immediately follows from the second set of conditions that the inverse of S is given by

$$S^{-1} = \begin{pmatrix} D^T & -B^T \\ -C^T & A^T \end{pmatrix}. \tag{A.4}$$

Let P and L be, respectively, a symmetric and an invertible $n \times n$ matrix. Then the block matrices

$$V_P = \begin{pmatrix} I & 0 \\ -P & I \end{pmatrix}, \quad U_P = \begin{pmatrix} -P & I \\ -I & 0 \end{pmatrix}, \quad M_L = \begin{pmatrix} L^{-1} & 0 \\ 0 & L^T \end{pmatrix} \tag{A.5}$$

all are symplectic. Moreover

- the set of all matrices J, M_L, V_P generates the symplectic group $\mathrm{Sp}(2n, \mathbb{R})$;
- the set of all matrices J, M_L, U_P generates the symplectic group $\mathrm{Sp}(2n, \mathbb{R})$.

It is often necessary to compose symplectic transformations with phase space translations; this motivates following definition.

Definition A.2. The semi-direct product $\mathrm{Sp}(2n, \mathbb{R}) \ltimes_s T(2n)$ of the symplectic group and the group of translations in $\mathbb{R}_z^{2n}$ is called the affine (or, inhomogeneous) symplectic group, and is denoted by $\mathrm{ASp}(n)$.

For practical calculations it is often useful to identify $\mathrm{ASp}(2n, \mathbb{R})$ with a matrix group.

The group of all matrices

$$[S, z_0] \equiv \begin{pmatrix} S & z_0 \\ 0_{1 \times 2n} & 1 \end{pmatrix}$$

is isomorphic to $\mathrm{ASp}(2n, \mathbb{R})$. (Here $0_{1 \times 2n}$ is the $2n$-column matrix with all entries equal to zero.)

A.2. The Unitary Group U(n)

One can associate a complex structure on $\mathbb{R}^{2n}$ to the standard symplectic matrix J; identifying $z = (x, p)$ with $x + ip$; it is defined by

$$(\alpha + i\beta)z = \alpha + \beta J z.$$

The unitary group $\mathrm{U}(n, \mathbb{C})$ acts in a natural way on $(\mathbb{R}^{2n}, \sigma)$ and that action preserves the symplectic structure.

Proposition A.3. The monomorphism $\mu : M(n, \mathbb{C}) \to M(2n, \mathbb{R})$ defined by $u = A + iB \mapsto \mu(u)$ with

$$\mu(u) = \begin{pmatrix} A & -B \\ B & A \end{pmatrix}$$

identifies the unitary group $\mathrm{U}(n, \mathbb{C})$ with the subgroup

$$\mathrm{U}(n) = \mathrm{Sp}(2n, \mathbb{R}) \cap O(2n, \mathbb{R}) \tag{A.6}$$

of $\mathrm{Sp}(n)$.

Proof. In view of (A.4) the inverse of $U = \mu(u)$, $u \in \mathrm{U}(n, \mathbb{C})$, is

$$U^{-1} = \begin{pmatrix} A^T & B^T \\ -B^T & A^T \end{pmatrix} = U^T,$$

hence $U \in O(2n, \mathbb{R})$ which proves the inclusion $U(n) \subset Sp(2n, \mathbb{R}) \cap O(2n, \mathbb{R})$. Suppose conversely that

$$U \in Sp(2n, \mathbb{R}) \cap O(2n, \mathbb{R}).$$

Then

$$JU = (U^T)^{-1}J = UJ$$

which implies that $U \in U(n)$ so that $Sp(2n, \mathbb{R}) \cap O(2n, \mathbb{R}) \subset U(n)$.   □

It is usual in the context of symplectic geometry to call $U(n)$ the "unitary group". Notice that it immediately follows from conditions (A.2), (A.3) that $A + iB \in U(n)$ if and only if the equivalent conditions

$$A^T B \text{ symmetric and } A^T A + B^T B = I, \tag{A.7}$$

$$AB^T \text{ symmetric and } AA^T + BB^T = I \tag{A.8}$$

are satisfied; these conditions follow from the identity

$$(A + iB)^*(A + iB) = (A + iB)(A + iB)^* = I$$

for the matrix $A + iB$ to be unitary.

In particular, taking $B = 0$ we see that the matrices

$$R = \begin{pmatrix} A & 0 \\ 0 & A \end{pmatrix} \text{ with } AA^T = A^T A = I \tag{A.9}$$

also are symplectic, and form a subgroup $O(n)$ of $U(n)$ which we identify with the rotation group $O(n, \mathbb{R})$. We thus have

$$O(n) \subset U(n) \subset Sp(2n, \mathbb{R}).$$

Appendix B

The Metaplectic Representation

The symplectic group $\mathrm{Sp}(2n,\mathbb{R})$ has a double covering which can be realized as a group of unitary operators on $L^2(\mathbb{R}^n)$, the metaplectic group $\mathrm{Mp}(2n,\mathbb{R})$. Each element of $\mathrm{Mp}(2n,\mathbb{R})$ is the product of two "quadratic Fourier transforms". This property allows the definition of the Maslov index on $\mathrm{Mp}(2n,\mathbb{R})$, which can be expressed in terms of the Leray index.

B.1. The Covering Groups $\mathrm{Sp}_q(2n,\mathbb{R})$

The symplectic group has covering groups of all orders $q = 1, 2, \ldots, +\infty$. By this we mean that there exist *connected* groups $\mathrm{Sp}_2(2n,\mathbb{R})$, $\mathrm{Sp}_3(2n,\mathbb{R}), \ldots, \mathrm{Sp}_\infty(2n,\mathbb{R})$ together with homomorphisms ("projections")

$$\Pi_q : \mathrm{Sp}_q(2n,\mathbb{R}) \longrightarrow \mathrm{Sp}(2n,\mathbb{R})$$

having the following properties: if $q < +\infty$, then:

- Π_q is onto and q-to-one: every $S \in \mathrm{Sp}(2n,\mathbb{R})$ is the image of exactly q elements $S_1, \ldots, S_q$ elements of $\mathrm{Sp}_q(2n,\mathbb{R})$; equivalently $\Pi_q^{-1}(I)$ consists of exactly q elements;
- Π_q is continuous, it is in fact a local diffeomorphism of $\mathrm{Sp}_q(2n,\mathbb{R})$ onto $\mathrm{Sp}(2n,\mathbb{R})$: every $S \in \mathrm{Sp}(2n,\mathbb{R})$ has a neighborhood $\mathcal{U}$ such that its coimage $\Pi_q^{-1}(\mathcal{U})$ is the disjoint union of neighborhoods $\mathcal{U}_1, \ldots, \mathcal{U}_k$ of $S_1, \ldots, S_q$.

In the case $q = +\infty$:

- $\mathrm{Sp}_\infty(2n, \mathbb{R})$ is a simply connected group and $\Pi_q^{-1}(I)$ is the additive integer group $(\mathbb{Z}, +)$.

The *existence* of the covering groups $\mathrm{Sp}_q(2n, \mathbb{R})$ follows from a standard argument from algebraic topology: one proves (see, e.g., [108]) that $\mathrm{Sp}(2n, \mathbb{R})$ is topologically the product $U(n) \times \mathbb{R}^{n(n+1)}$, and hence contractible to $U(n)$. It follows that the first homotopy group $\pi_1(\mathrm{Sp}(2n, \mathbb{R}))$ and $\pi_1(U(n, \mathbb{C}))$ are the same. Since $\pi_1(U(n, \mathbb{C}))$ is the integer group $\mathbb{Z}$ it follows that $\mathrm{Sp}(2n, \mathbb{R})$ has covering groups $\mathrm{Sp}_q(2n, \mathbb{R})$ which are in one-to-one correspondence with the quotient groups $\mathbb{Z}_q = \mathbb{Z}/q\mathbb{Z}$ for $q < \infty$, and that $\mathrm{Sp}_\infty(2n, \mathbb{R})$ is simply connected (it is the universal covering group of $\mathrm{Sp}(2n, \mathbb{R})$).

Among all these covering groups, there is one which plays a privileged role for us. It is the double covering $\mathrm{Sp}_2(2n, \mathbb{R})$; it can be represented as a group of unitary operators acting on the space $L^2(\mathbb{R}_x^n)$ of square integrable functions on configuration space. This "realization" of $\mathrm{Sp}_2(2n, \mathbb{R})$ is called the *metaplectic group*, and it is denoted by $\mathrm{Mp}(2n, \mathbb{R})$. We will see that $\mathrm{Mp}(2n, \mathbb{R})$ is generated by a set of operators that are closely related to the usual Fourier transform, but where the product $x \cdot x'$ is replaced by the generating function of free symplectic transforms.

In terms of representation theory $\mathrm{Mp}(2n, \mathbb{R})$ is thus a unitary representation of the double cover $\mathrm{Sp}_2(2n, \mathbb{R})$ in the square integrable functions. One can show that this representation is reducible, but that its sub-representations on the spaces $L_{\mathrm{odd}}^2(\mathbb{R}^n)$ and $L_{\mathrm{even}}^2(\mathbb{R}^n)$ consisting of odd and even square integrable functions, respectively, are irreducible. The proof consists in showing that each of these spaces is invariant under the action of the metaplectic group (the latter is easily achieved using the generators of $\mathrm{Mp}(2n, \mathbb{R})$ we will exhibit in a moment); see [83, Chapter 4].

B.2. Quadratic Fourier Transforms

We begin by defining the "quadratic Fourier transforms" associated with free symplectic matrices (or, rather, their generating functions).

The unitary Fourier transform $\widehat{J}$ is defined by

$$\widehat{J}\psi(x) = \left(\frac{1}{2\pi i \hbar}\right)^{n/2} \int e^{-\frac{i}{\hbar} x x'} \psi(x') d^n x' \tag{B.1}$$

(we assume that ψ in the Schwartz space $\mathcal{S}(\mathbb{R}^n)$). In this formula the argument of i is $\pi/2$, so the normalizing factor in front of the integral should be interpreted as

$$\left(\frac{1}{2\pi i \hbar}\right)^{n/2} = \left(\frac{1}{2\pi \hbar}\right)^{n/2} e^{-in\pi/4}. \tag{B.2}$$

In formula (B.1) $\hbar$ will be identified with Planck's constant h divided by 2π when doing physics, but mathematically speaking it just plays here the role of a parameter that we are free to fix as we like. The inverse $\widehat{J}^{-1}$ of $\widehat{J}$ is given by

$$\widehat{J}^{-1}\psi(x) = \left(\frac{i}{2\pi \hbar}\right)^{n/2} \int e^{\frac{i}{\hbar} x x'} \psi(x') \, d^n x', \tag{B.3}$$

that is

$$\widehat{J}^{-1}\psi = (\widehat{J}\psi^*)^*, \tag{B.4}$$

where the star * denotes complex conjugation. Both $\widehat{J}$ and $\widehat{J}^{-1} = J^\dagger$ are unitary Fourier transforms, in the sense that

$$||\widehat{J}\psi||_{L^2} = ||\psi||_{L^2}, \tag{B.5}$$

where $||\psi|| = \langle \psi | \psi \rangle$ is the usual norm on $L^2(\mathbb{R}^n_x)$, associated to the scalar product

$$\langle \psi | \phi \rangle = \int \psi^*(x)\phi(x) d^n x.$$

Let now S_W be a free symplectic matrix, with $W = (P, L, Q)$. Recall that P and Q are symmetric and that

$$\det(-W''_{xx'}) = \det L \neq 0,$$

where $W''_{xx'}$ is the matrix of mixed second derivatives of W (formula (2.75)). To S_W we associate the *two* "quadratic Fourier transforms" defined, for ψ in the Schwartz space $\mathcal{S}(\mathbb{R}^n_x)$, by the formula

$$\widehat{S}_{W,m}\psi(x) = \left(\frac{1}{2\pi i}\right)^{n/2} \Delta(W) \int e^{iW(x,x')} \psi(x') \, d^n x', \tag{B.6}$$

where

$$\Delta(W) = i^m \sqrt{|\det L|}. \tag{B.7}$$

The two possible choices for the integer m are defined by the condition

$$\arg \det L = m\pi \mod 2\pi. \tag{B.8}$$

Thus there exists an integer k such that

$$m = \begin{cases} 2k & \text{if } \det L > 0, \\ 2k + 1 & \text{if } \det L < 0. \end{cases} \tag{B.9}$$

Formula (B.8) can be written as

$$m = \arg \det(-W''_{xx'}). \tag{B.10}$$

The formulae above motivate the following definition: A choice of an integer m satisfying (B.9) is called a *Maslov index* of the quadratic form $W = (P, L, Q)$. Thus, exactly two Maslov indices modulo 4 are associated to each W, namely m and $m+2$. The Maslov index of a quadratic Fourier transform $\widehat{S}_{W,m}$ is then, by definition, the integer m; we write $m(\widehat{S}_{W,m}) = m$.

In particular,

$$m(\widehat{J}) = 0, \quad m(\widehat{J}^{-1}) = n \tag{B.11}$$

since we obviously have $\widehat{J} = \widehat{S}_{(0,I,0),0}$ and $\widehat{J}^{-1} = \widehat{S}_{(0,-I,0),n}$.

B.3. Definition and Properties

We will denote by $\aleph$ the set of all pairs (W, m), where m is one of the two integers modulo 4 defined by (B.8). We can now *define* the metaplectic group.

Definition B.1. The metaplectic group $\text{Mp}(2n, \mathbb{R})$ is the set of all products

$$\widehat{S} = \widehat{S}_{W_1,m_1} \cdots \widehat{S}_{W_k,m_k}$$

of a finite number of quadratic Fourier transforms.

Notice that it is not at all clear from this definition that $\text{Mp}(2n, \mathbb{R})$ really is a group! That $\text{Mp}(2n, \mathbb{R})$ is a semi-group is clear (it is closed under multiplication, and it contains the identity operator I, because $I = \widehat{J}\widehat{J}^{-1}$), but one does not immediately see why the inverse of a product of two quadratic Fourier transforms should also be such a product.

We will actually see that:

- $\mathrm{Mp}(2n, \mathbb{R})$ is a Lie group;
- there exists a "two-to-one" homomorphism

$$\prod : \mathrm{Mp}(2n, \mathbb{R}) \longrightarrow \mathrm{Sp}(2n, \mathbb{R})$$

("the projection") whose kernel $\Pi^{-1}(I)$ consists of the two elements $\pm I$ of $\mathrm{Mp}(2n, \mathbb{R})$.

It turns out that $\mathrm{Mp}(2n, \mathbb{R})$ is also connected; we will not prove this fact here; see [184] or [96].

Proof of Theorem 53. We will be rather sketchy in the proof of the first part of the proposition (the reader wanting to see the complete argument is referred to [96]). The obvious idea, if one wants to define $\Pi(\widehat{S})$ for arbitrary $\widehat{S}$ is to write $\widehat{S}$ as a product of quadratic Fourier transforms:

$$\widehat{S} = \widehat{S}_{W_1, m_1} \cdots \widehat{S}_{W_k, m_k}$$

and then to simply define the projection of $\widehat{S}$ by the formula

$$\prod(\widehat{S}) = S_{W_1} \cdots S_{W_k}.$$

There is however a true difficulty here, because one has to show that the right-hand side does not depend on the way we have factored $\widehat{S}$ (the factorization of an element of $\mathrm{Mp}(2n, \mathbb{R})$ is never unique: for instance, the identity operator can be written in infinitely many ways as $\widehat{S}_{W,m}(\widehat{S}_{W,m})^{-1} = \widehat{S}_{W,m}\widehat{S}_{W^*,m^*}$). However, once this is done, showing that Π is a homomorphism is straightforward. Let us show in detail the last part of the theorem, namely that Π is onto and two-to-one. Let $S = S_{W_1} \cdots S_{W_k}$ be an arbitrary element of $\mathrm{Sp}(2n, \mathbb{R})$. Then, for any choice of the (W_j, m_j) $(1 \leq j \leq k)$, we have

$$\prod(\widehat{S}_{W_1, m_1} \cdots \widehat{S}_{W_k, m_k}) = S_{W_1} \cdots S_{W_k}$$

and hence Π is onto. Let us next prove that Π is two-to-one. For this it suffices to show that $\mathrm{Ker}(\Pi) = \{\pm I\}$. Now, the inclusion $\mathrm{Ker}(\Pi) \supset \{\pm I\}$

is rather obvious. In fact, $\Pi(I) = I$ and

$$\prod(-I) = \Pi(\widehat{S}_{W,m}\widehat{S}_{W^*,m^*}) = S_W S_{W^*} = I.$$

Let us prove the opposite inclusion $\mathrm{Ker}(\Pi) \subset \{\pm I\}$ by induction on the number k of terms in the factorization $\widehat{S} = \widehat{S}_{W_1,m_1} \cdots \widehat{S}_{W_k,m_k}$. We first note that if $\prod(\widehat{S}) = I$, then we must have $k \geq 2$, because the identity is not a free symplectic matrix. Suppose next that

$$\prod(\widehat{S}_{W,m}\widehat{S}_{W',m'}) = I.$$

Then, either $S_W S_{W'} = I$, or $S_W S_{W'} = -I$, depending on whether $m' = m^*$ or $m' = m^* + 2$. This establishes the result when $k = 2$. Suppose now that we have proven the implication

$$\left.\begin{array}{r}\widehat{S} = \widehat{S}_{W_1,m_1} \cdots \widehat{S}_{W_k,m_k} \\[2mm] \prod(\widehat{S}) = I\end{array}\right\} \Rightarrow \widehat{S} = \pm I,$$

and let us prove that we then have

$$\left.\begin{array}{r}\widehat{S}' = \widehat{S}_{W_1,m_1} \cdots \widehat{S}_{W_{k+1},m_{k+1}} \\[2mm] \prod(\widehat{S}') = I\end{array}\right\} \Rightarrow \widehat{S}' = \pm I.$$

Since $\prod$ is a group homomorphism, we can write

$$\prod(\widehat{S}') = \Pi(\widehat{S})\Pi(\widehat{S}_{W'_{k+1},m_{k+1}}) = S S_{W'_{k+1}}$$

and it then follows from the induction assumption that we must have either $\widehat{S} = \widehat{S}_{W^*_{k+1},m^*_{k+1}}$ or $\widehat{S} = -\widehat{S}_{W^*_{k+1},m^*_{k+1}}$. But this means that we have either

$$\widehat{S}' = \widehat{S}_{W^*_{k+1},m^*_{k+1}} \widehat{S}_{W_{k+1},m_{k+1}} = I$$

or

$$\widehat{S}' = \widehat{S}_{W^*_{k+1},m^*_{k+1}} \widehat{S}_{W^*_{k+1},m^*_{k+1}} = -I,$$

which concludes the proof of the inclusion $\mathrm{Ker}(\Pi) \subset \{\pm I\}$. $\square$

B.4. The Structure of $\mathrm{Mp}(2n, \mathbb{R})$

Every symplectic matrix is the product of two free symplectic matrices, and a similar result holds for operators in $\mathrm{Mp}(2n, \mathbb{R})$ replacing the phrase "free symplectic matrix" by "quadratic Fourier transform".

Proposition B.2. Every $\widehat{S} \in \mathrm{Mp}(2n, \mathbb{R})$ is the product of two quadratic transforms $\widehat{S}_{W,m}$ and $\widehat{S}_{W',m'}$.

Proof. Let $\widehat{S}$ be an arbitrary element of $\mathrm{Mp}(2n, \mathbb{R})$, $S = \Pi(\widehat{S})$ its projection. Set $S = S_W S_{W'}$ and let $\widehat{S}_{W,m}$, $\widehat{S}_{W',m'}$ be two quadratic Fourier transforms with projections S_W and $S_{W'}$:

$$\Pi(\widehat{S}_{W,m}) = S_W \text{ and } \Pi(\widehat{S}_{W',m'}) = S_{W'}.$$

Then $\widehat{S} = \widehat{S}_{W,m}\widehat{S}_{W',m'}$ or $\widehat{S} = -\widehat{S}_{W,m}\widehat{S}_{W',m'} = \widehat{S}_{W,m+2}\widehat{S}_{W',m'}$. Either way $\widehat{S}$ can be written as the product of two quadratic Fourier transforms. $\quad\square$

Corollary B.3. The projections of the operators $\widehat{J}, \widehat{M}_{L,m}$ and $\widehat{V}_p$ are

$$\Pi(\widehat{J}) = J, \quad \Pi(\widehat{M}_{L,m}) = M_L, \quad \Pi(\widehat{V}_P) = V_P. \tag{B.12}$$

Proof. The fact that $\Pi(\widehat{J}) = J$ is obvious because $\widehat{J} = \widehat{S}_{(0,I,0),0}$. To prove that $\Pi(\widehat{M}_{L,m}) = M_L$ it suffices to note that the equality

$$\Pi(\widehat{M}_{L,m}\widehat{J}) = \Pi(\widehat{M}_{L,m})\Pi(\widehat{J})$$

implies that we have $\Pi(\widehat{M}_{L,m}) = \Pi(\widehat{M}_{L,m}\widehat{J})J^{-1}$ and hence

$$\Pi(\widehat{M}_{L,m}) = \begin{pmatrix} 0 & L^{-1} \\ -L^T & 0 \end{pmatrix} \begin{pmatrix} 0 & -I \\ I & 0 \end{pmatrix} = \begin{pmatrix} L^{-1} & 0 \\ 0 & L^T \end{pmatrix}.$$

The equality $\Pi(\widehat{V}_P) = V_P$ is proven in a similar fashion, using for instance the equality $\Pi(\widehat{V}_{-P}\widehat{J}) = \Pi(\widehat{V}_{-p})J$. $\quad\square$

B.5. Changing the Value of Planck's Constant

The metaplectic group $\mathrm{Mp}(2n, \mathbb{R})$ endowed with the projection Π is a twofold covering group of $\mathrm{Sp}(2n, \mathbb{R})$. However, a covering group can be "realized" in many different ways. Instead of choosing Π as a projection, we could as well have chosen any other mapping $\mathrm{Mp}(2n, \mathbb{R}) \to \mathrm{Sp}(2n, \mathbb{R})$ obtained from Π by composing it on the left with an inner automorphism of $\mathrm{Mp}(2n, \mathbb{R})$, or on the right with an inner automorphism of $\mathrm{Sp}(2n, \mathbb{R})$, or both. The essential point is here that the diagram

$$
\begin{array}{ccc}
\mathrm{Mp}(2n,\mathbb{R}) & \xrightarrow{\widehat{J}} & \mathrm{Mp}(2n,\mathbb{R}) \\
\Pi \downarrow & & \downarrow \Pi' \\
\mathrm{Sp}(2n,\mathbb{R}) & \xrightarrow{G} & \mathrm{Sp}(2n,\mathbb{R})
\end{array}
$$

is commutative, that is, we have $\Pi' \circ \widehat{J} = G \circ \Pi$, because for all such Π' we will then have

$$
\mathrm{Ker}(\Pi') = \{\pm I\}
$$

and Π' will be another honest covering mapping. We find it particularly convenient to define a new projection as follows. Set, for $\lambda > 0$:

$$
\widehat{M_\lambda} = \widehat{M_{\lambda I,0}} \tag{B.13}
$$

($\widehat{M_\lambda} \in \mathrm{Mp}(2n,\mathbb{R})$ is thus a "scaling operator" acting on functions on configuration space); we denote by M_λ ($= M_{\lambda I,0}$) its projection on $\mathrm{Sp}(2n,\mathbb{R})$. Let now η be a constant > 0, and define

$$
\widehat{S}^\eta = \widehat{M}_{1/\sqrt{\eta}} \widehat{S} \widehat{M}_{\sqrt{\eta}} \tag{B.14}
$$

for $\widehat{S} \in \mathrm{Mp}(2n,\mathbb{R})$. The projection of $\widehat{S}^\eta$ on $\mathrm{Sp}(2n,\mathbb{R})$ is then given by

$$
\Pi(\widehat{S}^\eta) = S^\eta = M_{1/\sqrt{\eta}} S M_{\sqrt{\eta}}. \tag{B.15}
$$

Now, we would like, for reasons that will become clear in a moment, to have a projection of $\mathrm{Mp}(2n,\mathbb{R})$ onto $\mathrm{Sp}(2n,\mathbb{R})$ that to $\widehat{S}^\eta$ associates, not S^η, but rather S itself. This can be achieved by defining the new projection

$$
\Pi^\eta : \mathrm{Mp}(2n,\mathbb{R}) \longrightarrow \mathrm{Sp}(2n,\mathbb{R})
$$

by the formula

$$
\Pi^\eta(\widehat{S}^\eta) = M_{\sqrt{\eta}}(\Pi(\widehat{S}^\eta)) M_{1/\sqrt{\eta}} \tag{B.16}
$$

which is of course equivalent to

$$
\Pi^\eta(\widehat{S}^\eta) = \Pi(\widehat{S}). \tag{B.17}
$$

Defining the "η-quadratic Fourier transform" $\widehat{S}^\eta_{W,m}$ associated with $\widehat{S}_{W,m}$ by

$$
\widehat{S}^\eta_{W,m} = \widehat{M}_{1/\sqrt{\eta}} \widehat{S}_{W,m} \widehat{M}_{\sqrt{\eta}}, \tag{B.18}
$$

we have explicitly

$$\widehat{S}^{\eta}_{W,m}\psi(x) = \left(\frac{1}{2\pi i\eta}\right)^{n/2} \Delta(W) \int e^{\frac{i}{\eta}W(x,x')}\psi(x')\,d^n x'. \qquad (B.19)$$

This is easily checked using the fact that W is homogeneous of degree two in (x, x'); the projection of $\widehat{S}^{\eta}_{W,m}$ on $\mathrm{Sp}(2n, \mathbb{R})$ is then the free symplectic matrix S_W:

$$\Pi^{\eta}(\widehat{S}^{\eta}_{W,m}) = S_W. \qquad (B.20)$$

When we use the covering mapping Π^{η} instead of Π, we will talk about the "metaplectic group $\mathrm{Mp}^{\eta}(2n, \mathbb{R})$". The reader should keep in mind that this is just a convenient way to say that we are using the projection Π^{η} instead of Π; of course $\mathrm{Mp}(2n, \mathbb{R})$ and $\mathrm{Mp}^{\eta}(2n, \mathbb{R})$ are identical as groups as is clear from formula (B.14)!

It follows from formula (B.18) that if η and η' are two positive numbers, then we have

$$\widehat{S}^{\eta}_{W,m} = M_{\sqrt{\eta'/\eta}}\widehat{S}^{\eta'}_{W,m}M_{\sqrt{\eta/\eta'}}.$$

This means, in the language of representation theory that the representations $\mathrm{Mp}^{\eta}(2n, \mathbb{R})$ and $\mathrm{Mp}^{\eta'}(2n, \mathbb{R})$ are equivalent. On the other hand it is possible to define $\mathrm{Mp}^{\eta'}(2n, \mathbb{R})$ also for $\eta' < 0$, but one can then show that $\mathrm{Mp}^{\eta}(2n, \mathbb{R})$ and $\mathrm{Mp}^{\eta'}(2n, \mathbb{R})$ are inequivalent representations (see [83, Chapter 4, Theorem 4.5.7]).

Appendix C

Born–Jordan Quantization

The main source is here our book [119], where we have listed many useful technical and historical references. Also see [117, 118].

C.1. Born's Canonical Commutation Relations

We assume $n = 1$. We denote by $\widehat{x}$ and $\widehat{p}$ two operators satisfying Born's canonical commutation relation (CCR)

$$[\widehat{x}, \widehat{p}] = \widehat{x}\widehat{p} - \widehat{p}\widehat{x} = i\hbar I;$$

for instance $\widehat{x}$ is multiplication by x and $\widehat{p} = -i\hbar(\partial/\partial x)$.

One easily proves by induction on the integers r and s that we have, more generally,

$$[\widehat{x}^r, \widehat{p}^s] = si\hbar \sum_{\ell=0}^{r-1} \widehat{x}^{r-1-\ell}\widehat{p}^{s-1}\widehat{x}^\ell, \tag{C.1}$$

$$[\widehat{x}^r, \widehat{p}^s] = ri\hbar \sum_{j=0}^{s-1} \widehat{p}^{s-1-j}\widehat{x}^{r-1}\widehat{p}^j, \tag{C.2}$$

$$[\widehat{x}^r, \widehat{p}^s] = \sum_{k=1}^{\min(r,s)} (i\hbar)^k k! \binom{r}{k}\binom{s}{k}\widehat{p}^{s-k}\widehat{x}^{r-k}, \tag{C.3}$$

$$[\widehat{x}^r, \widehat{p}^s] = -\sum_{k=1}^{\min(r,s)} (-i\hbar)^k k! \binom{r}{k}\binom{s}{k}\widehat{x}^{r-k}\widehat{p}^{s-k}. \tag{C.4}$$

C.2. Quantization of Monomials

Here are three examples of possible quantizations of the monomial $x^r p^s$:

- *the symmetric ordering*:

$$\mathrm{Op}_{\mathrm{sym}}(x^r p^s) = \frac{1}{2}(\widehat{p}^s \widehat{x}^r + \widehat{x}^r \widehat{p}^s);$$

- *the Weyl ordering*:

$$\mathrm{Op}_{\mathrm{W}}(x^r p^s) = \frac{1}{2^s} \sum_{\ell=0}^{s} \binom{s}{\ell} \widehat{p}^{s-\ell} \widehat{x}^r \widehat{p}^{\ell};$$

- *the Born–Jordan ordering*:

$$\mathrm{Op}_{\mathrm{BJ}}(x^r p^s) = \frac{1}{s+1} \sum_{\ell=0}^{s} \widehat{p}^{s-\ell} \widehat{x}^r \widehat{p}^{\ell}.$$

All three rules coincide when $s + r \leq 2$, but they are different as soon as $s \geq 2$ and $r \geq 2$. We now focus on the similarities and differences of the Weyl and Born–Jordan rules.

Using the generalized CCR (C.1)–(C.4) one shows that

$$\mathrm{Op}_{\mathrm{W}}(x^r p^s) = \sum_{\ell=0}^{\min(r,s)} (-i\hbar)^{\ell} \binom{s}{\ell} \binom{r}{\ell} \frac{\ell!}{2^{\ell}} \widehat{x}^{r-\ell} \widehat{p}^{s-\ell},$$

$$\mathrm{Op}_{\mathrm{BJ}}(x^r p^s) = \sum_{\ell=0}^{\min(r,s)} (-i\hbar)^{\ell} \binom{s}{\ell} \binom{r}{\ell} \frac{\ell!}{\ell+1} \widehat{x}^{r-\ell} \widehat{p}^{s-\ell};$$

these formulas give the expression of $\mathrm{Op}_{\mathrm{W}}(x^r p^s)$ and $\mathrm{Op}_{\mathrm{BJ}}(x^r p^s)$ in "normal form". For instance, in the usual choice $\widehat{x} = x$ and $\widehat{p} = -i\hbar(\partial/\partial x)$ this amounts writing the derivatives on the right side.

Introducing the "Shubin ordering"

$$\mathrm{Op}_{\tau}(x^r p^s) = \sum_{\ell=0}^{s} \binom{s}{\ell} (1 - \tau)^{\ell} \tau^{s-\ell} \widehat{p}^{s-\ell} \widehat{x}^r \widehat{p}^{\ell}, \tag{C.5}$$

where τ is an arbitrary real number, one sees that the choice $\tau = \frac{1}{2}$ immediately yields Weyl's rule; the choices $\tau = 0$ and $\tau = 1$ lead to the normal and antinormal rules $\mathrm{Op}_{\mathrm{N}}(x^r p^s) = \widehat{x}^r \widehat{p}^s$ and $\mathrm{Op}_{\mathrm{AN}}(x^r p^s) = \widehat{p}^s \widehat{x}^r$, respectively. Now, if we integrate the right-hand side of (C.5) for τ going from 0 to 1 we recover the Born–Jordan rule

$$\mathrm{Op_{BJ}}(x^r p^s) = \frac{1}{s+1} \sum_{\ell=0}^{s} \widehat{x}^\ell \widehat{p}^s \widehat{x}^{r-\ell};$$

this easily proven using the formula

$$\int_0^1 \tau^{s-\ell}(1-\tau)^\ell d\tau = \frac{(s-\ell)!\ell!}{(s+1)!} \tag{C.6}$$

familiar from the theory of the gamma and beta functions. This remark leads to the definition of Born–Jordan quantization in the general case.

C.3. Born–Jordan Quantization of an Observable

Let us write the Weyl operator $\widehat{A} = \mathrm{Op_W}(a)$ in integral form:

$$\widehat{A}\psi(x) = \left(\frac{1}{2\pi\hbar}\right)^n \iint e^{\frac{i}{\hbar}p(x-y)} a(\tfrac{1}{2}(x+y),p)\psi(y)d^n p\, d^n y. \tag{C.7}$$

We assume that $\psi \in \mathcal{S}(\mathbb{R}^n)$ and that the symbol a is a tempered distribution, thus viewing the integral above as a distributional bracket. Let τ be a real number as above. The Shubin operator $\widehat{A}_\tau - OP_\tau(a)$ is, by definition, given by

$$\widehat{A}_\tau\psi(x) = \left(\frac{1}{2\pi\hbar}\right)^n \int e^{\frac{i}{\hbar}p(x-y)} a((1-\tau)x + \tau y, p)\psi(y)d^n y\, d^n p. \tag{C.8}$$

One can show that when $a(x,p) = x^r p^s$ then $\mathrm{Op}_\tau(x^r p^s)$ is given by formula (C.5). As in the monomial case, the choice $\tau = \frac{1}{2}$ leads to the Weyl operator (C.7). This suggests the following definition.

Definition C.1. The Born–Jordan quantization rule associates to a symbol a the operator $\widehat{A}_{\mathrm{BJ}} = \mathrm{Op_{BJ}}(a)$ defined by

$$\widehat{A}_{\mathrm{BJ}} = \int_0^1 \mathrm{Op}_\tau(a)d\tau.$$

The formula above is not easy to use in practice. However, it leads to a remarkable harmonic analysis representation. Recall that the Weyl operator (C.7) can be expressed in terms of the symplectic Fourier transform

$$a_\sigma(z) = \left(\frac{1}{2\pi\hbar}\right)^n \int e^{-\frac{i}{\hbar}\sigma(z,z')} a(z') d^{2n}z$$

of the symbol as follows:

$$\widehat{A}\psi = \left(\frac{1}{2\pi\hbar}\right)^n \int a_\sigma(z)\chi_{\mathrm{BJ}}(z)\widehat{T}(z)\psi d^{2n}z, \qquad (\mathrm{C}.9)$$

where $\widehat{T}(z)$ is the Heisenberg–Weyl operator.

Theorem C.2. The Born–Jordan operator $\widehat{A}_{\mathrm{BJ}} = \mathrm{Op}_{\mathrm{BJ}}(a)$ is given by

$$\widehat{A}_{\mathrm{BJ}}\psi = \left(\frac{1}{2\pi\hbar}\right)^n \int a_\sigma(z)\chi_{\mathrm{BJ}}(z)\widehat{T}(z)\psi d^{2n}z, \qquad (\mathrm{C}.10)$$

where the function χ_{BJ} is defined by

$$\chi_{\mathrm{BJ}}(z) = \mathrm{sinc}(px/2\hbar). \qquad (\mathrm{C}.11)$$

For a proof of this result see [119, Proposition 2, Chapter 10]. One shows that, as is the case for the Weyl operator $\widehat{A} = \mathrm{Op}_{\mathrm{W}}(a)$, the operator $\widehat{A}_{\mathrm{BJ}}$ is a continuous $\mathcal{S}(\mathbb{R}^n) \to \mathcal{S}'(\mathbb{R}^n)$ for every $a \in \mathcal{S}'(\mathbb{R}^{2n})$.

Appendix D

Twisted Product and Convolution

Let us first prove the composition formulas (4.61)–(4.62) in Chapter 4. We are thereafter following closely our exposition in [112, Chapter 10].

We want to prove that when the product $\widehat{C} = \widehat{A}\widehat{B}$ of two Weyl operators with Weyl symbols a and b is defined, then $\widehat{C}$ is a Weyl operator with symbol c given by

$$
c(z) = \left(\frac{1}{4\pi\hbar} \right)^{2n} \iint e^{\frac{i}{2\hbar}\sigma(z',z'')} a\left(z + \frac{1}{2}z' \right) b\left(z - \frac{1}{2}z'' \right) d^{2n}z' d^{2n}z'',
$$

(D.1)

and that the symplectic Fourier transform of c is given by

$$
c_\sigma(z) = \left(\frac{1}{2\pi\hbar} \right)^n \int_{\mathbb{R}^{2n}} e^{\frac{i}{2\hbar}\sigma(z,z')} a_\sigma(z - z') b_\sigma(z') dz'.
$$

(D.2)

Proof of (D.1). Assume that the Weyl symbols a, b of $\widehat{A}$ and $\widehat{B}$ are in the Schwartz space $\mathcal{S}(\mathbb{R}^{2n})$. The kernel of the product $\widehat{A}\widehat{B}$ is given by

$$
K(x,y) = \left(\frac{1}{2\pi\hbar} \right)^{2n} \iiint e^{\frac{i}{\hbar}((x-\alpha)p + (\alpha-y)p)}
$$
$$
\times a\left(\frac{1}{2}(x+\alpha), \zeta \right) b\left(\frac{1}{2}(x+y), \xi \right) d^n\alpha d^n\zeta d^n\xi.
$$

The Weyl symbol c being given by

$$
c(x,p) = \int e^{-\frac{i}{\hbar}pu} K\left(x + \frac{1}{2}u, x - \frac{1}{2}u \right) d^n u
$$

we thus have

$$c(z) = \left(\frac{1}{2\pi\hbar}\right)^{2n} \iiiint e^{\frac{i}{\hbar}Q} a\left(\frac{1}{2}\left(x + \alpha + \frac{1}{2}u\right), \zeta\right)$$
$$\times b\left(\frac{1}{2}\left(x + \alpha - \frac{1}{2}u\right), \xi\right) d^n\alpha d^n\zeta d^n u d^n\xi$$

the function Q being given by

$$Q = (x - \alpha + \tfrac{1}{2}u)\zeta + (\alpha - x + \tfrac{1}{2}u)\xi - up$$
$$= (x - \alpha + \tfrac{1}{2}u)(\zeta - p) + (\alpha - x + \tfrac{1}{2}u)(\xi - p).$$

Setting $\zeta' = \zeta - p$, $\xi' = \xi - p$, $\alpha' = \frac{1}{2}(\alpha - x + \frac{1}{2}u)$ and $u' = \frac{1}{2}(\alpha - x - \frac{1}{2}u)$, we have

$$d^n\alpha d^n\zeta d^n u d^n\xi = 2^{2n} d^n\alpha' d^n\zeta' d^n u' d^n\xi'$$

and

$$Q = 2\sigma(u', \xi'; \alpha', \zeta').$$

It follows that

$$c(z) = \left(\frac{1}{\pi\hbar}\right)^{2n} \iiiint e^{\frac{2i}{\hbar}\sigma(u', \xi'; \alpha', \zeta')}$$
$$\times a(x + \alpha', p + \zeta')b(x + u', p + \xi')d^n\alpha' d^n\zeta' d^n u' d^n\xi';$$

formula (D.1) follows setting $z' = 2(\alpha', \zeta')$ and $z'' = -2(u', \xi')$.

Proof of (D.2). Writing the operators $\widehat{A}$ and $\widehat{B}$ in the usual form

$$\widehat{A} = \left(\frac{1}{2\pi\hbar}\right)^n \int_{\mathbb{R}^{2n}} a_\sigma(z_0)\widehat{D}(z_0)dz_0,$$

$$\widehat{B} = \left(\frac{1}{2\pi\hbar}\right)^n \int_{\mathbb{R}^{2n}} b_\sigma(z_1)\widehat{D}(z_1)dz_1,$$

we have, using the property

$$\widehat{D}(z_0 + z_1) = e^{-\frac{i}{2\hbar}\sigma(z_0, z_1)}\widehat{D}(z_0)\widehat{D}(z_1)$$

of the displacement operators,

$$\widehat{D}(z_0)\widehat{B} = \left(\frac{1}{2\pi\hbar}\right)^n \int b_\sigma(z_1)\widehat{D}(z_0)\widehat{D}(z_1)d^{2n}z_1$$

$$= \left(\frac{1}{2\pi\hbar}\right)^n \int e^{\frac{i}{2\hbar}\sigma(z_0,z_1)}b_\sigma(z_1)\widehat{D}(z_0+z_1)d^{2n}z_1,$$

and hence

$$\widehat{A}\widehat{B} = \left(\frac{1}{2\pi\hbar}\right)^{2n} \iint_{\mathbb{R}^{4n}} e^{\frac{i}{2\hbar}\sigma(z_0,z_1)}a_\sigma(z_0)b_\sigma(z_1)\widehat{D}(z_0+z_1)d^{2n}z_0d^{2n}z_1.$$

Setting $z = z_0 + z_1$ and $z' = z_1$, this formula can be rewritten

$$\widehat{A}\widehat{B} = \left(\frac{1}{2\pi\hbar}\right)^{2n} \int \left(\int e^{\frac{i}{2\hbar}\sigma(z,z')}a_\sigma(z-z')b_\sigma(z')d^{2n}z'\right) \widehat{D}(z)d^{2n}z$$

hence (D.2).

Let us now prove that the formal adjoint of the Weyl operator $\widehat{A} = \mathrm{Op}^{\mathrm{W}}(a)$ is the operator $\widehat{A}^\dagger = \mathrm{Op}^{\mathrm{W}}(a^*)$ (Proposition 59 of Chapter 4).

We have

$$\langle\psi|\widehat{A}\phi\rangle = \int a(z)W(\psi,\phi)(z)d^{2n}z,$$

where $W(\psi,\phi)$ is the cross-Wigner function. Let us write

$$\langle\psi|\widehat{A}^\dagger\phi\rangle = \int b(z_0)W(\psi,\phi)(z)d^{2n}z.$$

Since $\langle\psi|\widehat{A}^\dagger\phi\rangle = \langle\widehat{A}^\dagger\phi|\psi\rangle^* = \langle\phi|\widehat{A}\psi\rangle^*$ we have

$$\langle\psi|\widehat{A}^\dagger\phi\rangle = \left(\int a(z)W(\phi,\psi)(z_0)d^{2n}z\right)^*$$

$$= \int a^*(z)W(\phi,\psi)^*(z)d^{2n}z.$$

Using the relation $W(\phi,\psi)^* = W(\psi,\phi)$ we thus have

$$\langle\psi|\widehat{A}^\dagger\phi\rangle = \int a^*(z)W(\phi,\psi)^*(z)d^{2n}z$$

and hence $b = a^*$.

Bibliography

1. ABBONDANDOLO, A. AND MATVEYEV, S. How large is the shadow of a symplectic ball? Preprint (2012), arXiv:1202.3614v1[math.SG].
2. ABRAHAM, R., MARSDEN J.E. AND RATIU, T. *Manifolds, Tensor Analysis, and Applications*, Applied Mathematical Sciences, Vol. 75, Springer, 1988.
3. ALDA, J. Paraxial optics, in *Encyclopedia of Optical Engineering*, pp. 1920–1931, Marcel Dekker, 2003.
4. ANANDAN, J. Geometric angles in quantum and classical physics, *Phys. Lett. A* 129(4), 201–207 (1988).
5. ARNOLD, V.I. *Mathematical Methods of Classical Mechanics*, 2nd edition, Graduate Texts in Mathematics, Springer-Verlag, 1989.
6. ARNOLD, V.I. First steps in symplectic topology, *Uspekhi Mat. Nauk* 41(6), 3–18 (1986); *Russian Math. Surveys* 41(6), 1–21 (1986).
7. ARTSTEIN-AVIDAN, S. AND MILMAN, V. The concept of duality in convex analysis, and the characterization of the Legendre transform, *Ann. of Math.* 169, 661–674 (2009).
8. ARTSTEIN-AVIDAN, S., MILMAN, V. AND OSTROVER, Y. The M-ellipsoid, symplectic capacities and volume, *Comm. Math. Helv.* 83(2), 359–369 (2008); Preprint (2006) arXiv:math.SG/0603411.
9. ARVIND, DUTTA, B., MUKUNDA, N. AND SIMON, R. The real symplectic groups in quantum mechanics and optics, *Pramana J. Phys.* 45(6), 471–497 (1995).
10. BALL, K.M. Ellipsoids of maximal volume in convex bodies, *Geom. Dedicata* 41(2), 241–250 (1992).
11. BANYAGA, A. Sur la structure du groupe des difféomorphismes qui préservent une forme symplectique, *Comm. Math. Helv.* 53, 174–227 (1978).
12. BAPAT, R. AND RAGHAVAN, T.E.S. *Nonnegative Matrices and Applications*, Cambridge University Press, 1997.
13. BARCLAY, D.T. Convergent WKB series, *Phys. Lett. A* 185(2), 169–173 (1994).

14. BARNETT, S.M. AND RADMORE, P.M. *Methods in Theoretical Quantum Optics*, Oxford University Press, New York, 1997; Reprinted 2002.

15. BARROW, J.D. *The Constants of Nature: From Alpha to Omega*, Pantheon books, New York, 2002.

16. BARROW, J.D. Varying constants, *Philos. Trans. R. Soc. A* 363, 2139–2153 (2005).

17. BARROW, J.D. AND MAGUEIJO, J. Varying-a theories and solutions to the cosmological problems, *Phys. Lett. B* 443, 104–110 (1998).

18. BELINFANTE, F.G. *A Survey of Hidden-Variables Theories*, Pergamon Press, Oxford, 1973.

19. BELL, J. *Speakable and Unspeakable in Quantum Mechanics*, Cambridge University Press, 1993.

20. BENENTI, G. Gaussian wave packets in phase space: The Fermi g_F function, *Amer. J. Phys.* 77(6), 546–551 (2009).

21. BENENTI, G. AND STRINI, G. Quantum mechanics in phase space: first order comparison between the Wigner and the Fermi function, *Eur. Phys. J. D* 57, 117–121 (2010).

22. BERENGUT, J.C., FLAMBAUM, V.V., ONG, A., WEBB, J.K., BARROW, J.D., BARSTOW, M.A., PREVAL, S.P. AND HOLBERG, J.B. Limits on the dependence of the fine-structure constant on gravitational potential from white-dwarf spectra, *Phys. Rev. Lett.* 111(1), 010801 (2013).

23. BERNDT, R. *An Introduction to Symplectic Geometry*, Graduate Studies in Mathematics, Vol. 26, American Mathematical Society, 2001.

24. BERRY, M.V. Quantal phase factors accompanying adiabatic changes, *Proc. Roy. Soc. London A* 392, 45–57 (1984).

25. BERRY, M.V. Classical adiabatic angles and quantal adiabatic phase. *J. Phys. A* 18, 15–27 (1985).

26. BINZ, E. AND SCHEMPP, W. Quantum hologram and relativistic hodogram: Magnetic resonance tomography and gravitational wavelet detection, in *Proceedings of the Second International Conference on Geometry, Integrability and Quantization,* June 7–15, 2000, Varna, Bulgaria, eds. I.M. Mladenov and G.L. Naber, pp. 110–150, Coral Press, Sofia, 2001.

27. BLANCHARD, P. AND BRÜNING, E. *Mathematical Methods in Physics: Distributions, Hilbert Space Operators, Variational Methods, and Applications in Quantum Physics*, Vol. 69, Birkhäuser, 2015.

28. BOHM, D. *Quantum Theory*, Prentice-Hall, New York, 1951.

29. BOHM, D. A suggested interpretation of the quantum theory in terms of "hidden" variables: Part I, *Phys. Rev.* 85, 166–179 (1952).

30. BOHM, D. A suggested interpretation of the quantum theory in terms of "hidden" variables: Part II, *Phys. Rev.* 85, 180–193 (1952).

31. BOHM, D. Hidden variables and the implicate order, *Quantum Implications: Essays in Honour of David Bohm,* eds., B.J. Hiley and D. Peat, pp. 33–45, Routledge, London, 1987.

32. BOHM, D. AND HILEY, B. *The Undivided Universe: An Ontological Interpretation of Quantum Theory*, Routledge, London, 1993.

33. BOPP, F. La mécanique quantique est-elle une mécanique statistique particulière? *Ann. Inst. H. Poincaré* 15, 81–112 (1956).

34. BORN, M. AND JORDAN, P. Zur quantenmechanik, *Zeits. Physik* 34, 858–888 (1925).

35. BRILLOUIN, L. The actual mass of potential energy: A correction to classical relativity, *Proc. Natl. Acad. Sci. USA* 53(3), 475–482 (1965).

36. BRÖCKER, T. AND WERNER, F. Mixed states with positive Wigner functions, *J. Math. Phys.* 36(1), 62–75 (1995).

37. BRISLAWN, C. Kernels of trace class operators, *Proc. Amer. Math. Soc.* 104(4), 1181–1190 (1988).

38. BROWN, M.R. AND HILEY, B.J. Schrödinger revisited: the role of Dirac's 'standard' ket in the algebraic approach, Preprint (2001).

39. DE BRUIJN, N.G. *Uncertainty principles in Fourier analysis*, in Inequalities, ed., O. Shisha, pp. 55–71, Academic Press, New York, 1967.

40. DE BRUIJN, N.G. A theory of generalized functions, with applications to Wigner distribution and Weyl correspondence, *Nieuw Arch. Wiskd.* 21, 205–280 (1973).

41. BUSCH, P., HEINONEN, T. AND LAHTI, P. Heisenberg's uncertainty principle, *Phys. Rep.* 452(6), 155–176 (2007).

42. BUSCH, P., LAHTI, P.J. AND MITTELSTAEDT, P. *The Quantum Theory of Measurement*, Springer, Berlin, 1996.

43. BUSLAEV, V.C. Quantization and the W.K.B method, *Trudy Mat. Inst. Steklov* 110, 5–28 (1978) [in Russian].

44. BUZANO, E. AND TOFT, J. Schatten–von Neumann properties in the Weyl calculus, *J. Funct. Anal.* 259(12), 3080–3114 (2010).

45. BUŽEK, V., ADAM, G. AND DROBNÝ, G. Reconstruction of Wigner functions on different observation levels, *Ann. Phys.* 245, 37–97 (1996).

46. CASTELLANI, L. Quantization rules and Dirac's correspondence, *Il Nuovo Cimento* 48A(3), 359–368 (1978).

47. CARROLL, L. *On the Emergence Theme of Physics*, World Scientific, Singapore, 2010.

48. CHORIN, A.J., HUGHES, T.J.R, MCCRACKEN, M.F. AND MARSDEN, J.E. Product formulas and numerical algorithms, *Comm. Pure Appl. Math.* 31(2), 205–256 (1978).

49. CIELIEBAK, K., HOFER, H., LATSCHEV, J. AND SCHLENK, F. Quantitative symplectic geometry, in *Dynamics, Ergodic Theory, and Geometry*, pp. 1–44, Mathematical Research Institute Publishing, Vol. 54, Cambridge University Press, 2007; arXiv:math/0506191v1 [math.SG].

50. COHEN, L. *The Weyl Operator and its Generalization*, Pseudo-Differential Operators, Birkhäuser Basel, 2013.

51. CORNBLEET, S. On the eikonal function. *Radio Sci.* 31(6), 1697–1703 (1996).

52. CORDERO, E., DE GOSSON, M. AND NICOLA, F. On the invertibility of Born–Jordan quantization, *J. Math. Pure Appl.* 105(4), 537–557 (2016).

53. CORDERO, E., DE GOSSON, M. AND NICOLA, F. Positivity of trace class operators and the Cohen class, in *Application to Born–Jordan Operators*, Preprint (2017).

54. DIAS, N.C. AND PRATA, J. The Narcowich–Wigner spectrum of a pure state, *Rep. Math. Phys.* 63(1), 43–54 (2009).

55. CREHAN, P. The parametrisation of quantisation rules equivalent to operator orderings, and the effect of different rules on the physical spectrum, *J. Phys. A* 811–822 (1989).

56. DAMOUR, T. AND DYSON, F. The Oklo bound on the time variation of the fine structure constant revisited, *Nucl. Phys. B* 480(1), 37–54 (1996).

57. D'ARIANO, G.M. Universal quantum observables, *Phys. Lett. A* 300, 1–6 (2002).

58. D'ARIANO, G.M., MACCHIAVELLO, C. AND PARIS, M.G.A. Detection of the density matrix through optical homodyne tomography without filtered back projection, *Phys. Rev. A* 50(5), 4298–4303 (1994).

59. DAUBECHIES, I. Continuity statements and counterintuitive examples in connection with Weyl quantization, *J. Math. Phys.* 24(6), 1453–1461 (1983).

60. DIAS, N.C. AND PRATA, J.N. The Narcowich–Wigner spectrum of a pure state, *Rep. Math. Phys.* 63(1), (2009) 43–54.

61. DIAS, N., DE GOSSON, M. AND PRATA, J. Maximal covariance group of Wigner transforms and pseudo-differential operators, *Proc. Amer. Math. Soc.* 142(9), 3183–3192 (2014).

62. DIRAC, P.A.M. Quantised singularities in the electromagnetic field, *Proc. Roy. Soc. A* 133(1931), 60–72 (2014).

63. DIRAC, P.A.M. A new basis for cosmology, *Proc. Roy. Soc. London A* 165(921), 199–208 (1938).

64. DIXMIER, J. *Les C*-algèbres et leurs représentations*, Gauthier-Villars, 1969.

65. DRAGT, A.J. Lie methods for nonlinear dynamics with applications to accelerator physics, Preprint (2014), http://www.physics.umd.edu/dsat/.

66. DU, J. AND WONG, M.W. A trace formula for Weyl transforms, *Approx. Theory Appl.* 16(1), 41–45 (2000).

67. DUBIN, D.A., HENNINGS, M.A. AND SMITH, T.B. *Mathematical Aspects of Weyl Quantization and Phase*, World Scientific, 2000.

68. DUFF, M.J., OKUN, L.B. AND VENEZIANO, G. Trialogue on the number of fundamental constants, *J. High Energy Phys.* 2002(03) 023 (2002).

69. DUFF, M.J. Comment on time-variation of fundamental constants Preprint (2002), arXiv:hep-th/0208093.

70. DUFF, M.J. How fundamental are fundamental constants? *Contemp. Phys.* 56(1), 35–47 (2015).

71. DYSON, F.J. The fundamental constants and their time variation, in *Aspects of Quantum Theory*, pp. 213–36, eds., J. E. Lannutti and E. P. Wigner, Cambridge University Press, Cambridge, 1972.

72. EMCH, G.G. Geometric dequantization and the correspondence problem, *Internat. J. Theoret. Phys.* 22(5), 397–420 (1983).

73. EKELAND, I. AND HOFER, H. Symplectic topology and Hamiltonian dynamics, I and II, *Math. Zeit.* 200, 355–378 and 203, 553–567 (1990).

74. ELIASHBERG, YA., KIM, S. AND POLTEROVICH, L. Geometry of contact transformations and domains: orderability versus squeezing, *Geom. Topol.* 10, 1635–1747 (2006).

75. ESPOSITO, G., MARMO, G., MIELE, G. AND SUDARSHAN, G. *Advanced Concepts in Quantum Mechanics*, Cambridge University Press, 2015.

76. FANO, U. Description of states in quantum mechanics by density matrix and operator techniques, *Rev. Mod. Phys.* 29, 71–93 (1957).

77. FEDAK, W.A. AND PRENTIS, J.J. The 1925 Born and Jordan paper "On quantum mechanics", *Amer. J. Phys.* 77(2), 128–139 (2009).

78. FEICHTINGER, H.G. On a new Segal algebra, *Monatsh. Math.* 92, 269–289 (1981).

79. FEICHTINGER, H.G. Modulation spaces on locally compact abelian groups, *Technical Report*, January 1983; in R. Radha, M. Krishna and S. Thangavelu, eds., *Proc. Internat. Conf. Wavelets and Applications*, pp. 1–56, Chennai, January 2002. New Delhi Allied Publishers, 2003.

80. FERMI, E. L'interpretazione del principio di causalità nella meccanica quantistica, *Rend. Lincei* 11, 980 (1930); reprinted in *Nuovo Cimento* 7, 361 (1930).

81. FEYNMAN, R.P. AND HIBBS, A.R. *Quantum Mechanics and Path Integrals*, McGraw-Hill, New York, 1965.

82. FLACK, R. AND HILEY B. Weak measurement and its experimental realisation, *J. Phys. Conf. Ser.* 504(Conference 1), (2014).

83. FOLLAND, G.B. *Harmonic Analysis in Phase Space*, Annals of Mathematics Studies, Princeton University Press, Princeton, NJ, 1989.

84. FOLLAND, G,B. AND SITARAM, A. The uncertainty principle: A mathematical survey, *J. Fourier Anal. Appl.* 3(3), 207–238 (1997).

85. GAMOW, G. *Mr Tompkins in Paperback.* Cambridge University Press, Cambridge, 1993.

86. GARCÍA-BULLÉ, M., LASSNER, W.L AND WOLF, K.B. The metaplectic group within the Heisenberg–Weyl ring, *J. Math. Phys.* 27(1), 29–36 (1986).

87. GEL'FAND, I.M. AND NAIMARK, M.A. On the imbedding of normed rings into the ring of operators on a Hilbert space, *Math. Sbornik* 12(2), 197–217 (1943).

88. GEFTER, A. Beyond space-time: Welcome to phase space, *New Scientist* 2824, 08 August 2011.

89. GIACHETTA, G., MANGIAROTTI, L. AND SARDANASHVILI, G. A. *Geometric and Algebraic Topological Methods in Quantum Mechanics*, World Scientific, Singapore, 2005.

90. GIBBS, J.W. *Elementary Principles in Statistical Mechanics*, Dover Publications, New York, 1960.

91. GIEDKE, G., EISERT, J., CIRAC, J.I. AND PLENIO, M.B. Entanglement transformations of pure Gaussian states, Preprint (2003), arXiv:quant-ph/0301038, v1.

92. GOLDSTEIN, H. *Classical Mechanics*, Addison-Wesley, 1950; 2nd edition, 1980.

93. DE GOSSON, C. AND DE GOSSON, M. The phase space formulation of time-symmetric quantum mechanics, *Quanta* 4(1), (2015).

94. DE GOSSON, M. Maslov indices on Mp(n), *Ann. Inst. Fourier (Grenoble)* 40(3), 537–55 (1990).

95. DE GOSSON, M. On half-form quantization of Lagrangian manifolds and quantum mechanics in phase space, *Bull. Sci. Math.* 121, 301–322 (1997).

96. DE GOSSON, M. *Maslov classes, metaplectic representation and Lagrangian quantization, Res. Notes Math.* 95, (1997).

97. DE GOSSON, M. The quantum motion of half-densities and the derivation of Schrödinger's equation. *J. Phys. A* 31(2), (1998).

98. DE GOSSON, M. On the classical and quantum evolution of Lagrangian half-forms in phase space, *Ann. Inst. H. Poincaré* 70(6), 547–573 (1999).

99. DE GOSSON, M. Lagrangian path intersections and the Leray index, in *Aarhus Geometry and Topology Conference*, Contemporary Mathematics, Vol. 258, pp. 177–184, American Mathematics Society, Providence, RI, 2000.

100. DE GOSSON, M. *The Principles of Newtonian and Quantum Mechanics. The Need for Planck's Constant ℏ*, With a Foreword by Basil Hiley, Imperial College Press, London, 2001.

101. DE GOSSON, M. The symplectic camel and phase space quantization, *J. Phys. A* 34, 10085–10096 (2001).

102. DE GOSSON, M. The "symplectic camel principle" and semiclassical mechanics, *J. Phys. A* 35(32), 6825–6851 (2002).

103. DE GOSSON, M. A class of symplectic ergodic adiabatic invariants, in *Foundations of Probability and Physics*, Vol. 2, pp. 151–158, Växjö University Press, Växjö, 2003.

104. DE GOSSON, M. Phase space quantization and the uncertainty principle, *Phys. Lett. A* 317(5–6), 365–369 (2003).

105. DE GOSSON, M. The optimal pure Gaussian state canonically associated to a Gaussian quantum state, *Phys. Lett. A* 330(3–4), 161–167 (2004).

106. DE GOSSON, M. Uncertainty principle, phase space ellipsoids, and Weyl calculus, in *Pseudo-Differential Operators and Related Topics*, Operator Theory: Advances and Applications, Vol. 164, pp. 121–132, Birkhäuser, Basel, 2006.

107. DE GOSSON, M. The adiabatic limit for multi-dimensional Hamiltonian systems, *J. Geom. Symmetry Phys.* 4, 19–44 (2005–2006).

108. DE GOSSON, M. *Symplectic Geometry and Quantum Mechanics*, Operator Theory: Advances and Applications, Vol. 166, Birkhäuser, Basel, 2006.

109. DE GOSSON, M. Metaplectic representation, Conley–Zehnder index, and Weyl calculus on phase space, *Rev. Math. Phys.* 19(8), 1149–1188 (2007).

110. DE GOSSON, M. Spectral properties of a class of generalized Landau operators. *Commun. Partial Differential Equations* 33(11), (2008).

111. DE GOSSON, M. The symplectic camel and the uncertainty principle: The tip of an iceberg? *Found. Phys.* 99, 194–214 (2009).

112. DE GOSSON, M. *Symplectic Methods in Harmonic Analysis; Applications to Mathematical Physics*, Pseudo-Differential Operators, Birkhäuser, 2011.

113. DE GOSSON, M. The symplectic egg in quantum and classical mechanics, *Amer. J. Phys.* 81(5), (2013).

114. DE GOSSON, M. Symplectic covariance properties for Shubin and Born–Jordan pseudo-differential operators, *Trans. Amer. Math. Soc.* 365(6), 3287–3307 (2013).

115. DE GOSSON, M. Quantum blobs, *Found. Phys.* 43(4), 440–457 (2013).

116. DE GOSSON, M. Born–Jordan quantization and the uncertainty principle, *J. Phys. A: Math. Theory* 46(44), 445301 (2013).

117. DE GOSSON, M. Born–Jordan quantization and the equivalence of the Schrödinger and Heisenberg pictures, *Found. Phys.* 44(10), 1096–1106 (2014).

118. DE GOSSON, M. From Weyl to Born–Jordan quantization: the Schrödinger representation revisited, *Phys. Rep.* 623, 1–58 (2016).

119. DE GOSSON, M. *Introduction to Born–Jordan Quantization*, Fundamental Theories of Physics, Springer, 2016.

120. DE GOSSON, M. *The Wigner Transform*, Advanced Textbooks in Mathematics, World Scientific, 2017.

121. DE GOSSON, M. AND DE GOSSON, S. Weak values of a quantum observable and the cross-Wigner distribution, *Phys. Lett. A* 376(4), 293–296 (2012).

122. DE GOSSON, M. AND DE GOSSON, S. The reconstruction problem and weak quantum values, *J. Phys. A* 45(11), (2012).

123. DE GOSSON, M. AND HILEY, B. Imprints of the quantum world in classical mechanics, *Found. Phys.* 41(9), (2011).

124. DE GOSSON, M. AND HILEY, B. Short-time quantum propagator and Bohmian trajectories, *Phys. Lett. A* 377(42), 3005–3008 (2013).

125. DE GOSSON, M. AND HILEY, B. Hamiltonian flows, short-time propagators and the quantum Zeno effect, *J. Phys.: Conf. Ser.* 504(Conference 1), (2014).

126. DE GOSSON, M., HILEY, B., AND COHEN, E. Observing quantum trajectories: from Mott's problem to quantum Zeno effect and back, *Ann. Phys.* 374, 190–211 (2016).

127. DE GOSSON, M. AND LUEF, F. Quantum states and Hardy's formulation of the uncertainty principle: a symplectic approach, *Lett. Math. Phys.* 80(1), 69–82 (2007).

128. DE GOSSON, M. AND LUEF, F. Remarks on the fact that the uncertainty principle does not characterize the quantum state, *Phys. Lett. A* 364, (2007).

129. DE GOSSON, M. AND LUEF, F. Quantum states and Hardy's formulation of the uncertainty principle: a symplectic approach, *Lett. Math. Phys.* 80, 69–82 (2007).

130. DE GOSSON, M. AND LUEF, F. Principe d'incertitude et positivité des opérateurs à trace; applications aux opérateurs densité, *Ann. Inst. Henri Poincaré* 9(2), 329–346 (2008).

131. DE GOSSON, M. AND LUEF, F. Symplectic capacities and the geometry of uncertainty: the irruption of symplectic topology in classical and quantum mechanics, *Phys. Rep.* 484, 131–179 (2009).

132. DE GOSSON, M. AND LUEF, F. Preferred quantization rules: Born–Jordan vs. Weyl: Applications to phase space quantization, *J. Pseudo-Differ. Oper. Appl.* 2(1), (2011).

133. GOTAY, M.J. AND ISENBERG, G.A. The symplectization of Science, *Gazette Math.* 54, 59–79 (1992).

134. GRACIA-BONDIA, J.M. AND VARILLY, J.C. Algebras of distributions suitable for phase-space quantum mechanics, I, *J. Math. Phys.* 29(4), 869–879 (1988).

135. GRACIA-BONDIA, J.M. AND VARILLY, J.C. Nonnegative mixed states in Weyl–Wigner–Moyal theory, *Phys. Lett.* A 128(1–2), 20–24 (1988).

136. GROMOV, M. Pseudoholomorphic curves in symplectic manifolds, *Invent. Math.* 82, 307–347 (1985).

137. GROENEWOLD, H.J. On the principles of elementary quantum mechanics, *Physica* 12, 405–460 (1946).

138. GROSSMANN, A., LOUPIAS, G. AND STEIN, E.M. An algebra of pseudo-differential operators and quantum mechanics in phase space, *Ann. Inst. Fourier (Grenoble)* 18(2), 343–368 (1968).

139. GUILLEMIN, V. AND STERNBERG, S. *Geometric Asymptotics*, Mathematical Surveys and Monographs, Vol. 14, American Mathematical Society, Providence RI, 1978.

140. GUILLEMIN, V. AND STERNBERG, S. *Symplectic Techniques in Physics*, Cambridge University Press, Cambridge, MA, 1984.

141. GUTH, L. Symplectic embeddings of polydisks, *Invent. Math.* 172(3), 477–489 (2008).

142. HALL, M.J.W. AND REGINATTO, M. Schrödinger equation from an exact uncertainty principle, *J. Phys.* A 35, 3289–3303 (2002).

143. HAMILTON, W.R. *Mathematical Papers* 2, Cambridge University Press, 1940.

144. HAMILTON, W.R. On a general method of expressing the paths of light, and of the planets, by the coefficients of a characteristic function, *Dublin Univ. Rev. Quart. Mag.* I, 795–826 (1833).

145. HANNAY, J.H. Angle variable holonomy in adiabatic excursion of an integrable Hamiltonian, *J. Phys.* A 18, 221–230 (1985).

146. HARDY, G.H. A theorem concerning Fourier transforms, *J. London Math. Soc.* 8, 227–231 (1933).

147. HEISENBERG, W. Über den anschaulichen Inhalt der quantentheoretischen *Kinematik und Mechanik, Z. Phys. A Hadrons Nuclei* 43(3–4), 172–198 (1927).

148. HEISENBERG, W. *The Physical Principles of the Quantum Theory*, Chicago University Press, 1930; reprinted by Dover, New York, 1949.

149. HILEY, B.J. The algebra of process, in *Consciousness at the Crossroads of Cognitive Science and Philosophy*, Maribor, August 1994, pp. 52–67, 1995.

150. HILEY, B.J. *Non-commutative geometry, the Bohm interpretation and the mind-matter relationship, AIP Conf. Proc.* 573(1), 77–88 (2001).

151. HILEY, B.J. Phase space descriptions of quantum phenomena, in *Proc. Int. Conf. Quantum Theory: Reconsideration of Foundations*, Vol. 2, ed. Khrennikov, A., pp. 267–286, Växjö University Press, Växjö, Sweden (2003).

152. HILEY, B.J. On the Relationship between the Wigner–Moyal and Bohm approaches to quantum mechanics: A step to a more general theory, *Found. Phys.* 40, 365–367 (2010).

153. HILEY, B.J. Weak values: Approach through the Clifford and Moyal algebras, Preprint (2011), arXiv:quant-ph/1111.6536.

154. HILEY, B.J. AND PEAT, F.D. *Quantum Implications: Essays in Honour of David Bohm* (Routledge & Kegan Paul, 1987).

155. HILEY, B.J. AND AZIZ MUFTI, A.H. *The Ontological Interpretation of Quantum Field Theory Applied in a Cosmological Context, Fundamental Problems in Quantum Physics*, eds., M. Ferrero and A. van der Merwe, pp. 141–156, Kluwer Academic Publishers, Dordrecht. (1995).

156. HILEY, B.J. AND CALLAGHAN, R.E. Delayed-choice experiments and the Bohm approach. *Phys. Scripta* 74, 336–348 (2006).

157. HILEY, B.J. AND CALLAGHAN, R.E. Maroney O.J.E., Quantum trajectories, real, surreal or an approximation to a deeper process? Preprint (2000), arXiv:quant-ph/0010020.

158. HILLERY, M., O'CONNELL, R.F., SCULLY, M.O. AND WIGNER, E.P. Distribution functions in physics: fundamentals, *Phys. Rep.* 106(3), 121–167 (1984).

159. HOFER, H. AND ZEHNDER, E. *Symplectic Invariants and Hamiltonian Dynamics*, Birkhäuser Advanced Texts, Basler Lehrbücher, Birkhäuser Verlag, 1994.

160. HOLLAND, P.R. *The Quantum Theory of Motion: An Account of the de Broglie–Bohm Causal Interpretation of Quantum Mechanics*, Cambridge University Press, 1993.

161. HOLLAND, P. Hamiltonian theory of wave and particle in quantum mechanics II: Hamilton–Jacobi theory and particle back-reaction. *Nuovo Cimento B* 116, 1143–1172 (2001).

162. HOLM, D.D. *Geometric Mechanics, Part I: Dynamics and symmetry*, 2nd edition, Imperial College Press, 2011.

163. HSIAO, F. Y. AND SCHEERES, D.J. Fundamental constraints on uncertainty relations in Hamiltonian systems, *IEEE Trans. Automat. Control* 52(4), 686–691 (2007).

164. HUDSON, R.L. When is the Wigner quasi-probability density non-negative? *Rep. Math. Phys.* 6, 249–252 (1974).

165. HUTCHIN, R.A. Experimental Evidence for Variability in Planck's Constant. *Optics and Photonics Journal* 6, 124–137 (2016).

166. IBORT, A. MAN'KO, V.I. MARMO, G. SIMONI, A. AND VENTRIGLIA, F. An introduction to the tomographic picture of quantum mechanics, *Phys. Scripta* 79, 065013 (2009).

167. JAMMER, M. *The Conceptual Development of Quantum Mechanics,* Inst. Series in Pure and Appl. Physics, McGraw-Hill Book Company, 1966.

168. JAUCH, J.M. *Foundations of Quantum Mechanics*, Addison-Wesley Series in Advanced Physics, Addison-Wesley, 1968.

169. JOHN, F. Extremum problems with inequalities as subsidiary conditions, in *Studies and Essays Presented to R. Courant on his 60th Birthday*, January 8, (1948), pp. 187–204, Interscience Publishers, Inc., New York, NY, 1948.

170. KASTLER, D. The C^*-algebras of a free Boson field, *Commun. Math. Phys.* 1, 14–48 (1965).

171. KATOK, A. Ergodic perturbations of degenerate integrable Hamiltonian systems, *Izv. Akad. Nauk. SSSR Ser. Mat.* 37, 539–576, (1973) [in Russian]. English translation: *Math. USSR Izv.* 7, 535–571 (1973).

172. KAUDERER, M. *Symplectic Matrices: First Order Systems and Special Relativity*, World Scientific, 1994.

173. KAUFFMANN, S.K. Unambiguous quantization from the maximum classical correspondence that is self-consistent: the slightly stronger canonical commutation rule Dirac missed, Preprint (2010), arXiv:0908.3024.

174. KENNARD, E.H. Zur Quantenmechanik einfacher Bewegungstypen, *Z. Phys. A Hadrons Nuclei* 44(4–5), 326–352 (1927).

175. KENTOSH, J. AND MOHAGEG, M. Global positioning system test of the local position invariance of Planck's constant, *Phys. Rev. Lett.* 108(11), 110801 (2012).

176. KENTOSH, J. AND MOHAGEG, M. Testing the local position invariance of Planck's constant in general relativity, *Phys. Essays* 28(2), 286–289 (2015).

177. KUBO, R. Wigner representation of quantum operators and its applications to electrons in a magnetic field, *J. Phys. Soc. Japan* 19(11), 2127–2139 (1964).

178. LAGRANGE, J.L. *Mécanique analytique*, Facsimilé de la troisième édition, Librairie Albert Blanchard, Paris, 1965.

179. LANDAU, L.D. AND LIFSHITZ, E.M. *Mechanics*, 3rd edition, Pergamon Press, 1976; translated from *Analyse Lagrangienne RCP 25*, Strasbourg Collège de France, (1976–1977).

180. LANDSMAN, N.P. Lecture notes on Hilbert spaces and quantum mechanics, Preprint (2006), http://courses.daiict.ac.in/pluginfile.php/15332/mod_resource/content/0/hilbert_space_why.pdf.

181. LANDSMAN, N.P. *Mathematical Topics Between Classical and Quantum Mechanics*, Springer Science & Business Media, 2012.

182. LEAVENS, C.R. Weak measurements from the point of view of Bohmian Mechanics, *Found. Phys.* 35, 469–491 (2005).

183. LEONHARDT, U. AND PAUL, H. Realistic optical homodyne measurements and quasiprobability distributions, *Phys. Rev. A* 48, 4598 (1993).

184. LERAY, J. *Lagrangian Analysis and Quantum Mechanics, A Mathematical Structure Related to Asymptotic Expansions and the Maslov Index*, The MIT Press, Cambridge, MA, 1981; translated from *Analyse Lagrangienne RCP 25*, Strasbourg Collège de France (1976–1977).

185. LERAY, J. The meaning of Maslov's asymptotic method the need of Planck's constant in mathematics, *Bull. Amer. Math. Soc.*, Symposium on the Mathematical Heritage of Henri Poincaré, 1980.

186. LERAY, J. *Complément à la théorie d' Arnold de l'indice de Maslov*, Convegno di geometrica simplettica et fisica matematica, Instituto di Alta Matematica, Roma (1973).

187. LIBERMANN, P. AND MARLE, C.M. *Symplectic Geometry and Analytical Mechanics*, D. Reidel Publishing Company, 1987.

188. LION, G. AND VERGNE, M. *The Weil Representation, Maslov Index and Theta Series*, Progress in Mathematics, Vol. 6, Birkhäuser, 1980.

189. LITTLEJOHN, R.G. The semiclassical evolution of wave packets, *Phys. Rep.* 138(4–5), 193–291 (1986).

190. LITTLEJOHN, R.G. The Van Vleck formula, Maslov theory, and phase space geometry, *J. Statist. Phys.* 68(1/2), 7–50 (1992).

191. LOUPIAS, G. AND MIRACLE-SOLE, S. C^*-algèbres des systèmes canoniques, I, *Commun. Math. Phys.* 2, 31–48 (1966).

192. LOUPIAS, G. AND MIRACLE-SOLE, S. C^*-Algèbres des systèmes canoniques, II, *Ann. Inst. Henri Poincaré* 6(1), 39–58 (1967).

193. LUO, S. Heisenberg uncertainty relation for mixed states, *Phys. Rev. A* 72, 042110–042112 (2005).

194. LUO, S. AND ZHANG, Z. An Informational characterization of Schrödinger's uncertainty relations, *J. Statist. Phys.* 114(516), 1557–1576 (2004).

195. LUO, S.L. Quantum versus classical uncertainty, *Theoret. Math. Phys.* 143(2), 681–688 (2005).

196. LVOVSKY, A.I. AND RAYMER, M.G. Continuous-variable optical quantum-state tomography, *Rev. Mod. Phys.* 8, 299–332 (2009).

197. MACKEY, G.W. *The Mathematical Foundations of Quantum Mechanics* Benjamin, Inc., New York, 1963.

198. MACKEY, G.W. *Unitary Group Representations*, The Benjamin/Cummings Publ. Co., Inc., Reading, MA, 1978.

199. MACKEY, G.W. *The Relationship Between Classical and Quantum Mechanics,* Contemporary Mathematics, Vol. 214, American Mathematical Society, Providence, RI, 1988.

200. MAKRI, N. AND MILLER, W.H. Correct short time propagator for Feynman path integration by power series expansion in Δt, *J. Chem. Phys. Lett.* 151, 1–8 (1988).

201. MAKRI, N. AND MILLER W.H. Exponential power series expansion for the quantum time evolution operator. *J. Chem. Phys.* 90, 904–911 (1989).

202. MANCINI, S. MAN'KO, V.I. AND TOMBESI, P. Symplectic tomography as classical approach to quantum systems, *Phys. Lett. A* 213, 1–6 (1996).

203. MANGANO, G. LIZZI, F. AND PORZIO, A. Inconstant Planck's constant, *Internat. J. Mod. Phys. A* 30(34), 1550209 (2015).

204. MAN'KO O. AND MAN'KO,V.I. Quantum states in probability representation and tomography, *J. Russian Laser Res.* 18(5), 407–444 (1997).

205. MARCIANO, W.J. Time variation of the fundamental "constants" and Kaluza–Klein theories, *Phys. Rev. Lett.* 52(7), 489–491 (1984).
206. MARUSKIN, J.M., SCHEERES, D.J. AND BLOCH, A.M. Dynamics of symplectic subvolumes, *SIAM J. Appl. Dynam. Syst.* 8(1), 180–201 (2009).
207. MASLOV, V.P. *Théorie des Perturbations et Méthodes Asymptotiques*, Dunod, Paris, 1972; translated from Russian [original Russian Edition 1965].
208. MASLOV, V.P. AND FEDORIUK, M.V. *Semi-classical Approximations in Quantum Mechanics*, Reidel, Boston, 1981.
209. MASOLIVER, J. AND ROS, A. From classical to quantum mechanics through optics, *Eur. J. Phys.* 31, 171–192 (2010).
210. MESSIAH, A. *Quantum Mechanics, I, II*, North-Holland, 1991; translated from the French; original title: *Mécanique Quantique*, Dunod, Paris, 1961.
211. MNEIMÉ, R. AND TESTARD, T. *Introduction à la Théorie des Groupes de Lie Classiques*, Collection Méthodes, Hermann, Paris, 1986.
212. MOSER, J. On the volume element of a manifold, *Trans. Amer. Math. Soc.* 120, 286–294 (1965).
213. MOTT, N.F. The wave mechanics of α-ray tracks, *Proc. Roy. Soc. A* 126, 79–84 (1929).
214. MOYAL, J.E. Quantum mechanics as a statistical theory, *Proc. Cambridge Phil. Soc.* 45, 99–124 (1947).
215. MURPHY, M.T., WEBB, J.K., FLAMBAUM, V.V., DZUBA, V.A., CHURCHILL, C.W., PROCHASKA, J.X., BARROW, J.D. AND WOLFE, A.M. Possible evidence for a variable fine-structure constant from QSO absorption lines: motivations, analysis and results. *Mon. Not. R. Astron. Soc.* 327, 1208–1222 (2001).
216. MURPHY, M.T., WEBB, J.K. AND FLAMBAUM, V.V. Further evidence for a variable fine-structure constant from Keck/HIRES QSO absorption spectra. *Mon. Not. R. Astron. Soc.* 345(2), 609–638 (2003).
217. NARCOWICH, F.J. AND O'CONNELL, R.F. Necessary and sufficient conditions for a phase-space function to be a Wigner distribution, *Phys. Rev. A* 34(1), 1–6 (1986).
218. NARCOWICH, F.J. Conditions for the convolution of two Wigner distributions to be itself a Wigner distribution, *J. Math. Phy.* 29(9), 2036–2041 (1988).
219. NARCOWICH, F.J. Distributions of $\hbar$-positive type and applications. *J. Math. Phys.* 30(11), 2565–2573 (1989).
220. NARCOWICH, F.J. Geometry and uncertainty, *J. Math. Phys.* 31(2) 354–364 (1990).
221. NARCOWICH, F.J. AND O'CONNELL, R.F.A unified approach to quantum dynamical maps and Gaussian Wigner distributions, *Phys. Lett. A* 133(4), 167–170 (1988).
222. NAZAIKIINSKII, V., SCHULZE, B.W. AND STERNIN, B. Quantization Methods in Differential Equations, Preprint, Potsdam (2000).
223. NELSON, E. Feynman integrals and the Schrödinger equation, *J. Math. Phys.* 5, 332–343 (1964).

224. NELSON, E. *Topics in Dynamics I: Flows, Mathematical Notes*, Princeton University Press, 1969.

225. NICOLA, F. Convergence in L^p for Feynman path integrals, *Adv. Math.* 294, 384–409 (2016).

226. VON NEUMANN, J. Wahrscheinlichkeitstheoretischer Aufbau der Quantenmechanik, *Göttinger Nachr.* 1, 245–272 (1927).

227. OZAWA, M. Universally valid reformulation of the Heisenberg uncertainty principle on noise and disturbance in measurement, *Phys. Rev.* A 67(4), 042105 (2003).

228. PARIS, M. AND REHÁČEK, J. eds. *Quantum State Estimation*, Lecture Notes in Physics, Vol. 649, Springer, Berlin, 2004.

229. PAO-KENG YANG. How does Planck's constant influence the macroscopic worlds? *Eur. J. Phys.* 37, 055406 (2016).

230. PARK, D. *Classical Dynamics and Its Quantum Analogues*, 2nd edition; 1st edition, Lecture Notes in Physics, Vol. 110, Springer, 1990.

231. PARTHASARATHY, K.R. What is a Gaussian state? *Commun. Stoch. Anal.* 4(2), 143–160 (2010).

232. PARTHASARATHY, K.R. *An Introduction to Quantum Stochastic Calculus,* Springer Science & Business Media, 2012.

233. PARTHASARATHY, K.R. AND SCHMIDT, K. *Positive Definite Kernels, Continuous Tensor Products, and Central Limit Theorems of Probability Theory,* Lecture Note in Mathematics, Vol. 272, Springer, Berlin, 1972.

234. PASSERONE, D. Computing the density of paths in complex systems, *J. Chem. Phys.* 124, 134103 (2006).

235. PAULI, W. *General Principles of Quantum Mechanics*, Springer Science & Business Media, 2012 [original title: *Prinzipien der Quantentheorie,* published in *Handbuch der Physik*, Vol. 5.1, 1958].

236. PEGG, D.T. Future variation of Planck's constant, *Nature* 267, 408–409 (1977).

237. PENROSE, R. *The Emperor's New Mind*, Oxford University Press, 1989.

238. PERES, A. *Quantum Theory, Concepts and Methods*, Kluwer Academic Publishers, 1995.

239. PHILIPPIDIS, C., DEWDNEY, C. AND HILEY, B.J. Quantum interference and the quantum potential, *Nuovo Cimento* 52B, 15–28 (1979).

240. POLTEROVICH, L. *The Geometry of the Group of Symplectic Diffeomorphisms*, Lectures in Mathematics, Birkhäuser (2001).

241. POTOČEK, V. AND BARNETT, S.M. On the exponential form of the displacement operator for different systems, *Phys. Scripta* 90, 065208 (2015).

242. PRIMAS, H. *Chemistry, Quantum Mechanics and Reductionism,* Springer, Berlin, 1983.

243. RADON, J. Über die Bestimmung von Funktionen durch ihre Integralwerte längsgewisser Mannigfaltigkeiten, *Sächs. Akad. Wiss. Leipzig, Math. Nat. Kl.* 69, 262–277 (1917).

244. ROBERTSON, H.P. The uncertainty principle. *Phys. Rev.* 34, 163–164 (1929).

245. ROYER, A. Wigner functions as the expectation value of a parity operator, *Phys. Rev. A* 15, 449–450 (1977).

246. SAMUEL REICH, E. How camels could explain quantum uncertainty (Why quantum uncertainty is all biblical geometry), *New Scientist* 2697, 12 (2009).

247. SCHEMPP, W. *Harmonic Analysis on the Heisenberg Nilpotent Lie Group*, Pitman Research Notes in Mathematics, Vol. 147, Longman Scientific & Technical, 1986.

248. SCHEERES, D.J., HSIAO, F.-Y., PARK, R.S., VILLAC, B.F. AND MARUSKIN, J.M. Fundamental limits on spacecraft orbit uncertainty and distribution propagation, *J. Astronaut. Sci.* 54, 505–523 (2006).

249. SCHEERES, D.J., DE GOSSON, M.A. AND MARUSKIN, J.M. Fundamental limits on orbit uncertainty, in *Information Fusion (FUSION), 2012: 15th International Conference*, 2012.

250. SCHILLER, R. Quasi-classical theory of the nonspinning electron, *Phys. Rev.* 125, 1100–1108 (1962).

251. SCHILLER, R. Quasi-classical transformation theory, *Phys. Rev.* 125, 1109–1115 (1962).

252. SCHLENK, F. Symplectic embeddings of ellipsoids, *Israel J. Math.* 138, 215–252 (2003).

253. SCHLENK, F. *Embedding Problems in Symplectic Geometry*, De Gruyter Expositions in Mathematics, Vol. 40, de Gruyter, Berlin, 2005.

254. SCHRÖDINGER, E. *The Interpretation of Quantum Mechanics: Dublin (1949–1955) and Other Unpublished essays*, ed. M. Bitbol, Ox Bow Press, Woodbridge, CT, 1995.

255. SCHRÖDINGER, E. Zum Heisenbergschen Unschärfeprinzip, *Berliner Berichte*, 296–303 (1930) [English translation: A. Angelow, M.C. Batoni, About Heisenberg uncertainty relation, *Bulgarian J. Phys.* 26 5/6 (1999) 193–203, and http://arxiv.org/abs/quant-ph/9903100.

256. SCHRÖDINGER, E. Quantisierung als Eigenwertproblem. *Ann. Physik* 79, 361 (1926). English Translation in Schrödinger, E. *Collected Papers on Wave Mechanics*, Chelsea Publishing Company, 1978.

257. SCHRÖDINGER, E. Quantisierung als Eigenwertproblem, *Ann. Physik* (1926), 1st communication: 79, 489–527, 2nd communication: 80, 437–490, 3rd communication: 81, 109–139.

258. SCHULMAN, L.S. *Techniques and Applications of Path Integrals*, Wiley, NY, 1981.

259. SEGAL, I.E. Foundations of the theory of dynamical systems of infinitely many degrees of freedom (I), *Mat. Fys. Medd., Danske Vid. Selsk.* 31(12), 1–39 (1959).

260. SEIFERT, H. Periodische Bewegungen mechanischer Systeme, *Math. Zeit.* 51, 197–216 (1948).

261. SHALE, D. Linear symmetries of free Boson fields, *Trans. Amer. Math. Soc.* 103, 149–167 (1962).

262. SHUBIN, M.A. *Pseudodifferential Operators and Spectral Theory*, Springer, 1987 [original Russian edition in Nauka, 1978].

263. SIBURG, K.F. Symplectic capacities in two dimensions, *Manuscripta Math.* 78, 149–163 (1993).

264. SIKORAV, J.C. Quelques propriétés des plongements lagrangiens (Some properties of Lagrangian embeddings), *Mém. Soc. Math. France* (*Nouv. Sér.*) 46, 151 (1991).

265. SIMON, B. *Trace Ideals and their Applications*, Cambridge University Press, Cambridge, 1979.

266. SIMON, R. MUKUNDA, N. AND DUTTA, N. Quantum noise matrix for multimode systems: U(n)-invariance, squeezing and normal forms. *Phys. Rev. A* 4, 15867–15839 (1994).

267. SIMON, R., SUDARSHAN, E.C.G. AND MUKUNDA, N. Gaussian–Wigner distributions in quantum mechanics and optics, *Phys. Rev. A* 36(8), 3868–3880 (1987).

268. SOURIAU, J.M. *Structure des Systèmes Dynamiques*, Dunod, Paris, 1970; [English translation by C.H. Cushman-de-Vries: *Structure of Dynamical Systems*, Birkhäuser, 1997].

269. STEIN, E.M. *Harmonic Analysis: Real Variable Methods, Orthogonality, and Oscillatory Integrals*, Princeton University Press, 1993.

270. STEWART, I. The symplectic camel, *Nature* 329, 17–18 (1987).

271. STONEY, G. On the physical units of nature, *Phil. Mag.* 11, 381–391 (1881).

272. SUDARSHAN, E.C.G., CHIU, C.B. AND BHAMATI, G. Generalized uncertainty relations and characteristic invariant for the multimode states, *Phys. Rev. A* 52(1), 43–54 (1995).

273. SRINIVAS, M.D. AND WOLF, E. Some nonclassical features of phase-space representations of quantum mechanics, *Phys. Rev. D* 11(6), 1477–1485 (1975).

274. TAKHTAJAN, L.A. *Quantum Mechanics for Mathematicians*, Vol. 95, American Mathematical Society, 2008.

275. TANNOR, D. *A Time-Dependent Perspective*, University Science Books, Sausalito, 2007.

276. THEKKADATH, G.S., GINER, L., CHALICH, Y., HORTON, M.J., BANKER, J. AND LUNDEEN, J.S. Direct measurement of the density matrix of a quantum system, *Phys. Rev. Lett.* 117, 120401 (2016).

277. TROTTER, H.F. On the product of semi-groups of operators. *Proc. Amer. Math. Soc.* 10, 545–551 (1959).

278. VAN VLECK, J.H. Quantum principles and line spectra, *Bull. Natl. Res. Council* 10(54), 1–316 (1926).

279. VAN VLECK, J.H. The correspondence principle in the statistical interpretation of quantum mechanics. *Proc. Natl. Acad. Sci. USA* 14(2), 178–188 (1989).

280. VAN HOVE, L. Sur le Problème des Relations entre les Transformations Unitaires de la Mécanique Quantique et les Transformations Canoniques de la Mécanique Classique, *Mém. Acad. Roy. Belg.* 26, 610 (1951).

281. VISSER, M. AND VAN VLECK, J.H. Determinants: Traversable wormhole spacetimes *Phys. Rev. D* 49, 3963–3980 (1994).

282. VOGEL, K. AND RISKEN, H. Determination of quasiprobability distributions in terms of probability distributions for the rotated quadrature phase, *Phys. Rev. A* 40(5), 2847–2849 (1989).

283. WALLACH, N. *Lie Groups: History, Frontiers and Applications*, Vol. 5: *Symplectic Geometry and Fourier Analysis*, Math Sci Press, Brookline, MA, 1977.

284. WANG, D. Some aspects of Hamiltonian systems and symplectic algorithms,*Physica D* 73, 1–16 (1994).

285. WANG, Y. AND XU, C. Density matrix estimation in quantum homodyne tomography, *Statistica Sinica* 953–973 (2015).

286. WEBB, J.K., MURPHY, M.T., FLAMBAUM, V.V., DZUBA, V.A., BARROW, J.D., CHURCHILL, C.W. AND WOLFE, A.M. Further evidence for cosmological evolution of the fine structure constant, *Phys. Rev. Lett.* 87(9), 091301 (2001).

287. WEBB, J.K., FLAMBAUM, V.V., CHURCHILL, C.W., DRINKWATER, M.J. AND BARROW, J.D. Search for time variation of the fine structure constant, *Phys. Rev. Lett.* 82(5), 884 (1999).

288. WEBB J.K., KING J.A., MURPHY M.T., FLAMBAUM V.V., CARSWELL, R.F. AND BAINBRIDGE M.B. Indications of a spatial variation of the fine structure constant, *Phys. Rev. Lett.* 107(19), 191101 (2011).

289. WEIERSTRASS, K. *Mahematische Werke*, Band I: pp. 233–246, Band II: pp. 19–44, Nachtrag: pp. 139–148, Mayer/Müller, Berlin, 1858.

290. WEIL, A. Sur certains groupes d'opérateurs unitaires, *Acta Math*, 111, 143–211 (1964); also in *Collected Papers*, Vol. III, pp. 1–69, Springer, Heidelberg, 1980.

291. WEINSTEIN, A. Periodic orbits for convex Hamiltonian systems, *Ann. Math.* 108, 507–518 (1978).

292. WEN-CHAO QIANG, AND SHI-HAI DONG. Proper quantisation rule. *Europhys. Lett.* 89, 10003 (2010).

293. WERNER, R. Quantum harmonic analysis on phase space, *J. Math. Phys.* 25(5), 1404–1411 (1984).

294. WEYL, H. Quantenmechanik und Gruppentheorie, *Z. Physik* 46, (1927).

295. WEYL, H. Invariants. *Duke Math. J.* 5(3), 489–502 (1939).

296. WEYL, M.W. *Transforms*, Springer-Verlag, 1989.

297. WHEELER, A. AND ZUREK, H.Z., eds. *Quantum Theory and Measurement*, Princeton Series in Physics, Princeton University Press, 2014.

298. WIGNER, E. On the quantum correction for thermodynamic equilibrium, *Phys. Rev.* 40, 799–755 (1932)

299. WIGNER, E.P. The unreasonable effectiveness of mathematics in the natural sciences, *Commun. Pure Appl. Math.* 13, 1–14 (1960).

300. WILLIAMSON, J. On the algebraic problem concerning the normal forms of linear dynamical systems, *Amer. J. Math.* 58, 141–163 (1936).

301. WILCZYNSKA, M.R., WEBB J.K., KING J.A., MURPHY M.T., BAINBRIDGE M.B. AND FLAMBAUM V.V. A new analysis of fine-structure constant measurements and modelling errors from quasar absorption lines, *Mon. Not. R. Astron. Soc.* 454(3), 3082–3093 (2015).

302. XIAO-YAN GU AND SHI-HAI DONG. The improved quantization rule and the Langer modification, *Phys. Lett. A* 372, 1972–1977 (2008). Also see: The theory of dynamical systems of infinitely many degrees of freedom (I), *Mat. Fys. Medd., Danske Vid. Selsk.* 31(12), 1–39 (1959).

303. YUEN, H.P. Multimode two-photon coherent states, in *International Symposium on Spacetime Symmetries*, eds., Y.S. Kim and W.W. Zachary, North-Holland, Amsterdam, 1989.

304. ZHONG-QI MA AND BO-WEI XU. Quantum correction in exact quantisation rules, *Europhys. Lett.* 69, 685 (2005).

305. ZHONG-QI MA AND BO-WEI XU. Exact quantisation rules for bound states of the Schrödinger equation, *Int. J. Mod. Phys. E* 14, 599–610 (2005).

Index

Printed in the United States
By Bookmasters